1st International Conference on Chemistry (ICCHEM 2018)

Journal of Physics: Conference Series
Volume 1156

Yogyakarta, Indonesia
28 - 30 September 2018

Editors:

Sri Handayani
Eli Rohaeti
Suwardi
Cahyorini Kusumawardhani

ISBN: 978-1-5108-8118-1
ISSN: 1742-6588

Printed from e-media with permission by:

Curran Associates, Inc.
57 Morehouse Lane
Red Hook, NY 12571

Some format issues inherent in the e-media version may also appear in this print version.

No changes have been made to the content of these proceedings. There may be changes to pagination and minor adjustments for aesthetics.

Printed by Curran Associates, Inc. (2019)

For permission requests, please contact the Institute of Physics at the address below.

Institute of Physics
Dirac House, Temple Back
Bristol BS1 6BE UK

Phone: 44 1 17 929 7481
Fax: 44 1 17 920 0979

techtracking@iop.org

Additional copies of this publication are available from:

Curran Associates, Inc.
57 Morehouse Lane
Red Hook, NY 12571 USA
Phone: 845-758-0400
Fax: 845-758-2633
Email: curran@proceedings.com
Web: www.proceedings.com

Table of contents

Volume 1156

International Conference of Chemistry 2018

28–30 September 2018, Yogyakarta, Indonesia

Accepted papers received: 10 December 2018
Published online: 6 February 2019

IOP Conf. Series: Journal of Physics: Conf. Series **1156** (2019) 011001 doi:10.1088/1742-6596/1156/1/011001

The 1st International Conference of Chemistry (ICCHEM)

Yogyakarta, Indonesia

28-30 September 2018

Editors:

Assoc. Prof. Dr. Sri Handayani

Department of Chemistry Education

Faculty of Mathematics and Natural Sciences

Universitas Negeri Yogyakarta, Indonesia

handayani@uny.ac.id

Assoc. Prof. Dr. Eli Rohaeti

Department of Chemistry Education

Faculty of Mathematics and Natural Sciences

Universitas Negeri Yogyakarta, Indonesia

eli_rohaeti@uny.ac.id

Dr. Suwardi

Departement of Chemistry Education

Faculty of Mathematics and Natural Sciences

Universitas Negeri Yogyakarta

suwardi@uny.ac.id

Dr. Cahyorini Kusumawardhani

Department of Chemistry Education

Faculty of Mathematics and Natural Sciences

Universitas Negeri Yogyakarta

cahyorini.k@uny.ac.id

IOP Conf. Series: Journal of Physics: Conf. Series **1156** (2019) 011001 doi:10.1088/1742-6596/1156/1/011001

PREFACE

The International Conference of Chemistry (ICCHEM) is an annual conference organized by Department of Chemistry Education, Faculty of Mathematics and Natural Sciences, Universitas Negeri Yogyakarta, Yogyakarta, Indonesia and successfully held from 28 to 30 September, 2018. The theme of the 1st ICCHEM is optimizing and empowering chemistry for sustainable development of science and technology. The conference was a forum for researchers, educators, students, policy makers, and practitioner to achieve the innovation and social development through research and education on chemistry, as it is accentuated by the theme of this conference. The scope of this Conference covers the area of chemistry and chemistry education. This proceeding contains 34 papers that have been carefully peer-reviewed and selected from 65 papers submitted to the conference.

We would like to express our gratitude to the reviewers of these manuscripts, who provided constructive criticism and stimulated comments and suggestions to the authors. We are extremely grateful as organizers, technical program committee and editors and extend our most sincere thanks to all the participants of the conference for their fruitful work and their excellent contribution to the development of this conference proceedings. Our sincere gratitude also goes to the IOP Publishing editors and managers for their helpful cooperation during the preparation of the proceedings.

On behalf of the Organizing Committee of the 1st ICCHEM
Assoc. Prof. Dr. Retno Arianingrum

IOP Conf. Series: Journal of Physics: Conf. Series **1156** (2019) 011001 doi:10.1088/1742-6596/1156/1/011001

List of Committees

International Advisory Committee

1. Prof. Dr. Muammer Calik (Chemistry Education, Trabzon University, Turkey)
2. Assoc. Prof. Dr. Khajornsak Buaraphan (Chemistry Education, Institute of Innovative Learning, Mahidol University, Thailand)
3. Assoc. Prof. Dr. Azlan Kamari (Chemistry Education, Sultan Idris Education University, Malaysia)
4. Prof. KH. Sugiyarto, Ph.D (Department of Chemistry Education, Universitas Negeri Yogyakarta, Indonesia)
5. Assoc. Prof. Dr. Dwi Siswanta (Department of Chemistry, Universitas Gadjah Mada, Indonesia)
6. Assoc. Dr. Hafsah binti Taha (Chemistry Education, Sultan Idris Education University, Malaysia)
7. Assoc. Prof. Dr. Tienthong Thongpanchang (Department of Chemistry, Mahidol University, Thailand)
8. Assoc. Prof. Dr. Muthuraaman Bhagavathi Achari (Madras University, India)
9. Dr. Gerhard Marini (MED-EL Institut Innsbruck, Austria)
10. Prof. Dr. Sri Atun (Department of Chemistry Education, Universitas Negeri Yogyakarta)
11. Dr. Harry Firman (Department of Chemistry Education, UPI, Bandung)

Organizing Committee

1. Assoc. Prof. Dr. Hartono
2. Assoc. Prof. Dr. rer.nat Senam
3. Assoc. Prof. Jaslin Ikhsan, Ph.D
4. Assoc. Prof. Suyanta
5. Assoc. Prof. Dr. Retno Arianingrum
6. Sukisman Purtadi, M.Pd
7. Erfan Priyambodo, M.Si
8. Prof. Dr. Endang Wijayanti, LFX
9. Dr. Antuni Wiyarsi
10. Anggiyani Retnaningtyas Eka N., M.Pd.
11. Dr. Dyah Purwaningsih
12. Sulistiyani, M.Si
13. Marfuatun, M.Sc.
14. Annisa Fillaeli. M.Si

IOP Conf. Series: Journal of Physics: Conf. Series **1156** (2019) 011002 doi:10.1088/1742-6596/1156/1/011002

Peer review statement

All papers published in this volume of *Journal of Physics: Conference Series* have been peer reviewed through processes administered by the proceedings Editors. Reviews were conducted by expert referees to the professional and scientific standards expected of a proceedings journal published by IOP Publishing.

IOP Conf. Series: Journal of Physics: Conf. Series **1156** (2019) 012001 doi:10.1088/1742-6596/1156/1/012001

Synthesis activated carbon of Screw-pine leaves by HNO_3 and its properties

A Fillaeli*, S Kristianingrum, E D Siswani, Sulistyani and S D Fatimah

Department of Chemistry Education, Faculty of Mathematics and Natural Sciences, Universitas Negeri Yogyakarta, Indonesia, Jl. Colombo 1, Yogyakarta 55281, Indonesia

*Email: annisa_fillaeli@uny.ac.id

Abstract. The leaves of Screw-pine (*Pandanusodorifer*) which contain natural cellulose were used as sources of carbon to prepare activated carbon. Contained 28-30% of water, the leaves were dried for 3-5 days by sunlight. The pyrolysis process of dry leaves yielded 25.7% of carbon that was then followed by grounding and sifting steps to 80 mesh of particle size. This uniform particles were mixed with 5% HNO_3 activator in batch system for 24 hours. Physical activation conducted at 700^0C and obtained 70.8% activated carbon. The characteristics of activated carbon were compared to the carbon base by using porosimeter, referencing to SNI 06-3730-1995 and finding the adsorption isotherm patterns. Activated carbon surface area was 337.9 m^2/g that is classified as micro-pore substance. In accordance to SNI requirements, the activated carbon fulfilled water content, ash content and adsorption capacity of I_2while the study of efficiency adsorption is satisfy at 92.5436%, 97.5707%, 97.9811% and 96.2066% of Cd, Cu, Zn and Fe respectively.

1. Introduction

Screw-pine (*Pandanus odorifer*) is a species of *pandanaceae* family that commonly can be found along the coastal area. This species has not been used maximally as it used as basic ingredients of making handy-craft that sometimes leave a lot of wastes. Actually, the tree also has abundant potential to be used in modern industry due to cellulose content of the leaves. Naturally, its untreated-leaves have cellulose content of about 35% and can be maximized into 70% of cellulose extract by enrichment process with a chemical [1].

The natural cellulose content can be used as precursor of making activated carbon because the cost is very cheap and the amount of leaves are abundant in nature. Additionally, the based carbon to the activated process is low cost but effective in producing a material that are easy to regenerate and selective to adsorb heavy metals [2].

The natural cellulose can be converted into carbon via carbonation process. In order to increase adsorption capacity, the raw carbon must be activated in chemical and physical activations. The enlarger capacity aims to remove dissolved organic compounds, inhibits tar formation, and increases adsorption capacity of heavy metal contaminants due to the large carbon surface area [3-4].

Chemical activation of carbon involves the use of activators in the form of base or acidic compounds. Basically, the use of acid activators will produce activated carbon with a better quality compared to alkaline activators [5]. HNO_3 solution is one of acid activators that can be used to treat the raw carbon because HNO_3 is strong acid that can properly dissolve organic impurities and inhibit tar formation, so that the pores on the surface of the activated carbon become larger [6]. According to Tao and Xiaoqin [7], the activation of carbon from peanut shell by HNO3 (20% (by mass)) produced adsorption efficiency

10.3 times greater than commercial activated carbon. All the acidic activation to the raw carbon were followed by physical activation by involving heating and carbon gasification at high temperatures to expand the carbon surface area to the maximum [8].

Escalating interest in activation with nitric acid from such leaves precursors has led to focus on the mechanism of the activated carbon synthesis with a good developed micro-porous structure and a high capacity adsorption. The activated carbon characterization will refer to SNI 06-3730-1995 because this ordinance regulates the use of activated carbon mainly for industrial/commercial purposes. Investigation of adsorption patterns in several metal solutions of Cu, Cd, Fe, Zn as the simulation of liquid waste will complete the physical properties measurements.

The paper presents the investigation results on character of Screw-pine leaves activated carbon which was activated by 5% HNO_3 and the adsorption properties in liquid waste systems. The concentration of HNO_3 chosen is purposed for finding the better character of activated carbon with lower concentration than Tao and Xiaoqin [7] did, because the lower acidic is the better for environment.

2. Experimental Section

2.1 Materials and Instruments

Material for making raw carbon was Screw-pine leaves (*P. odorifer*) from the coastal area of Trisik, Kulonprogo. The chemicals were 5% HNO_3, 0.1M I_2, 0.1M $Na_2S_2O_3$, $Cu(NO_3)_2$, $CdSO_4$, $Fe(NO_3)_3$, $Zn(NO_3)_2$, amylum, distilled-water and completed with filter paper and pH paper. All the chemicals were provided from various commercial distributors with pure analysis (p.a.) grade and used without further purification. The characterization were done by Surface Area Analyzer (SAA) and Atomic Absorption Spectrophotometer (AAS), supported with blowing furnace, 80 mesh strainer, analytical scale, magnetic stirrer, oven and laboratory glass wares.

2.2. Synthesis of carbon

The Screw-pine leaves were washed and then cut into small pieces before dried under the sun until completely dry (got constant weighing). The leaves then being carbonated in 500^0C for 6 hours using a furnace. The obtained carbon was smoothed to reduce the particle size and strained with 80 mesh sieve to make it more uniform.

2.3 Activation of Carbon

Five percent of HNO_3 was mixed with homogeneous raw carbon (in size) in a batch system for 24 hours at ambient temperature. Followed by filtering the liquid, the activated carbon product was being washed with distilled-wateruntil neutral in pH in accordance with pH of distilled water. The next step was drying the neutral carbon with oven at 100°C for 3 hours or until the carbon weight has not changed when carbon underwent in further drying process. The yielded carbon then was heated at 700°C for 2 hours.

2.4 Characterization

The size of porous were identified by porosimeter for both raw and activated carbons. According to SNI 06-3730-1995, the activated carbon was tested in finding of the content of water, ash level, volatile degree, and the capacity adsorption of activated carbon to I_2.The properties of adsorption were investigated by interacting activated carbon with solutions of Cu, Cd, Fe and Zn in a ratio of 1:50 (g/mL) for 24 hours in medium pace respectively to find the capacity, the efficiency and the isotherm models.

3. Result and Discussion

Synthesis of activated carbon from natural cellulose has a lot of advantages due to its renewable source characteristic. It opens the possible commercialization of product because the activated carbon capable to achieve prominent adsorption level mainly in removing hazard substances. Higher degree of adsorption, larger benefit of using activated carbon will be gained. The quality of activated carbon depends on the making process and the choices of appropriate chemical to activate carbon.

IOP Conf. Series: Journal of Physics: Conf. Series **1156** (2019) 012001 doi:10.1088/1742-6596/1156/1/012001

The Screw-pine leaves were selected as source of carbon due to cellulose level of both lignocellulose and hemicellulose [9]. It can easily be found in the south coastal of Yogyakarta of Trisik, Kulonprogo. To start the activated carbon process, the selected leaves of Screw-pine were dried under the sunlight for 3-5 days depend on the weather until it really dried. A significant weighing reduction showed that the water content of the leaves is 28-30%.

The dried leaves were ready to place in the furnace at 500^{0}C for 6 hours to get raw carbon. The yielded raw carbon presented in the ratio of 1:5 to the wet leaves. The raw carbon was treated in further steps of activation both chemically and physically. It aimed at getting high capacity of adsorption that will be meaningful if the activated carbon has a great surface area. According to Alfiany et al [6], the size of the carbon particles affect the surface area of carbon produced. The smaller size of the carbon particles, the greater the carbon surface area so that during activation process, the larger pores will be formed on each carbon particle.

Chemical activation process using 5% HNO_3 solution was done to the raw carbon of Screw-pine leaves in batch system for 24 hours by solid-liquid interaction in constant stirring. This acid solution can clean the pore surface of the impurity compound, limit the formation of tar, degrade organic molecules in subsequent heating processes, and rearrange interchangeable atoms. This 5% HNO_3 solution is then removed from activated carbon by washing it using distilled-water to neutral pH compared to pH of distilled-water. After that, the carbon were activated physically at temperature 700°C for 2 hours. Heating at high temperatures will influence the carbon structure to be more organized and evaporate water trapped in the pore of carbon. It aims to open the pore surface of activated carbon with the maximum porosity. Growing and opening of these carbon surface pores are expected to increase the adsorption capacity and efficiency of activated carbon so that it is more effective in heavy metal adsorption of the wastewater solution.

3.1 Characterization of activated carbon

The characterization results of the activated carbon of Screw-pine leaves are shown in table 1. The characterization including moisture content, ash content, volatile substance, bond carbon content, and capacity adsorption to I_2 were done to fulfill SNI 06-3730-1995 requirements. On the other hand, the characterization with porosimeter were completed the SNI measurement.

Table 1. Characterization of activated carbon with 5% HNO_3

Name	Water content (%)	Ash content (%)	Volatile substance content (%)	Bond carbon content (%)	Capacity adsorption of I_2 (mg/g)	Single point surface area (m^2/g)	BET surface area (m^2/g)
Parameter according to SNI 06-3730-1995	Max. 15	Max. 10	Max. 25	Min. 65	Min. 750	-	-
Carbon	7.46	21.62	70.43	0.49	2157,5546	3.7719	3.5756
Activated carbon	1.02	6.44	65.58	26.96	2251.2530	337.9532	318.1830

Findings showed that water content before and after activation process have accordance with SNI requirement that were less than 10%. The purpose of this characterization was to know the hygroscopic nature of activated carbon. Low water levels of carbon indicate that the water content has evaporated much during the process of drying, carbonation and physical activation.

Ash content before activation process was more than 10% while ash content after activation process was less than 10%. It meets SNI 06-3730-1995 of activated carbon requirements. The purpose of characterization of ash content was to find the remaining minerals left behind in the carbon after the activation process. The results examined that the level of ash after activation was smaller than initially because the impurities and minerals that attach to the carbon were much lost during the activation process. High ash content can reduce the adsorption capacity of activated carbon because the pores are blocked by minerals that still lag in carbon [10].

The volatile carbon content before and after the activation process were greater than 25% so it do not accordance SNI 06-3730-1995. The purpose of characterization of volatile substances was to determine the volatile compounds contained in activated carbon. The high levels of these volatile substances indicated that many non-carbon compounds such as H atoms and O atoms are still strongly bonded to carbon atoms. The amount of these non-carbon compounds can clog pores of the carbon surface so it reduces the adsorption capacity of activated carbon to heavy metals in waste water [11].

The adsorption capacity of activated carbon to iodine has fulfilled the quality of activated carbon of SNI 06-3730-1995 that is more than 750 mg/g. Determination of activated carbon adsorption to iodine was done by iodometric titration. This procedure aimed at investigating of adsorption ability of activated carbon to metal ion with a relatively small molecular weight. The larger the iodine number the greater the adsorption capacity of the active carbon will be achieved [6].

Because the characterization result of water content, ash content, volatile content of carbon before and after activation were under the SNI 06-3730-1995 standard, the resulting carbon content is not accordance with SNI 06-3730-1995 that is at least 65%. The carbon content bond before activation is 0.49% while after activation process is 26.96%. It is far less than SNI 06-3730-1995 requirements. It indicates the carbon shape tend to create single component than making amorphous structure that appropriates to adsorb molecules.

The characterization data was also completed by porosimeter measurement. Chemical and physical process applied on carbon of Screw-pine leaves will influence carbon surface structure. After carbonization, the raw carbon needed to increase the pore volume and to enlarge the pore diameter. The activation process causes the breakdown of the hydrocarbon bonds in the carbon structure so that the surface area of the pore will increase [12]. It is proved by this study of surface area comparison between before and after activation of raw carbon which became greater. The surface area of carbon prior to activation was 3.7719 m^2/g. That area is classified into meso-pore carbon category with surface area in the range of 2-100 m^2/g. The function of pore with meso-pore size will be meaningful for transportation purposes. However, for reducing liquid chemical pollution the meso-pore must be reduced to micro-pore level. This level requires surface area size of $\geq$91 m^2/g. It was gained by treatment in combination of chemical and physical activations. The porosimeter output of surface area of activated carbon measurement is 337.9532 m^2/g. This result is an evidence that the activated carbon belongs to a micro-pore of its surface area. The activation process alters the meso-pore structure of carbon into micro-pore structures with the helping of acidic chemical solutions and heating at high temperatures. Carbon with micro-pore size can adsorb small molecules on its surface [13].

3.2 Adsorption of wastewater solution properties

The heavy metal adsorption process wastewater in batch system. The ratio of activated carbon and the wastewater solution was 1:50 (gram/mL). The contact time between activated carbon and the waste water solution was 24 hours [14].

This research purposes to find the efficiency, capacity, and adsorption isotherm patterns of activated carbon. The instrument to measure the concentration of Cu, Cd, Zn and Mg before and after adsorption process was Atomic Absorption Spectrophotometer (AAS).

Adsorption of Cd^{2+}, Cu^{2+}, Zn^{2+} and Fe^{3+} with activated carbon showed the various adsorption efficiency and capacity levels. All the adsorption efficiency measurement showed that the use of activated carbon is very significant to reduce the metal solutions content of samples due to higher efficiency than 90%. Additionally, levels of adsorption capacity are very randomly each other because of undefined relation between pore and ion particle size.

Table 2. Adsorption capacity and efficiency of activated carbon with 5% HNO_3

Type of ions	Adsorption capacity (mg/g)	Adsorption efficiency (%)
Cd^{2+}	0.1170	92.5436
Cu^{2+}	0.6858	97.5707
Zn^{2+}	0.2471	97.9811
Fe^{3+}	0.5116	96.2066

Variations of concentrations of wastewater showed a linear relationship between the initial concentration and the adsorbed concentration. The results showed that the greater the concentration of metal ions in the wastewater, the greater amount of metal ions adsorbed as shown in table 2. This is consistent with the theory that the greater the solute concentration in the solution, the larger amount of solvated substances are adsorbed to reach a certain equilibrium.

To find the adsorption characteristics, two treatments were applied to activated carbon in the relation between log Xm/m vs log Ce. Xm/m describes the adsorbed metal concentration every gram of activated carbon and Ce is the final concentration after adsorption process. These data will examine Langmuir and Freundlich models of adsorption isotherm. This differentiation is to find the pattern of mono or multi-layer of adsorption. The analysis of adsorption pattern of Cd, Cu, Fe and Zn can be concluded by looking at the determination coefficient of Freundlich and Langmuir adsorption isotherm curves at 5% significant level. Thus, it should be about 0.95 of R^2 coefficient for each adsorption. Determination the adsorption isotherm graph of Freundlich and Langmuir were investigated by the data in table 3.

Table 3. The relation between Log Xm/m vs Log Ce of activated carbon by 5% HNO_3

Cd		Cu		Fe		Zn	
x	y	x	Y	x	y	X	y
log Xm/m	log Ce	log Xm/m	log Ce	log Xm/m	log Ce	log Xm/m	log Ce
-0.4944	-1.8182	0.5801	-0.70509	0.43364	-0.5439	0.04254	-
-0.1848	-1.2299	0.89244	-0.71265	0.75581	-0.5216	0.29747	-1.7865
0.1017	-0.7525	1.07148	-0.67572	0.94098	-0.5675	0.57346	-0.5937
0.3109	-0.449	1.19655	-0.5583	1.06965	-0.584	0.75396	-0.4881
0.4023	-0.4638	1,29354	-0.4661	1.16863	-0.5924	0.87422	-0.2886
0.5173	-0.239	1,37132	-0.31283	1.24943	-0.6191	0.98512	-0.4727

By put the relation of log Xm/m and log Ce, it resulted graph of Freundlich and Langmuir adsorption patterns. The graph of Freundlich model of Cd, Cu, Fe and Zn are shown in figure 1. Based on the figure 1, it is clearly seen that there were various linearity numbers. From the plotting of linearity formula, the R^2 of determination coefficient are 0.9194, 0.6895, 0.7032 and 0.7647 for Cd, Cu, Fe, Zn respectively so that none of R^2 measurement of Freundlich model at figure 1 exceed the linearity requirement at least 0.95.

IOP Conf. Series: Journal of Physics: Conf. Series **1156** (2019) 012001 doi:10.1088/1742-6596/1156/1/012001

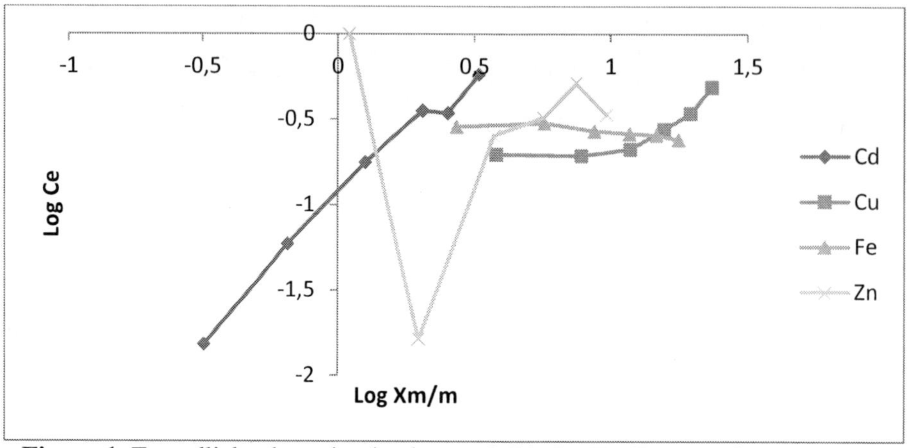

Figure 1. Freundlich adsorption isotherm of Cd, Cu, Fe, Zn on activated carbon.

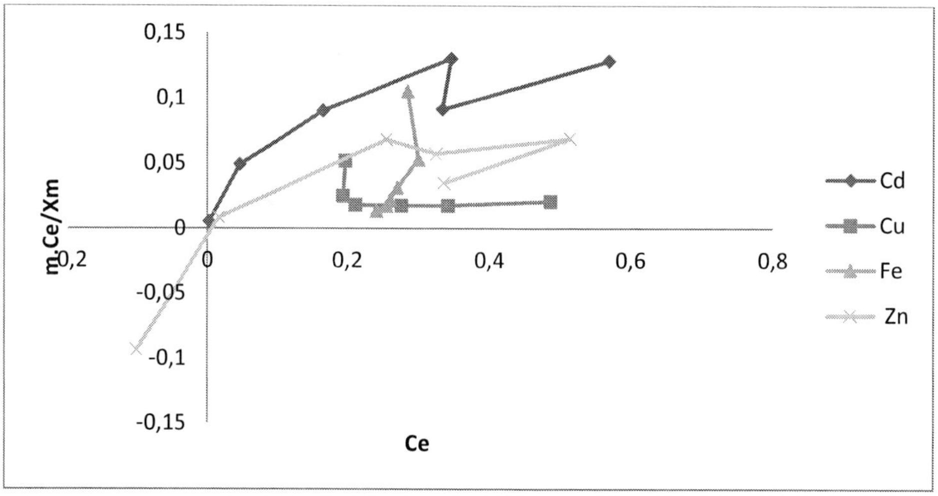

Figure 2. Langmuir adsorption isotherm of Cd, Cu, Fe, Zn on activated carbon.

On the other hand, plotting data of Langmuir model is given by figure 2. It is seen that the R^2 of determination coefficient are 0.7737, 0.1578, 0.7406, 0.5092 of Cd, Cu, Fe, Zn respectively. Compared to the previous isotherm models, the Langmuir are unacceptable to describe the adsorption pattern because of poorer determination coefficient than Freundlich. Hence, the adsorption pattern cannot be concluded to follow neither Langmuir nor Freundlich models.

All in all, the study of adsorption properties to wastewater solution only indicated the high efficiency of using activated carbon to reduce heavy metal level. Even though the capacity adsorption should be recalculated in further research, the efficiency adsorption gives high benefits in lowering level of Cd, Cu, Fe, and Zn in the wastewater.

4. Conclusion

The activated carbon produced from Screw-pine leaves remained a low quality of the properties of ash content, volatile content and carbon bond character of activated carbon based on SNI 06-3730-1995. However, water content, ash content, and absorption capacity of I_2 at activated carbon by 5% HNO_3 have accordance with the SNI standard. Additionally, the chemical and physical activation influence in making the surface area greater to 100 times porosity size of carbon pores. Thus, efficiency adsorptions are satisfy at 92.5436%, 97.5707%, 97.9811% and 96.2066% while adsorption capacities are 0.1170mg/g, 0.6858mg/g, 0.2471mg/g, and 0.5116mg/g of Cd, Cu, Zn and Fe respectively.

5. References

[1] Owolabi and Megat-Yusoff 2018 *J. Material Sci. Eng.* **7**(1) 1000425
[2] Wan Ngah W S and Hanafiah M A K M 2008 *Bioresource Technology* ***99*** 3935–48
[3] Winarno F 2015 *Kelapa Pohon Kehidupan* (Yogyakarta: Gramedia Pustaka Utama)
[4] Laeliocattleya R A, Muchlisyiyah J and Putri W D 2017 *Kimia Fisik Pangan* (Malang: Universitas Brawijaya Press)
[5] Sivaraj R, Rajendran V and Gunalan G S 2010 *E journal of chemistry* ***7***(4) 1314–19
[6] Alfiany H, Bahri S and Nurakhirawati 2013 *Journal of Natural Science* ***2***(3) 75–86
[7] Tao X U and Xiaoqin L I U 2008 *Chin. J.Chem.Eng.* **16**(3) 401–6
[8] Anirudhan T S and Sreekumari 2011 *J.Environ. Sc.* ***23***(12) 1989–98
[9] Gusmailina 2010 *Jurnal Penelitian Hasil Hutan 28*(**I**) 66–76
[10] Pari G, Sofyan K, Syafii W and Buchari 2005 *Jurnal Penelitian Hasil Hutan* **23**(3) 207-18
[11] Prasetyo I, Rochmadi, Teguh A and R Yunanto 2013 *Indo. J. Chem.* **13**(2) 95-100
[12] Meisrilestari Y, Khomaini R, and Wijayanti H 2013 *Konversi* ***2***(1) 46–51
[13] Smisek M and Cerny S 1970 *Active Carbon* (II) (Amsterdam: Elsevier Publishing Co.)
[14] Kristianingrum S, Siswani E D and Fillaeli A 2014 *Jurnal Sains Dasar* ***3***(1) 48–55

Acknowledgment
Thanks to Rector of UniversitasNegeri Yogyakarta who funded this research by DIPA scheme for Mathematics and Sciences faculty.

Preparation and characterization adsorbent based on zeolite from Klaten, Central Java, Indonesia

D W Astuti[1,2*], Mudasir[1] and N H Aprilita[1]

[1]Chemistry Department, Faculty of Mathematics and Natural Sciences, Gadjah Mada University, Yogyakarta 55281, Indonesia
[2]Medical Laboratory Technology Department, STIKES Guna Bangsa Yogyakarta, 55283, Indonesia

*E-mail: dian_wuriastuti@mail.ugm.ac.id

Abstract. The aim of this study was to prepare and characterize a potential adsorbent based on zeolite from Klaten, Central Java, Indonesia. The pre-treatment of natural zeolite (ZA) consists of chemical activation using HCl (ZAA) and NaCl (ZAT). Cationic surfactant benzalkonium chloride was used for modification of the zeolitic surface. ZA, ZAA, ZAT, and modified zeolites were characterized by FTIR, XRD, and AAS. The cation exchange capacity (CEC) of the natural zeolite measured by 1 M NH_4OAc method. Results show that the higher Si/Al ratio was achieved when natural zeolite was activated by the strong acid of HCl, but acid activation did not change the structure of the zeolite itself. The result of XRD analysis supports the AAS analysis result in which the degrees of crystallinity of zeolite increased after acid activation. The studies by FTIR and XRD analysis indicated that the zeolite structure remained unchanged after the modification with the cationic surfactant. Thus the activated and modified zeolites create a new material adsorbent to reduce pollutants in the environment.

1. Introduction

Indonesia has a large number of minerals including natural zeolite. Zeolites are crystalline hydrated aluminosilicate material with three-dimensional frameworks that based on TO4 tetrahedra, where T is an aluminium or silicon atom (phosphorus in aluminophosphates) arranged regularly in which each oxygen atom is divided between two tetrahedra [1]. Zeolites have a great potential as effective sorbents for water treatment applications, such as dyes, heavy metals, and other pollutants from wastewater and natural waters. Zeolite modification extends obviously the zeolites application. The surface properties of zeolites may be greatly modified with cationic surfactant by ion-exchange reactions [2].

To improve ion-exchange capacities, it is important to modify their chemical structure and composition by applying a suitable pretreatment. In this paper, the natural zeolite (ZA) was activated with different activating agents. The objectives of this study are determining the Si/Al ratio of the zeolite after HCl or NaCl pretreatment. HCl is frequent reagent used to activate the natural zeolite. The ability of HCl-activated natural zeolite was higher than the natural zeolite without activation that is used to remove of rhodamine B from aqueous solution [3]. Ion exchange with sodium chloride solution is widely employed because sodium ions can readily remove other cations and therefore enhancing the adsorption capacity.

This research was aimed to characterize the zeolites by using of NaCl and HCl to activate the natural zeolite. A negatively charged framework and its hydrophilic properties cause the zeolite to be inappropriate for the binding of anionic and hydrophobic species [4]. However, adsorption of surfactants can modify the properties of the zeolite surface [5]. Therefore, this adsorbent has a good affinity toward inorganic pollutants, but little or no ability to remove organic pollutants. The surface properties of zeolites may be greatly modified with cationic surfactant by ion-exchange reactions. Currently, many research studies have investigated the modified zeolite using cetyl trimethyl ammonium bromide (CTAB), hexadecyl ammonium bromide (HDTMA), and sodium dodecyl sulfate (SDS) [6-7]. By the addition of cationic surfactants, the zeolite surface can be modified to improve the anion exchange capacity and, to an extent, the potential to eliminate cations, anions, and organic compounds [8-10].

For this purpose, modification of the zeolite surface was performed using cationic surfactant benzalkonium chloride, in which each molecule is composed of a hydrophilic and positively charged head group and a hydrophobic tail. Benzalkonium chloride may form monolayers and/or bi-layers on the zeolite surface [11].

2. Experimental Section

2.1. Materials and reagents

Natural zeolite (ZA) from Klaten Central Java Indonesia was used as starting material. Reagents of analytical purity NaCl, HCl, benzalkonium chloride were used on the activation and modification, respectively. Benzalkonium chloride supplied by Sigma. Aquabides was used to the solvent for making the chemicals solution. The chemical structure of quaternary ammonium salt benzalkonium chloride is shown in figure 1 was used as a cationic surfactant.

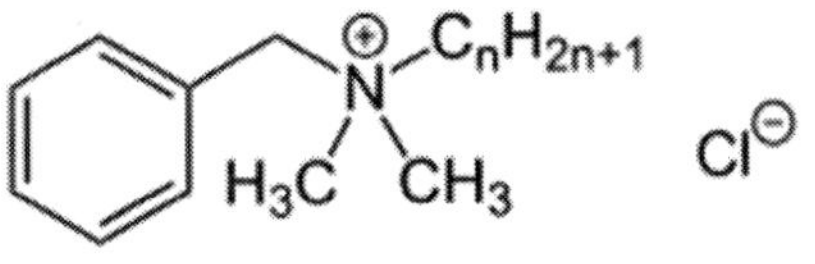

n=8,10,12,14,16,18

Figure 1. Chemical structure of benzalkonium chloride

2.2. Preparation and activation of the natural zeolite

ZA was crushed and sieved to yield particle 250 mesh. Some 50 g of ZA was weighed, then washed and dried in an oven at 80 °C for 24 hours. Washing and drying step were carried out prior to characterization of ZA chemical properties. Atomic Absorption Spectroscopy (AAS) was utilized for revealing the composition of the cation oxide chemical compound in the ZA [12]. The result of AAS analysis was used to calculate the Si/Al ratio inside the ZA. X-Ray Diffraction (XRD) powder diffraction was used to confirm the composition analysis result. Characterization of functional groups of ZA was used Fourier Transform Infrared Spectroscopy (FTIR). Activation of the zeolite was carried out by contact of the powdered material with an aqueous solution of NaCl and HCl. Solutions of HCl 3 M and NaCl 0.1 M were prepared as an activating agent.

A known quantity of ZA (1 g) was mixed with 100 mL of NaCl solution 1 M for 24 hours, at room temperature. Then, the suspension was filtered and the material washed with aquabides. The wet activated material was dried at 105 °C for 24 hours and used in the modification step. The product of this process is called the activated zeolite (ZAT).

Afterward, The ZA was submerged in 150 mL HCl solution 3 M and dried for 30 minutes at 72 °C. The mixture was filtered and washed with aquabides until neutral (pH=7), dried in an oven at 110-120 °C for 3 hours. Then, the product was diluted with 100 mL of NaCl solution 1 M and stirred for 24 hours at room temperature. The suspension was filtered with aquabides and dried at 105°C. The product of this process is called activated zeolite (ZAA).

Studies of cationic exchange capacity (CEC) of the ZA, ZAT, and ZAA used the reagent of analytical purity NH_4OAc. The mineralogical compositions of the zeolite were determined by X-ray diffraction analyses (XRD) with an automated Rigaku multiflex diffractometer with Cu anode using Co Kα radiation at 32 kV and 20 mA over the range (2θ) of 2-65° with a scan time of 4 °/min. The crystalline phases present in the zeolitic materials were identified with the help of ICDD (International Centre for Diffraction Data).

2.3. Modification of the activated zeolite

The modification of the ZAT and ZAA was performed contacting the powdered material samples with benzalkonium chloride (BC). Thus, 1 g of the ZAT was contacted with 100 mL of the BC solution equivalents of twice as of its CEC for 24 hours at room temperature. The suspension was filtered, washed with aquabidest, and dried at 60°C for 24 hours [11]. The similar process was repeated by substitution of ZAT with ZAA. The obtained products denoted as ZATB and ZAAB. Both ZATB and ZAAB will be examined by XRD and FTIR analysis.

FTIR spectra of the natural zeolite, activated zeolite, and modified zeolite was recorded in the range of 400-4000 cm^{-1} using a Shimadzu FTIR-8201 PC. The zeolitic materials were dispersed in KBr and compressed into disks.

3. Results and Discussion

3.1 Characterization by Atomic Absorption Spectroscopy (AAS)

The chemical composition of the ZA, ZAT, and ZAA determined by AAS is shown in table 1. Natural zeolite (ZA), besides contain the negative charge compensator such as alkali and alkaline earth cations, it also contains the undesired component such as oxides of metals. The metal oxides are trapped in the frame of zeolite when it is formed naturally. The existence of those oxide contaminants could decrease pore sizes and thereby decrease the ability of zeolite matrices as an adsorbent. By activation, those contaminants would be discarded the zeolite matrices.

Table 1. Chemical composition of the sample.

Components	ZA (%)	ZAT (%)	ZAA (%)
Si	27.96	26.30	28.34
Al	9.20	9.00	8.52
K	0.75	0.50	0.55
Na	1.19	1.70	1.80
Ca	0.83	0.70	0.60
Si/Al ratio	3.04	2.92	3.33

The characterization results of the samples with different activating agent and concentration show a significant change after the activation. The AAS analysis results have been used to calculate the Si/Al ratio as reported in table 1.

The Si/Al ratio of the ZAA increase was also followed by the increase of adsorptivity because the releasing of Al from the active site Si-O-Al caused the transformation of the active site into Si-O-Si in which the Si was coming from the outer structure. But, keep in mind that the zeolite tends to be a hydrophobic adsorbent when the Si/Al ratio increased. The formation of the localized electrostatic pole between cation and anion resulted in hydrophilic characteristic to adsorb polar pollutant well. The releasing of aluminium (dealumination) would reduce the number of cation inside the zeolite structure and reduced the hydrophilic characteristic of the zeolite. The removal of the Al atom from ZA framework increases the Si/Al ratio of ZAA. Characterization of ZA, ZAT, and ZAA using AAS (Table 1) shows that the Si/Al ratio after the acid activation increase from 3.04 to 3.33 for ZA and

ZAA respectively. The Si/Al ratio for zeolites is associated with the cation exchange capacity. ZA showed low values of Si/Al ratio (92.80 me/100 g) indicating materials with high cation exchange capacity [13]. The CEC values of the ZAT (99.09 me/100 g) were higher than ZAA (54.40 me/100 g). This result indicated that negative charge of zeolite decrease thus capability to exchange the cation decrease too.

3.2. Characterization by Fourier Transform Infrared Spectroscopy
The natural zeolite and activated zeolite were characterized by FTIR spectroscopy, and their IR spectra are shown in figure 2. The features positioned at 3441.0 cm^{-1} is characteristic bands of the mordenite connected to the vibration of the hydrogen-bonding hydroxyl groups. A band at 1635.64 cm^{-1} is associated with the deformation vibration of the absorbed water, whereas the band at 1049.28 cm^{-1} is related to the asymmetric valence vibration in the SiO_4 tetrahedra [11]. No variations in the frequency of these bands in the spectra of ZA, ZAT, and ZAA samples were observed, confirming that differential activation of zeolite does not affect the zeolite structure.

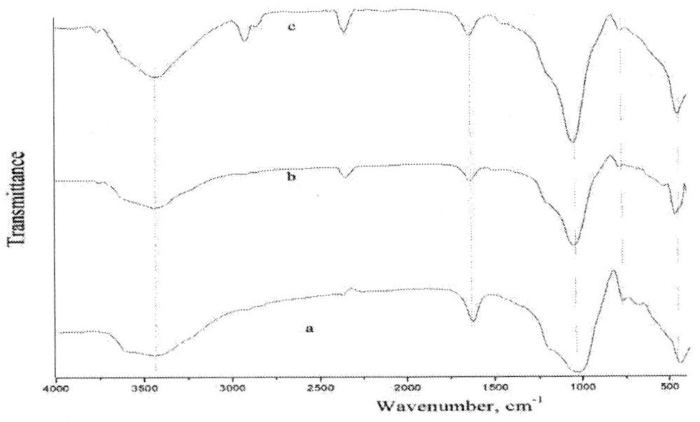

Figure 2. FTIR spectra of ZAT (a), ZA (b), ZAA (c)

However, three new bands appear, implying the presence of BC on the composite, two bands assigned to the C-H stretching vibrations of the hydrocarbon chain, 2924.09 and 2854.65 cm^{-1}, and a third one band corresponding to the C-H bending of the methyl and methylene groups at 1465.9 cm^{-1}. The IR spectra for modified zeolites are shown in figure 3.

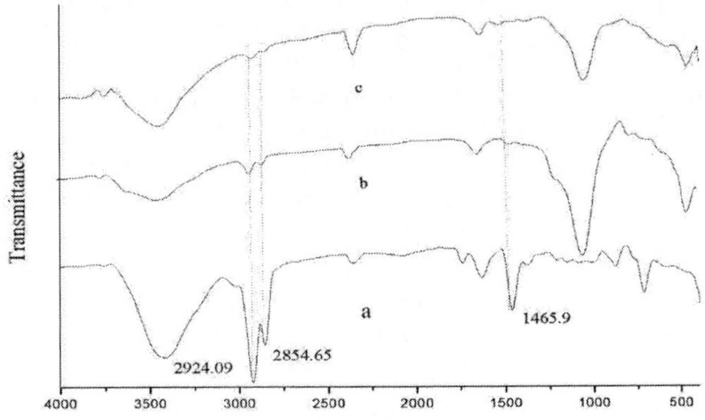

Figure 3. FTIR spectra of BC (a), ZATB (b), ZAAB (c)

No relevant variations in the frequency of the band assigned to the ZAA and ZAT after the treatment with the BC were observed, which indicates that the zeolite structure remains unaltered after the modification.

3.3. *Characterization by X-ray Diffraction (XRD) analysis*
Diffractogram of natural zeolite results in table 2 shows the intensity at an angle 2θ = 25.66 Å, 22.24 Å and 26.68 Å. There is a maximum peak at an angle of 2θ = 25.66 with a distance of 3.47Å. These results provide diffraction pattern shown in the display figure 4(c).

Table 2. Results of X-ray Diffraction analysis of natural zeolite

2θ	d(Å)	I/I1	Intensity (Counts)
9.66	9.15	65.6	223
22.24	3.99	74.4	253
25.66	3.47	100	340
26.68	3.34	71.2	242
27.74	3.21	52.6	179

X-ray diffractograms of natural zeolite, activated zeolites, and modified zeolites are shown in figure 4.

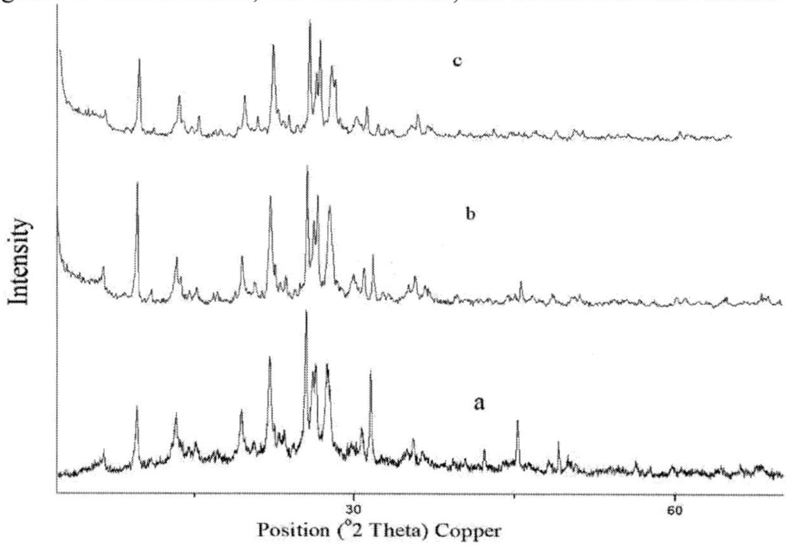

Figure 4. XRD graph of zeolite; (a) ZAA, (b) ZAT, (c) ZA

Figure 4 showed that the natural zeolite and activated zeolite composed of crystalline structures. The powder contains a mixture of zeolite mineral groups and the inert minerals. Mordenite was one of zeolite forming minerals found in the zeolitic stone spread in Central Java Indonesia. From diffractogram resulted from XRD analysis, this mineral had the peak value of d 13.89 Å, 9.15 Å, 6.61 Å, 3.47 Å, 3.38 Å and 3.21 Å. Mordenite had such physical property as the orthorhombic structure with 28% void volume. Other zeolite forming minerals existed in diffractogram were albite ($NaAlSi_3O_8$), muscovite ($KAl_2Si_3AlO_{10}(OH)_2$), pyrite (FeS_2), chlorite ($(Mg,Al)_6(Si,Al)_4O_{10}(OH)_8$), and quartz ($SiO_2$). Based on the XRD analysis, mordenite zeolite type was the major component in the zeolite. Mordenite zeolite type is a common zeolite found in silica-rich rocks [14].

The pretreatment by HCl and NaCl sharpened the peak in the spectra as shown in figure 4. Thus, this process can increase the crystallinity of zeolite. Application of HCl and NaCl for zeolite activation cause the angle of the peak to become sharper (the angle got smaller). The figure 4 shows that there was no significant change in the XRD patterns before and after activation.

4. Conclusions
Natural zeolites from Klaten Central Java Indonesia have, in its crystalline structure of mordenite. Activated and modified zeolite mordenite with BC is prepared without altering the starting zeolite structure. The adsorption potential of the Klaten natural zeolites can be studied by HCl and NaCl activation. The acid activation of ZA could increase the Si/Al ratio and decrease the CEC which correspond to the low hydrophilic characteristic of the zeolites.

Acknowledgment
The authors are grateful for the financial from BUDI-DN. More, the authors thank all colleagues for supporting this study at the Department of Chemistry Gadjah Mada University and STIKES Guna Bangsa Yogyakarta.

References
[1] Vyas R K, Shashi and Kumar S 2004 *Indian J. Chem Techn* **11** 704-709
[2] Tharcila C R, Juliana de C I, Raquel R A, Lucas C G, Denise A F 2015 *Acta Velit* **1** 79
[3] J Lin, Y Zhan, Z Zhu and Y Xing 2011 *J. Hazard. Mater.* **193** 102-111
[4] Haynes W M 2012 *CRC Handbook of Chemistry and Physics* (Florida: CRC Press) pp 2143 - 2184
[5] H Guan, E Bestland, C Zhu, H Zhu, D Albertsdottir, J Hutson, C T Simmons, M Ginic-Markovic, X Tao, A V Ellis 2010 *J. Hazard. Mater.* **183** 616
[6] Huijuan Wei, Sujuan Xie, Kefeng Liu, Wenjie Xin, Xiujie Li, Shanhe Liu, Shunming Gu, Shenglin Liu, Longya Xu 2015 Chinese J. Catal. **36** 1766-1776
[7] Xiaoying Jin, Ming-qin Jiang, Xaio-quan Shan, Zhi-guo Pei, Zuliang Chen 2008 J. Colloid Interf. Sci. **328** 243-247
[8] Apreutesei R E, Catrinescu C, Teodosiu C 2008 Environ. Eng. Manag. J. **7** 149-161
[9] Gupta P, Khanday W A, Majid S A, Kushwa V, Tomar S S, Tomar R 2013 J. Environ. Chem. Eng. **1** 510-515
[10] Misaelides P 2011 Microporous Mesoporous Mater. **144** 15-18
[11] Jevtić S, Grujic S, Hrenovic J and Rajic N 2012 *Micropor. Mesopor. Mat.* **159** 30–35
[12] Deer W A, Howie R A and Zussman J 1992 *An Introduction to The Rock-forming Minerals* (New York: John Wiley)
[13] J C Izidoro, D A Fungaro F S, Santos and S Wang 2012 *Fuel Process. Technol.* **97** 38-44
[14] Ostrooumov M and Corona-Chavez P 2003 *Rev. Mex. Cienc. Geol.* **20** 133-138

IOP Conf. Series: Journal of Physics: Conf. Series **1156** (2019) 012003 doi:10.1088/1742-6596/1156/1/012003

Adsorption of HA (humic acid) using sulfuric acid-crosslinked chitosan/pectin polyelectrolyte complex film

D Siswanta*, F Farida, D Zunaim and N H Aprilita

Department of Chemistry, Universitas Gadjah Mada, Sekip Utara, Yogyakarta, Indonesia

*E-mail: dsiswanta@ugm.ac.id

Abstract. A sulphuric acid-crosslinked chitosan/PEC film had been synthesized and applied for adsorption of HA from aqueous solution. The parameters for adsorption study were an optimum ratio of chitosan/pectin composition, contact time, the effect of the solution pH, and HA concentration. The model for adsorption kinetics and isotherm adsorption as well as desorption study were also conducted. The result showed that the film was durable in water at pH 2 to 12. The analysis using FTIR indicated a shift of absorption peak at 1527 cm^{-1} (bending vibration N–H in $-NH_3^+$). Analysis with SEM showed the morphological differences of the film before and after adsorption and after desorption. Adsorption reached its optimum capacity in the film with a ratio of chitosan and pectin composition 70:30, at pH 6, contact time of 480 mins and HA concentration of 450 mg L^{-1}. The kinetic model of HA adsorption could be explained by the second-order kinetic model of Ho & McKay and the adsorption isotherm follows the Langmuir isotherm model with a maximum adsorption capacity of 50.0 mg g^{-1}, the Langmuir constant (K_L) of 1.49×10^5 L mol^{-1} and the adsorption energy was found to be 29.7 kJ mol^{-1}. The highest desorption was 82.5% using 1 M NaOH.

1. Introduction

Water is a very important component of human survival and is needed sustainably. Efforts were required to provide clean water that meets health requirements such as removal of compounds that cause water to become smelly, taste and color. According to the World Health Organization [1], the color in raw water for drinking water is usually due to the presence of colored organic matter (mainly humic acid (HA) and fulvic acid (FA)) which is combined with the humus fraction on the soil. The density of colors derived from humic compounds can also indicate a high tendency to produce side products from the disinfection process.

Trihalomethane is an example of Disinfection By-Products (DBPs) as a result of the reaction of humic substances with chlorine in water treatment. Humic substances also play a significant role in increasing the transport of hydrophobic or heavy metal organic contaminants which bind to HA, causing bacterial growth in water distribution systems due to their role as providers of food sources and causing unpleasant and colored taste in drinking water[2].The presence of HA certainly can reduce the quality of water to be consumed and endanger the health. Its effects on human health must be prevented by removing organic compounds from water before being supplied for the chlorination process [3].

Various techniques for removing HA in water have been carried out such as coagulation [4], photocatalysis [5-6] and adsorption using activated carbon [7]. The disadvantage of the coagulation technique is that it produces large volumes of mud and requires high operational costs [2]. Photocatalytic techniques require photocatalysts whose prices are relatively expensive and relatively simple equipment.

Another technique is Fenton oxidation and photocatalysis with TiO_2 which is also efficient in decomposing HA but requires the addition of several Fenton reagents or nanomaterial into the water to initiate chemical reactions that might provide secondary pollution as micro-pollutants of water sources [8].

The adsorption process is considered the best method because of its simplicity, high efficiency and ease of operation [9]. Adsorption also does not produce DBPs or large volumes of mud so that it can be used as an alternative technique for removing HA from the water. Adsorption with activated carbon which has been recommended by the USA Environmental Protection Agency turned out to be inefficient and practical for large molecular organic compounds such as natural organic substances [3]. The adsorbent is quite expensive and for higher quality, the cost is also greater. Therefore, to date, a number of studies have been conducted to develop cheaper and more effective adsorbents including natural polymers such as chitosan [10].

Chitosan is a natural polymer and is obtained from deacetylation of chitin which is the second most abundant biopolymer in nature. Chitosan shows chemical stability and high reactivity, but has several disadvantages such as swelling on acidic media, tends to agglomerate, low specific surface area, poor mechanical properties [11], less hydrophilic and low degradability [12] thus limiting its application as an adsorbent. However, the weakness of chitosan (low hydrophilicity and degradability) and pectin (very good solubility in water) can be overcome through the formation of PEC so that it can be used for various applications. Furthermore, the modification of chitosan membranes by adding crosslinking agents can increase swelling property and mechanical strength in the application[13].

In this study, we developed a sulfuric acid-crosslinked chitosan/pectinPEC film as an adsorbent for HA in a batch system. Good hydrophilicity and free protonated amine groups that can interact with the adsorbate become the main active sites so that the adsorption capacity increases. The parameters studied in the film-making process are the mechanical properties and stability of the film in the acid-base medium. The adsorption study includes the effect of chitosan/pectin composition ratio, contact time, solution pH and HA concentration on the adsorption ability of chitosan/pectinPEC films. The desorption aspect studied was the effect of NaOH concentration on percent desorption of films.

2. Experimental Section

2.1. Material

The materials used in this study were pro analytical quality including chitosan with 87% deacetylation (produced by Bogor Agricultural University) and pectin from Chemix Yogyakarta.CH_3COOH solution, 96% H_2SO_4 solution, NaOH, KCl, NaCl, Na_2HPO_4, KH_2PO_4 (Merck), HAsodium salt (Sigma Aldrich) were used as received.

2.2. Instrumentation

FTIR spectrophotometer (Shimadzu FTIR PRESTIGE 21) and Scanning Electron Microscope/SEM (JEOL JSM-6510LA) were used for characterization of the film. UV-Vis Spectrophotometer (Thermo Scientific Evolution 201) was used for determination of HA concentration.Other instruments include pH meter (BNQ PH-777) and a sonicator (Branson 3210).

2.3. Procedure

2.3.1 Synthesis of the film. Pectin was dissolved with a mass of 0.03 g using 4 mL of distilled water and stirred to obtain an homogeneous solution. About 0.07 g of chitosan was added and stirred vigorously until the chitosan was dispersed in the solution, then an acetic acid solution (0.4 M, 16 mL) was added to the mixture with stirring. The solution was sonicated to ensure that chitosan and pectin were mixed homogeneously. The solution was poured into a petri dish and dried in an oven at 70 °C overnight. The film formed was immersed in 1 M NaOH solution for 4 hours until it was peeled off, washed using distilled water and phosphate buffer. The film was then immersed in 0.1 M H_2SO_4 solution for 2 hours and dried. The same procedure was performed for film synthesis with the ratio of 80:20 chitosan-pectin (0.08 g chitosan and 0.02 g pectin) and 90:10 (0.09 g chitosan and 0.01 g pectin).

2.3.2 Characterization of chitosan/pectin PEC films crosslinked sulfuric acid. Functional groups in the film were characterized by FTIR spectrophotometer at 4000-400 cm^{-1}. Surface morphology of the film was analyzed using SEM for the film before and after adsorption and after desorption. The swelling property of the films was examined by weighing the film after immersed in water for 2 h, and % swelling was calculated relative to the weight of the dry film. The resistance of films in acid and the alkaline medium was tested by soaking the film in the solution of various pH (2-12) for 24 h.

2.3.3 Adsorption study. Films (0.04 g) were put into a 30 mL of HA solution (100 mg L^{-1}) and shaken out for 6 h. The film was separated from the solution and the remaining concentration of the HA solution was measured using a UV-Vis spectrophotometer at its maximum wavelength. The variables examined were the mass composition of chitosan/pectin, optimum contact time, optimum pH, and HA concentration.

2.3.4 Desorption Study. The desorption process was studied using NaOH solution of 1.0, 0.1 and 0.05 M. The film that had adsorbed HA was soaked in NaOH solution at various concentrations with a volume of 30 mL and was shaken for 6 h. The HA concentration in the filtrate obtained was determined by a UV-Vis spectrophotometer to estimate the concentration of the remaining HA in the film.

3. Discussion

3.1. Film Characterization

Qualitative analysis using FTIR indicating functional groups of chitosan, pectin, and chitosan/pectinPEC films crosslinked sulfuric acid are shown in figure 1. The main absorption peak of chitosan is shown at 1651 cm^{-1} (C=O amide) and 1605 cm^{-1}(N–H). For pectin, the absorption peaks at 1743 cm^{-1} (C=O carboxylate methyl ester) and 1635 cm^{-1} (–COO^-) were observed. After crosslinking using H_2SO_4, some shifted peaks were observed at 1527 cm^{-1}(N–H of–NH_3^+), 1628 cm^{-1} (–COO^-) and 1720 cm^{-1} (C=O carboxylate methyl ester).

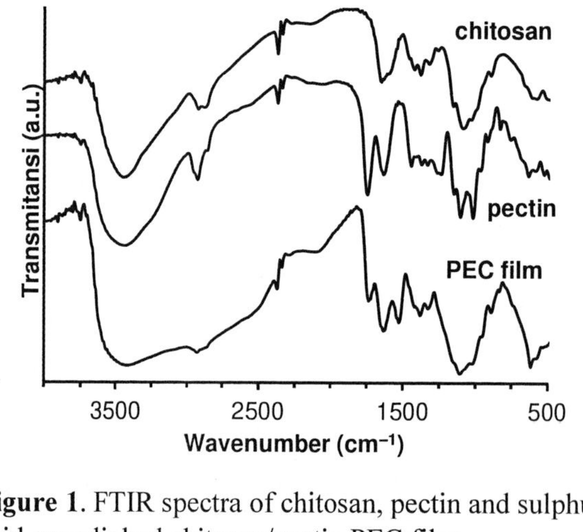

Figure 1. FTIR spectra of chitosan, pectin and sulphuric acid-crosslinked chitosan/pectin PEC film

IOP Conf. Series: Journal of Physics: Conf. Series **1156** (2019) 012003 doi:10.1088/1742-6596/1156/1/012003

The shift is in accordance with the results of Chen et al. (2010) that showed the peaks of 1564 cm^{-1} (chitosan) and 1611 cm^{-1} (pectin) shifted to 1598 cm^{-1} on chitosan/pectin composite material. The peak shift was explained as the result of a strong ionic interaction of the amine group from chitosan with the carboxyl group of pectin so the C = O peak to be covered. Peaks at 1103 and 617 cm^{-1} were the contributions of SO_4^{2-} ions that confirm the effectiveness of the crosslink reactions and are consistent with previous studies [14-15].

3.2 The surface morphology of chitosan/pectinPEC films crosslinked sulfuric acid

The morphological analysis of the surface of the film using SEM was carried out on films that before and after HA adsorption as well as after the desorption process. The results of the SEM analysis of the film are presented in Figure2. The surface of the film after the adsorption was smoother than before the adsorption which indicates that the film has successfully absorbed HA. The film after HA desorption (Figure 2.c) contains particles scattered almost throughout the surface of the film that probably was NaOH crystals formed after the desorption process.

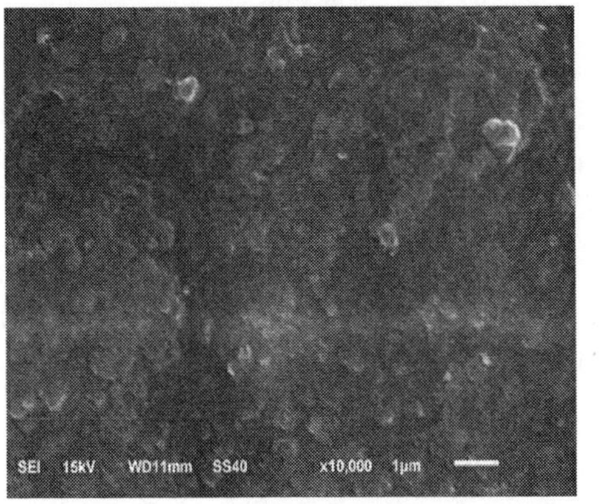

(a)

(b)

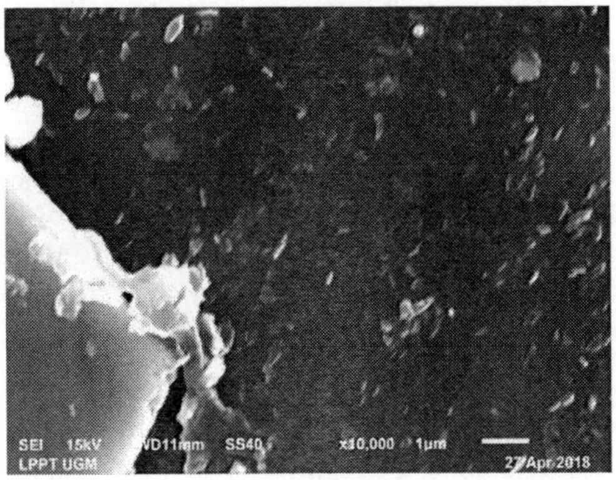

(c)

Figure 2. SEM images of chitosan/pectin PEC film crosslinked sulfuric acid with a magnification of 10,000 times (a) before adsorption, (b) after adsorption and (c) after desorption.

3.3 Swelling of sulfuric acid-crosslinked chitosan/pectinPEC films

The swelling test was conducted to determine the hydrophilicity properties of the film based on the ratio of the composition of chitosan and pectin and the results are presented in figure 3. The result shows that the greater the composition of chitosan the lower the water absorption ability of the film as the hydrophilicity of the film is also lower.

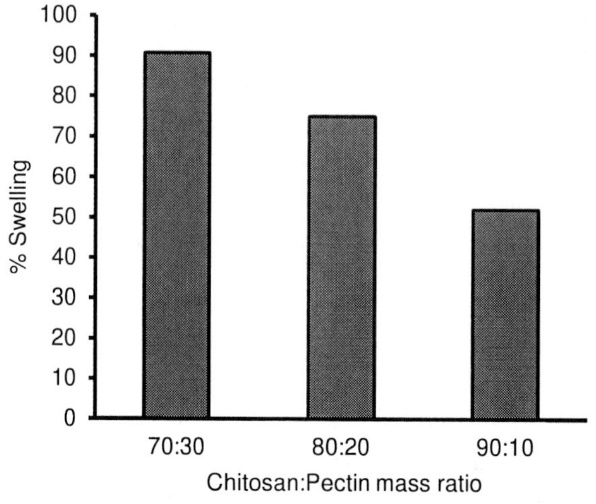

Figure 3. Effect of the ratio of film composition on the results of the film development test

Pectin plays an important role related to film hydrophilicity because of the abundant content of the carboxyl group ($-COO^-$) [12], that may be responsible for binding of water molecules. Therefore, the water absorption capacity in sulfuric acid-crosslinked chitosan/pectinPEC films is mainly influenced by the presence of pectin in the film.

3.4 The resistance of films in acid-base medium

Testing of the acid to alkaline medium aims to determine the stability of the film. Overall, sulfuric acid-crosslinked chitosan/pectin PEC films are resistant to acid-base medium. The chitosan amine group ($-NH_2$) will be protonated to be $-NH_3^+$ group at pH <6.4 [16] and will interact electrostatically with the pectin carboxyl group ($-COO^-$) to form a strong PEC. According to Ngah et al. Ionic bonds formed between chitosan protonated amines ($-NH_3^+$) and SO_4^{2-} ions improve chemical stability [15]. This makes the film stable at acidic pH. At pH 7-10, even though the protonated amine group is in a lesser amount, the film is still stable. However, at pH 11-12, the film start to partially dissolve because of weakening the interaction between $-NH_3^+$ and $-COO^-$ groups.

3.5. Adsorption Study of Films against HA Solutions

3.5.1 Determination of film composition ratio. The effect of the ratio of the composition of chitosan and pectin of the film on the adsorption of HA was examined and the result is shown in Figure 4. The highest adsorption capacity of 32.8 mg g^{-1} was observed for the film with a composition ratio of chitosan : pectin 70:30. The result proves that the more mass of chitosan or the less mass of pectin, the lower the adsorption capacity of the film. The effect of the chitosan/pectin mass ratio on the film on the adsorption of HA can be attributed to its ability to absorb water. The adsorption process of HA from an aqueous solution into an adsorbent was assumed to consist of 3 steps, namely migration of HA molecules from

IOP Conf. Series: Journal of Physics: Conf. Series **1156** (2019) 012003 doi:10.1088/1742-6596/1156/1/012003

solution to the surface of the adsorbent, diffusion through the boundary layer to the inside of the adsorbent (interior) and adsorption site [17]. The high composition of pectin causes more anionic groups and more water absorption capacity [18] that will facilitate the process of diffusion of HA into the adsorbent. Therefore, there is a positive correlation between water absorption (swelling) and HA adsorption capacity.

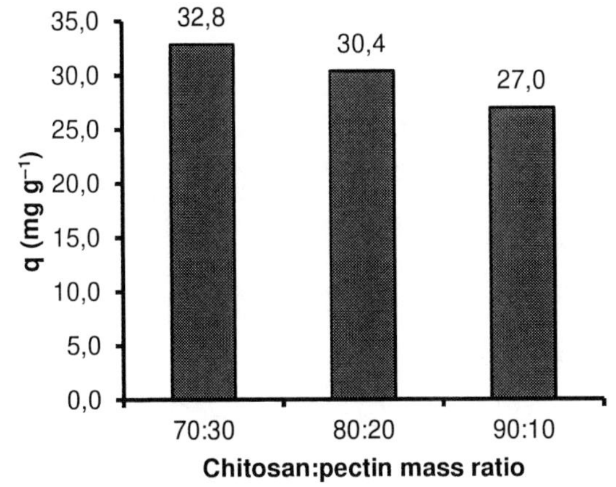

Figure 4. Effect of the mass ratio of the chitosan/pectin composition to the HA adsorption capacity.

3.5.2 Determination of optimum contact time and adsorption kinetics. The optimum contact time between the adsorbate and the adsorbent was recorded to determine the kinetic model of HA adsorption onto the sulfuric acid-cross linked chitosan/pectinPEC film. The results are presented in Figure 5.The main active sites ($-NH_3^+$) may be responsible for the most adsorption of HA through electrostatic interaction with the $-COO^-$ group of HA. In addition, weak interaction such as physisorption may also contribute to the HA adsorption.The data were examined for the pseudo-first-order and pseudo-second-order kinetic models and the result was summarized in Table 1. Based on R^2 value, the pseudo-second-order kinetic model that obtained by plotting the t versus t/q_t showed the best linearity and the calculated q_e was comparable to experimental value (33.1 vs 33.7 mg g^{-1}). This result supports the assumption that the adsorption of HA in chitosan/pectinPEC films should be a chemisorption.

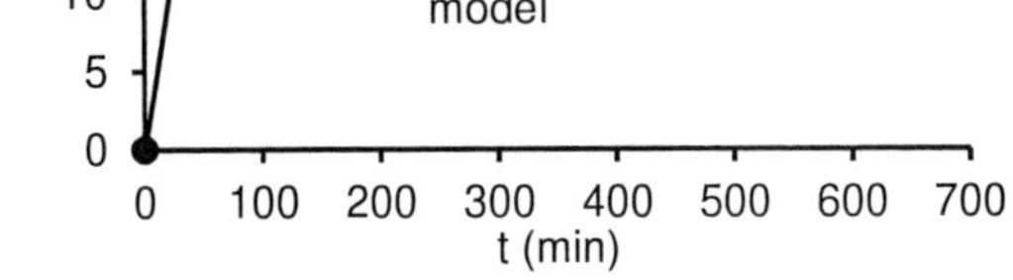

Figure 5. Effect of contact time on the adsorption of HA.

Table 1. Parameters of the adsorption kinetics model.

Kinetic Model	Parameter	
Pseudo-first order	R^2	0.4908
	k_1' (min^{-1})	2.40×10^{-3}
	q_e (mg g^{-1})	5.96
Pseudo-second order	R^2	0,9917
	k_2' (mg g^{-1} min)	1.64×10^{-3}
	q_e (mg g^{-1})	33.1

3.5.3 Determination of optimum pH. The HA adsorption on the sulfuric acid-crosslinked chitosan/pectinPEC films was examined in a pH range of 4 to 10 and the results are presented in Figure 6. The highest adsorption capacity of 37.1 mg g^{-1} was observed at pH 6. The adsorption capacity of HA solution increases at acidic pH and decreases at neutral to basic pHs. A solution with a low pH (3–6) produces a protonation of the amine group on which produce strong electrostatic interaction between the $-NH_3^+$ group of the chitosan and the $RCOO^-$ group of HA, and this resulted in a significant increase in the amount of HA adsorbed. The optimum pH value for adsorption of HA on chitosan/pectin film occurs at pH 5–7.5. Meanwhile, the carboxylic group begins to dissociate at pH 4-6 so it suggests that at pH 6 the electrostatic attraction between the deprotonated carboxylic group from HA and the protonated amine group from chitosan is the strongest so that the maximum adsorption capacity of HA is observed at that pH value [10].

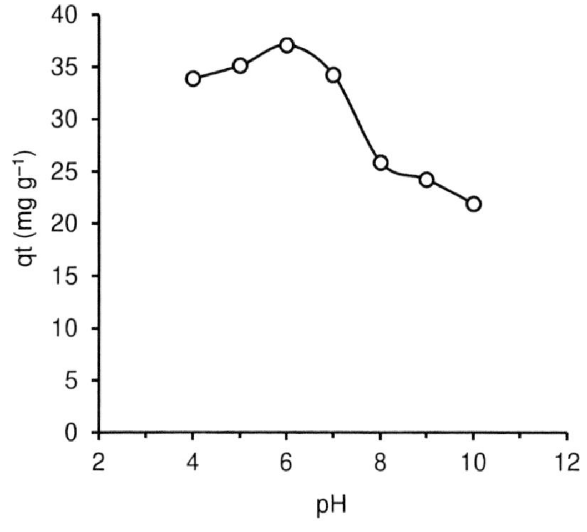

Figure 6. Effect of pH on the adsorption of HA onto sulfuric acid-crosslinked chitosan/pectin PEC films.

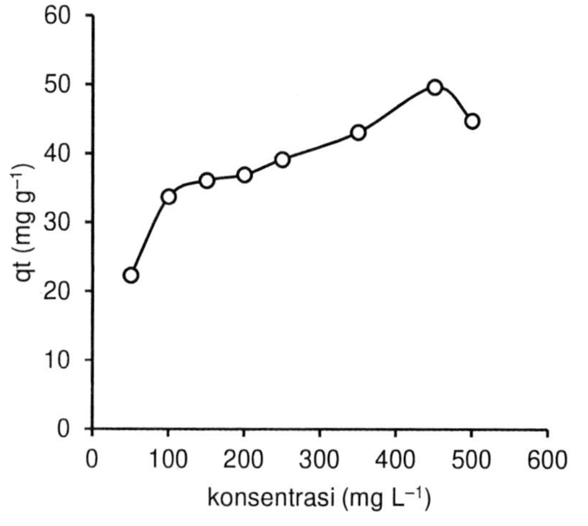

Figure 7. Effect of initial concentration of HAon the adsorptionof HA onto sulfuric acid-crosslinked chitosan/pectinPEC films.

3.5.4 Determination of optimum HA concentration. The effect of the initial HA concentration on the HA adsorption capacity of the sulfuric-acid crosslinked chitosan/pectin PEC film was examined to determine the appropriate adsorption isotherm model. The result is shown in Figure 7 that indicated the increase of the adsorption capacity until it reaches its optimum point. The saturation point occurs when all the active sites of the adsorbent have bonded to the adsorbate and the adsorption capacity reaches its optimum point. Adsorption isotherms are related to how HA interacts with chitosan/pectin PEC films crosslinked with sulfuric acid. Table 2 shows the Freundlich and Langmuir isotherm models for the data obtained in this experiment. As the value of R^2 in the Langmuir isotherm model is greater than that of the Freundlich isotherm model, the HA adsorption the sulfuric-acid crosslinked chitosan/pectinPEC film may follow the Langmuir isotherm model. The chitosan/pectin PEC film adsorbent has an homogeneous adsorption site ($-NH_3^+$). and adsorption of HA on this adsorbent is assumed to be a monolayer adsorption as the result of the interaction of the protonated amine group ($-NH_3^+$) from the chitosan and carboxylic group ($-COO^-$) from HA.

Table 2. Parameters of the Freundlich and Langmuir adsorption isotherm models.

Isotherm Model	Parameter	
Langmuir	R^2	0.9861
	q_{maks} (mg g^{-1})	50.0
	K_L (L mol^{-1})	1.49×10^5
	E (kJ mol^{-1})	29.7
	R_L	7.48×10^{-5}
Freundlich	R^2	0.9329
	K_F (L g^{-1})	11.6
	n	4.25

The maximum adsorption capacity obtained in this study is compared with the results of previous studies and presented in Table 3. The data show that the sulfuric acid-crosslinked chitosan/pectinPEC film exhibited a high HA adsorption capacity compared to the other reported adsorbent. The film has the potential as a good performance adsorbent for HA removal applications in waters.

Table 3. Comparison of the adsorption capacity of HA on different adsorbents

Adsorbent	q_{max}(mg g^{-1})	Reference
Activated carbon from husk ash	45.5	(Daifullah et al., 2004)
Chitosan-epichlorohydrin bead	44.84	(Ngah et al., 2008)
Activated sludge	2.78	(Jonglertjunya and Lertchutimakul, 2012)
ZnO nano particles	0.82	(Oskoei et al., 2016)

3.5.5. Desorption study. Desorption test is related to the film regeneration to be reused as an adsorbent. Good desorption performance of an adsorbent is important in relation to the potential for practical application [16]. The desorption process at this stage was done by interacting the chitosan/pectinPEC film crosslinked sulfuric acid that had adsorbed HA under optimum conditions with NaOH solution at various concentration. The result on the effect of NaOH concentration on percent desorption is presented in Figure 8. The highest percent desorption of 82.5% was observed when 1 M NaOH solution was used as a desorbant solution. Lin and Zhan [17] reported that the low adsorption capacity of chitosan/zeolite against HA solution at high pH implies that HA adsorbed on the adsorbent might be de-absorbed in alkaline solutions. The desorption efficiency is relatively high which might be related to the fact that in alkaline solutions, the positive charge in the amine group is deprotonated and the attraction between chitosan/zeolite and HA molecules becomes weaker.

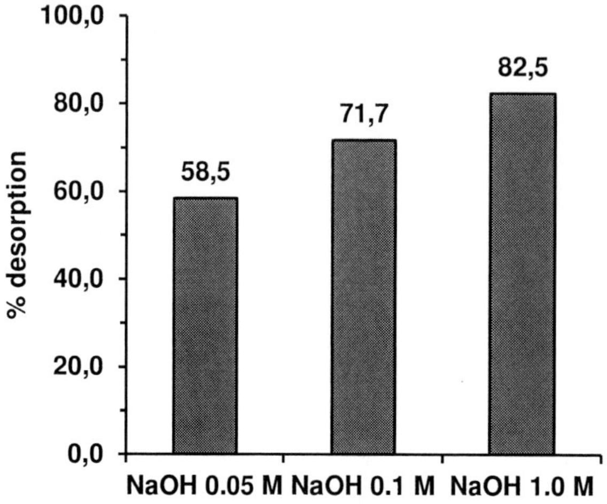

Figure 8. Fffect of NaOH concentration on desorption of HA.

4. Conclusion

The sulfuric acid-crosslinked chitosan/pectinPEC film showed good stability in water at wide range medium acidity (pH 2-10). The best film composition was films with chitosan to pectin mass ratio of (70:30). The adsorption isotherm pattern suitable for HA adsorption in chitosan/pectinPEC film crosslinked sulfuric acid is a Langmuir isotherm model with maximum adsorption capacity value of 50.0 mg g^{-1}, Langmuir constant (K_L) 1.49×10^5 L mol^{-1} and the adsorption energy is 29.7 kJ mol^{-1}.Desorption of HA from thechitosan/pectinPEC film can be achieved by using a 1.0 M NaOH solution with a percent desorption value of 82.5%.

References

[1] World Health Organization 2017 *Guidelines for Drinking Water Quality: Fourth Edition Incorporating the First Addendum* World Health Organization Geneva 224–230
[2] Kim H, Park S, Lee C, Han Y, Park J. and Kim S 2009 *Environ. Eng. Res.* **14** 9–12
[3] Jonglertjunya W and Lertchutimakul T 2012 *Songklanakarin J. Sci. Technol.* **34** 669–677
[4] Jin P, Song J, Yang L, Jin X and Wang X C 2018 *Environ. Pollut.* **233** 290–298
[5] Oskoei V, Dehghani M H, Nazmara S, Heibati B, Asif M, Tyagi I, Agarwal S and Gupta V K 2016 *J. Mol. Liq.* **213** 374–380
[6] Rajca M and Bodzek M 2013 *Sep. Purif. Technol.* **120** 35–42
[7] Daifullah A A M, Girgis B S and Gad H M H 2004 *J. Col. Surfa.* **235** 1–10
[8] Li Y, Qu G, Zhang L, Wang T, Sun Q, Liang D and Hu S 2017 *Sep. Purif. Technol.* **180** 36–43
[9] Derakhshani E and Naghizadeh A 2018 *J. Mol. Liq.* **259** 76–81
[10] Ngah WSW, Hanafiah MAKM and Yong SS 2008 *J. Col. Surf. B.* **65** 18–24
[11] Salih S S and Ghosh T K 2018 *Int. J. Biol. Macromol.* **106** 602–610
[12] Chen P, Kuo T, Kuo J, Tseng Y, Wang D, Lai J and Hsieh H 2010 *Carbohydr. Polym.* **82** 1236–1242
[13] Cui Z, Xiang Y, Si J, Yang M, Zhang Q and Zhang T 2008 *Carbohydr. Polym.* **73** 111–116
[14] Fideles T B, Santos J L, Tomás H, Furtado G T F S, Lima D B, Borges S M P and Fook M V L

2018 *OALib. J.* **5** 1–13
[15] Ngah W S W, Fatinathan S and Yosop N A 2011 *J. Desal.* **272** 293–300
[16] Dong C, Chen W, Liu C, Liu Y and Liu H 2014 *Col. Surface. A Physicochem. Eng. Asp.* **446** 179–189
[17] Lin J and Zhan Y 2012 *Chem. Eng. J.* **200–202** 202–213
[18] da Costa M P M, Ferreira I L de M and Cruz M T de M 2016 *Carbohydr. Polym.* **146** 123–130

Application of copper(II) oxide of electrocoagulation products of electroplating waste water as ceramic glaze dyes

R T Padmaningrum*, S Marwati, Sunarto and Sulistyani

Department of Chemistry Education, Faculty of Mathematics and Natural Sciences, Yogyakarta State University, Jl Colombo 1, Yogyakarta 55281 Indonesia

*E-mail: regina_tutikp@uny.ac.id

Abstract. This study aims to determine the effect of combustion temperatures and copper(II) oxide concentrations on ceramic glaze color. Electrocoagulation process was carried out using optimum conditions: pH 8, electrocoagulation time of 120 minutes, Al-Fe electrodes, and current density of 1.25 mA/cm^2. Experiment variables include combustion temperatures (999, 1060, 1101 °C) and copper(II) oxide concentrations (0.88; 1.76, 2.64, 3.53%). The results of the coagulant characterization with XRF showed copper(II) content of (22.04 ± 0.73)%. Besides that it also contains of other metals such as Cr, Cd, and Pb however below the detection limit. Coagulant is a metal hydroxide which is dried and calcined formed the oxide. Then, it was applied as a glaze ceramic dye. Based on the analysis of homogeneity using chromameter, the brightness value (L*) increased at the combustion temperature of 999 and 1060 °C, but at a temperature of 1101 °C it decreased. The green color value (a*) decreases when the combustion temperature is raised but the yellow value (b*) increased. When the metal oxide concentration was increased, L* and a* values decreased. While the b* value increased at a concentration of 0.88 to 1.76%, but at a concentration of 2.64 to 3.53% decreased.

1. Introduction

Electroplating waste water of the silver handicraft industry contains heavy metals such as Ag^+, Hg_2^{2+}, Pb^{2+}, Hg^{2+}, Bi^{3+}, Cu^{2+}, Co^{2+}, Al^{3+}, Cr^{3+}, Fe^{2+}, Mn^{2+}, Ni^{2+}, and Zn^{2+} [1]. Electroplating waste water that is discharged directly into the sewer without being treated first will cause environmental pollution. Water contaminated with electroplating liquid waste will endanger health if it used for daily needs [2]. The impact of the metal pollution varies, for example if someone is exposed to copper(II) exceeds the threshold, the excess of copper(II) in the body can cause stomach upset, nausea, and within a certain period of time causing the liver and kidney disease [3]. Therefore, this electroplating wastewater treatment needs to be done so as not to pollute the environment.

Electroplating wastewater treatment in Kotagede has been carried out by neutralization method with lime and precipitation with alum [1]. Another method is precipitation with hydroxides, sulfides, and carbonates from several precipitation agents [4-5]. In addition, electroplating liquid waste can be used as a metal coating material by recovery method or electrolysis technique [6-7]. Waste processing has advantages and disadvantages in each process and condition. In this study, we chosen the electrocoagulation method it is an easy, cheap, and without the addition of reagent. The method produced coagulants (floc) so that electroplating waste water could be utilized. The coagulant is expected to contain a maximum amount of heavy metal ions so that it can be used for purposes that are more useful and of economic value.

IOP Conf. Series: Journal of Physics: Conf. Series **1156** (2019) 012004 doi:10.1088/1742-6596/1156/1/012004

Electrocoagulation is a coagulation process using electric current in the direction of electrodes into the waste. Electrocoagulation is a combination of electrolysis and flocculation [1]. In this method there are no additional reagents so that new species that cause pollution will not form [8]. In addition, floc of coagulation contain heavy metal ions can be used as a mixture of oxides. Then, the oxide mixture from the electrocoagulation process can be used as a ceramic glaze material by calcination process [7].

Optimum conditions of electrocoagulation process encompass: pH 8, current density 1.25 mA/cm^2, combination of Al-electrode Fe (cathode-anodes), and electrocoagulation time for 120 minutes [9]. The electrocoagulation process using the optimum conditions of copper(II) ions allows maximum floc or coagulant deposition. Coagulant is expected that in the form of $Cu(OH)_2$ mixed with other metal hydoxides including $Pb(OH)_2$, $Zn(OH)_2$, $Cr(OH)_3$, $Cr(OH)_6$, $Cd(OH)_2$, dan $Fe(OH)_2$. The coagulant will pass through the drying and calcination process so that the oxide form will be applied as a glaze ceramic dye. The color of a ceramic glaze is influenced by several factors including the combustion temperature, dye concentration, combustion conditions, the composition of the glassy material, and the type of metal oxide as coloring [10]. In this study, the effect of dye concentration and combustion temperature on the final color of ceramic glaze will be studied. Other factors that influence the color of ceramic glassware have been controlled.

2. Experimental Section

2.1 Materials

Electroplating liquid waste from Kotagede, Yogyakarta, taken on August 22, 2017, at 11.00 WIB, the buffer solution was pH 4 and 7, Fe and Al Electrodes were 15 x 3 x 0.03 cm, acetone, 5 M NH_4OH solution, transparent soft glaze (TSG), zircon oxide (ZrO_2), ceramic biscuits, plastic wrap, distilled water, whatman filter paper no. 42. All of the materials were purchased from various commercial sources. The grade of them is pure analysis (pa) and used without further purification.

2.2 Instruments

XRF (X-Ray Fluorecense), AAS (Atomic Absorption Spectroscopy) AA-7000, Pyrometer dan Thermocouple, Pyrometric Cone, Chromameter, DC Power Supply (Cimarec dan Faithful), furnace, oven, pH meter (Gemmy), digital scales, Hot plate-magnetic stirrer, and laboratory glassware.

2.3 Characterization of the physical-chemical properties of the sample

As many as 1.5 liters of homogenized samples were analyzed at the Yogyakarta Health Laboratory Center (BLK). The parameters tested were physical properties such as odor, turbidity, color, temperature and chemical properties such as anion content, heavy metals, DO, and pH. In this study, electroplating wastewater samples are estimated to contain dangerous heavy metal ions.

2.4 Precipitation of metal ions by electrocoagulation method

As many as 500 mL of electroplating liquid waste samples were put into a 1 liter beaker and a 4 cm magnetic bar was also included. Then the electrocoagulation process was carried out using optimum conditions for copper(II) metal deposition including pH 8, electrocoagulation time of 120 minutes, combination of Al-Fe electrodes, current density of 1.25 mA/cm^2.

2.5 Characterization of sample coagulants with XRF

5 grams of coagulant were analyzed using X-Ray Fluoresence at BATAN.

2.6 Application of copper(II) oxides as ceramic glaze coloring with combustion temperature variations and copper(II) oxide concentrations

As many as 21.74 grams of TSG glassware, 3.26 grams of ZrO_2, and 2 grams of copper(II) oxide were mixed with water little by little. Then the mixture is mashed using pestle and mortar to form a slurry-like mixture. The mixture of the glassy formula is filtered using a filter mesh measuring 120 mesh and accommodated in a small bowl. Ceramic coating is then carried out by dipping for 5 seconds into a glassy formula mixture. Furthermore, the ceramic is burned using a furnace to reach temperatures of 999, 1060 and 1101°C.

The effect of variations concentration of copper(II) oxide was studied by repeating all treatments and changing the concentration of Cu(II) oxide to 0.88; 2.64; and 3.53% or equivalent to 1, 3 and 4 grams of coagulant. Furthermore, the ceramic is burned using a furnace at a temperature of 1060°C.

2.7 Analysis of ceramic glaze color homogeneity using a chromameter

As many as 2 pieces of ceramic glaze from each combustion temperature variation and variation of copper(II) oxide concentration were analyzed by using chromameter konica minolta CR-400 at the UGM Faculty of Agricultural Engineering.

3. Discussion

Electroplating liquid waste sample characterization aims to determine the content of heavy metals. Characterization results of electroplating liquid waste showed samples containing several dangerous metal ions such as Cu^{2+}, Cd^{2+}, Cr^{3+}, Pb^{2+}, and Zn^{2+}. The highest level of metal ions in waste, namely copper(II) is 178.4510 ppm.

Metal ions in the sample are then precipitated using electrocoagulation method at optimum conditions. The sample electrocoagulation process was carried out 31 times. The results of sedimentation in the form of colored coagulants obtained as much as 19.266 grams. The coagulants obtained are still in the form of wet hydroxide. In this study, the coagulant applied and acts as a dye for ceramic glaze is a form of metal oxide. Wet metal hydroxide coagulant will be dried using an oven until it turns into dry hydroxide. The calcination process in the process of burning ceramic glaze will convert hydroxide to oxide. Coagulant drying was carried out at 110°C for 6 hours. Before the coagulant is applied as a dye, it needs to be characterized to determine the composition of the elements in the coagulant qualitatively and quantitatively. The color of the glaze is determined by the largest metal content. So, the characterization to determine the composition of the metal content in the coagulant needs to be done. Based on the results of the characterization, it is known that the largest metal content in the coagulant is copper(II) of 22.04±0.73%, while other metals such as Cr, Cd, and Pb are below the detection limit of the XRF device. Then the copper(II) coagulant was used as a coloring material in a glassy formula. The application of copper(II) oxides as dyes in combustion temperature variations and concentrations results in green ceramic glaze, as in figures 1, 2 and 3.

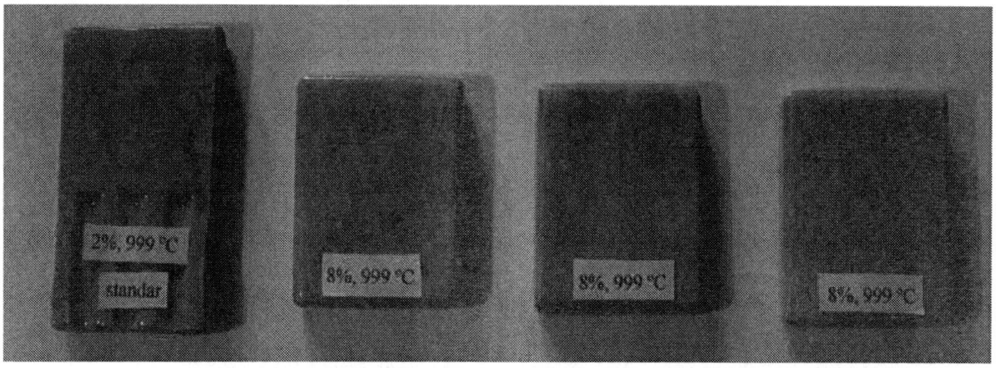

Figure 1. Ceramic tiles produced by combustion at a temperature of 999°C.

Figure 2. Ceramic tiles produced by combustion at a temperature of 1060°C.

Figure 3. Ceramic tiles produced by combustion at a temperature of 1101°C.

Based on a visual analysis of the color of the emerging ceramic glaze, it has a typical copper metal oxide color at all temperature variations. In general, the typical color of copper metal oxide in all ceramic glazes is almost the same, namely green. A striking difference is the 999°C ceramic glaze, which is less visible green color. The green color that appears in the form of bluish green and uneven on the surface. This is due to the fact that there are still many coagulants that have not melted, which is marked by the appearance of many black dots on the surface of the ceramic.

Furthermore, the ceramic glaze with a heating temperature of 1060°C, the color is light green and more evenly distributed on the surface of ceramic glaze. That is, copper(II) oxides on ceramic glaze with this combustion temperature have fused more, so that the black dots begin to fade compared to the previous temperature. One of the reasons is that the combustion temperature of 1060°C is the temperature that is closest to the melting point of the copper metal which is around 1038°C [11]. At a temperature of 1101°C, the color of the ceramic glaze is dark green. This dark green color appears more evenly on the surface of the ceramic glaze. Metal oxide that has not melted on the surface of the glaze looks less than ceramic with the previous combustion temperature. That means the coagulant has fused more maximally. According to the theory, the sparkling color that appears for copper(II) oxide is light green or Turkish [10]. So visually, ceramics with a combustion temperature variation that has been carried out successfully produces ceramic colors that are close to the color of the theory. The higher the temperature of combustion of ceramic glaze, the lower the brightness and green color of Cu(II) oxide.

Based on figure 4, it can be seen that the L* value increases at the combustion temperature of 999 and 1060°C, ie from 60.79 to 61.52, while at the combustion temperature of 1101°C it shows a decrease to 56.67. The increase and decrease in the L* value indicate the brightness level on the surface of the ceramic glaze. Meanwhile, the higher the combustion temperature, the a* value decreases to -7.45; -6.35; and -5.64. Decreasing the value of a* indicates a reduction in green on the surface of ceramic glaze. The value of b * is directly proportional to the increase in the temperature of burning ceramic glaze. Increased b * value is 3.95; 7.59; and 9.28 indicates increased yellow color, although visually less visible.

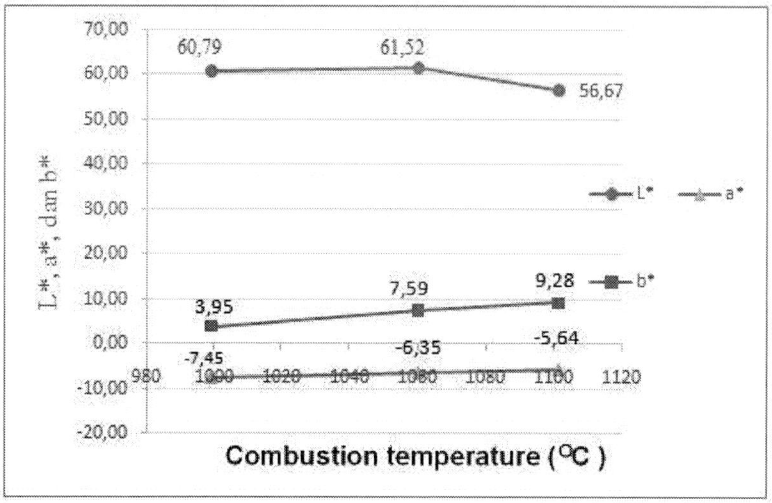

Figure 4. Graph of the relationships of L*, a*, and b* to the combustion temperature.

Based on a visual analysis of the color of glassy ceramics, all variations of copper(II) oxide (impure) concentration produce typical colors of copper(II) oxide. Smoothness in each color of ceramic glaze has similarities, but brightness decreases with increasing concentration. The typical color of copper(II) oxide seen in all ceramics is green. However, the green color of the ceramic glaze that appears has a difference in each variation of concentration. The difference is seen in the ceramic glaze with 0.88% copper(II) oxide concentration in the form of bluish green with dark green dots scattered on the glassy surface. Ceramic glaze with 1.76% copper(II) oxide concentration, light green color spread evenly on the surface of ceramic glaze. That is, the coagulant on glaze ceramic has fused more at this combustion temperature, so that black dots begin to fade compared to previous concentrations. Ceramic glaze with a concentration of copper(II) oxide 2.64%, the resulting color is a rather dark green. Ceramic glaze with 3.53% copper(II) oxide concentration, the color is darker green than before with black dots covering evenly on the surface of ceramic glaze.

The results of the glassy ceramic variation of copper(II) oxide concentration can be seen in figures 5, 6, 7, and 8.

Figure 5. Ceramic glaze with a concentration of Cu(II) oxide 0.88%.

IOP Conf. Series: Journal of Physics: Conf. Series **1156** (2019) 012004 doi:10.1088/1742-6596/1156/1/012004

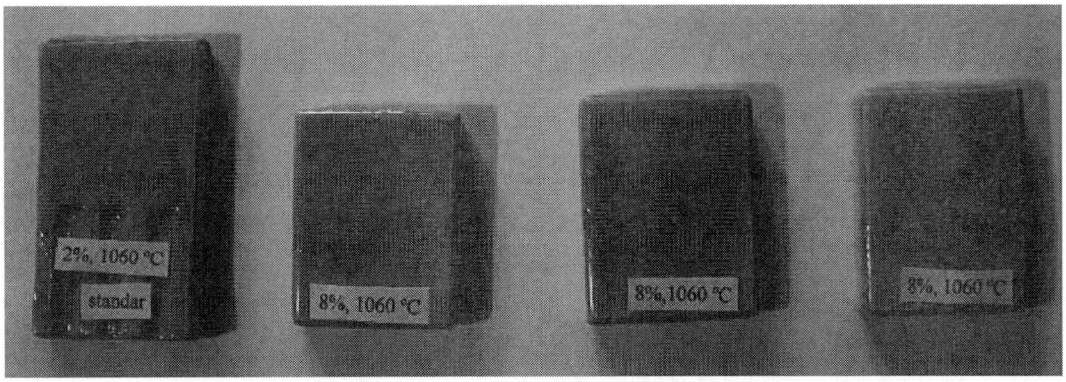

Figure 6. Ceramic glaze with a concentration of Cu(II) oxide 1.76%.

Figure 7. Ceramic glaze with a concentration of Cu(II) oxide 2.64%.

Figure 8. Ceramic glaze with a concentration of Cu(II) oxide 3.53%.

Figure 9 shows that the higher the copper(II) oxide concentration, the lower the L* and a* values. The decrease in L* value is 70.36; 61.52; 47.96; and 43.03. This shows the brightness on the surface of the ceramic glaze is decreasing. Meanwhile, a decrease in the value of a* is -9.09; -6.35; -3.71; and -0.03. Decreasing the value of a* shows a reduction in green on the surface of ceramic glaze. The value of b* is increasing at a concentration of 0.88 and 1.76%, which is from 0.34 to 7.59. However, at concentrations of 2.64 and 3.53%, the value of b* decreased to 6.65 and 2.67. The increase and decrease in b* value shows an indication of increasing or decreasing yellow color although visually less visible.

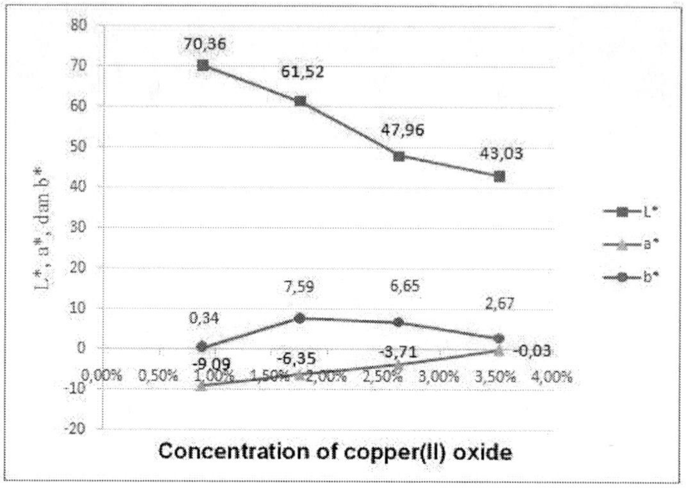

Figure 9. Graph of the relationship of L*, a*, and b* to the concentration of copper(II) oxide.

4. Conclusions

Based on the research that has been done, it can be concluded that the higher the temperature of combustion of ceramic glaze, the brightness of the ceramic color will increase, the green color of copper(II) oxide will decrease, but the resulting yellow color increases. Meanwhile, the greater the concentration of copper(II) oxide causes the less brightness and green color to be produced, while the yellow color increases.

References

[1] S Marwati, R T Padmaningrum and Marfuatun 2008 *Proc Seminar Penelitian, Pendidikan dan Penerapan MIPA* FMIPA UNY
[2] Goyal N, Jaina S C and Banerjeeb U C 2003 *J. Adv. Eniron. Res.* 7 311-319
[3] Agency for Toxic Substance and Disease Registry (ATSDR) 2004 *Toxicological profile for copper U.S.* (Atlanta: Department of Health and Humans Services, Public Health Humans Services, Centers for Diseases Control)
[4] Naim R, Kisay L, Park J, Qaisar Z, Noshin M and Janil K 2010 *Int. J. Environ. Res* 4(**4**)
[5] Weltrowski M, Martel B and Morcellet M 1996 *J. Appl. Poly. Sci.* 59
[6] Fu F and Qi W 2011 *J Environ Manag* 92
[7] S Marwati, R T Padmaningrum, and Marfuatun 2009 *Jurnal Penelitian Saintek* 14(**1**)
[8] Hanum F, Tambun R, Ritonga M Y and Kasim W W 2015 *Jurnal Teknik Kimia* 4(**4**) 13-17
[9] F D Santos 2017 *Optimasi kondisi proses elektrokoagulasi ion logam tembaga(II) dalam limbah cair elektroplating* Skripsi (Yogyakarta: FMIPA UNY)
[10] Budiyanto and Wahyu Gatot, Kriya Keramik 2008 *Kriya keramik* Book 3th ed (Jakarta: Direktorat Pembinaan Sekolah Menengah Kejuruan)
[11] Svehla G 1990 *Buku teks analisis anorganik kualitatif makro dan semimikro* 5th ed (Jakarta: PT Kalman Media Pustaka)

IOP Conf. Series: Journal of Physics: Conf. Series **1156** (2019) 012005 doi:10.1088/1742-6596/1156/1/012005

Identification of activated NaOH carbon of synthesis of sea pandanus leaves (*P. odorifer*) for Fe^{3+} and Cu^{2+} ions adsorption

Sulistyani*, S Kristianingrum, E D Siswani, and A Fillaeli

Department of Chemistry Education, Faculty of Mathematics and Natural Sciences, Yogyakarta State University, Jl Colombo 1, Yogyakarta 55281 Indonesia

*E-mail: sulistyani@uny.ac.id

Abstract. This research aims to characterize activated NaOH carbon of sea pandanus leaves (*P. odorifer*) which include water content, ash content, volatile content, carbon content and adsorption capacity of iodine. Further, we investigate adsorption capacity of activated NaOH carbon to Fe^{3+} and Cu^{2+} ions, and its adsorption type. Sea pandanus leaves (*P. odorifer*) was taken from Trisik Beach in Kulonprogo. The carbonation process is carried out at a temperature of 500 °C, followed by chemical activation using 1% NaOH solution and physical activation with a temperature of 700 °C while carbon dioxide gas flowed for 2 hours. Identification of capacity of activated carbon to heavy metals is conducted in batch. The characterization results show that water content and iodine adsorption capacity of activated NaOH carbon complied with SNI 06-3730-1995, respectively of 5.38% and 2463.63 mg/g. While the volatile content (59.09%), ash content (17.88%), carbon content (17.66%) did not fulfill SNI 06-3730-1995. The adsorption capacity of activated NaOH carbon to heavy metal ions of Fe^{3+} and Cu^{2+} respectively is 0.895 and 0.996 mg/g.

1. Introduction

At present, the industry has grown rapidly. The development of the industry may have a negative impact on the heavy metal content contained in the waste. The content of heavy metals in industrial waste water is very dangerous when disposed directly into the environment. One of the fastest growing industries in Indonesia is the metal coating industry. In metallic coating industry using electroplating method, produces colored waste water and odor [1].

There are various methods that can be used to reduce the pollution of heavy metals in the environment such as chemical precipitation, ion exchange, adsorption, membrane filtration, coagulation, flotation and electrochemical methods [2-4]. A common method for heavy metal adsorption is the adsorption method because it can use a variety of adsorbents at relatively low cost such as active clay [5], activated carbon [6], and zeolite [7].

The adsorbent can also be synthesized from the raw material of sea pandanus leaves [8]. Carbon of sea pandanus leaves can be activated with acid or base solution to increase its adsorption ability. In previous studies, the activator substance to activate carbon of sea pandanus leaves are acid solutions such as HCl and H_2SO_4 [8], HNO_3 [9]. The kind of activator used in the process of carbon activation will greatly affect the ability of the adsorbent [6]. In addition, activator concentration and activation techniques also affect the ability of the adsorbent produced.

IOP Conf. Series: Journal of Physics: Conf. Series **1156** (2019) 012005 doi:10.1088/1742-6596/1156/1/012005

Adsorbents are generally synthesized from abundant and under utilized material, such as egg shells [10], and sea pandanus leaves (*P. odorifer*) [8]. In previous research, synthesis of activated carbon of sea pandanus leaves using acid solution as activator. In this study, we will synthesis activated carbon of sea pandanus leaves using alkaline activator, i.e. sodium hidroxide (NaOH). We want to investigate the ability of alkaline solutions to activate carbon from the ingredients of sea pandan leaves. The resulting activated carbon is then used as an adsorbent for heavy metal ions such as Fe^{3+} and Cu^{2+}.

2. Experimental Section

2.1 Materials

Sea pandanus leaves (*P. odorifer*) from Trisik Beach in Kulonprogo, 1% (b/v) NaOH, $Pb(NO_3)_2$, $Fe(NO_3)_3$, $Cr(NO_3)_3$, $Ni(NO_3)_2$, $CdSO_4$, $Cu(NO_3)_2$, $Zn(NO_3)_2$, 0.1 M I_2, 0.1 M $Na_2S_2O_3$, indicator of amylum, distilled water, filter paper, pH paper. All of the materials were purchased from various commercial sources. The grade of them is pure analysis (pa) and used without further purification.

2.2 Instruments

Surface Area Analyzer (SAA), Atomic Absorption Spectrophotometer (AAS), blowing furnace, 80 mesh sieve, analytical scales, oven, magnetic stirrer, pumpkin powder, and laboratory glass ware.

2.3 Synthesis of carbon

The sea pandanus leaves are cleaned and then washed thoroughly. Then, it cut into small pieces and then dried under the sun to dry. The dried sea pandanus leaf is then carbonated with a furnace at a temperature of 500 °C for 4 hours, then mashed and sieved with an 80 mesh sieve.

2.4 Activation of carbon

200 grams of 80-mesh carbon are added with 600 mL of 1% NaOH solution while stirring for 24 hours. The activated carbon is then filtered and washed with distilled water until the pH is neutral. Then, activated carbon is dried in an oven at 100 °C for 3 hours. Furthermore, the physically activation was done by inserting carbon in the furnace with a temperature of 700 °C while carbon dioxide gas flowed for 2 hours.

2.5 Adsorption of activated carbon to heavy metal ions (Fe^{3+} and Cu^{2+})

Identification of activated carbon capacity to heavy metals is conducted in batch. Preparing as many as 6 erlenmeyer, each added 1 gram of activated carbon of sea pandanus leaves. Then, each erlenmeyer is added 50 mL of Fe^{3+} ion solution with various concentration (3-18 mg/L for Fe^{3+} and 4-24 mg/L for Cu^{2+}). The mixture is allowed to stand for 24 hours while stirring. Then, the mixture was filtered and the filtrate was taken for AAS testing.

3. Discussion

Activated carbon has been widely known as a good absorbent because it has a high adsorption capacity. The type of activator, activation conditions and time of the activation process affect the physical properties of activated carbon [11]. The activation process has the goal of enlarging pore size by breaking hydrocarbon bonds or oxidizing surface molecules, so that carbon can undergo changes both physically and chemically [9].

Chemical activation was performed using 1% NaOH solution. The filtrat color of its process is brown because NaOH solution is a strong base and corrosive. Thus causing a lot of ash to be released by carbon [12]. Chemical activation is continued with physical activation to optimize the pores of activated carbon. The physical activation is carried out by heating the carbon at a temperature of 700 °C while flowed by carbon dioxide gas.

3.1 Characterization of activated NaOH carbon of sea pandanus leaves

Characterization of activated NaOH carbon of sea pandanus leaves is carried out to determine the quality of the carbon that has been synthesized and compared with the standard quality based on SNI 06-3730-1995.

Characters that are seen at table 1 are water content, ash content, volatile content, carbon content, and absorption capacity of iodine. The result of characterization obtained then compared with SNI 06-3730-1995.

Table 1. Characters of carbon from sea pandanus leaves.

	Water Content (%)	Ash Content (%)	Volatile Content (%)	Carbon Content (%)	Absorption capacity of iodine (mg/g)
SNI	Maks 15	Maks 10	Maks 25	Min 65	Min 750
Carbon before activation	5.42%	19.62%	70.50%	4.46%	2157.55
Carbon after activation	5.38%	17.88%	59.09%	17.66%	2463.64

Generally, characters of carbon from sea pandanus leaves after activation is better than carbon before activation. The value of water content meets the standard which means it has value less than the maximum standard water content that has been established SNI 06-3730-1995. The absorption capacity of iodine also meets the standards which means it has a value of more than a minimum standard of iodine absorption capacity. The percentage of carbon content obtained has not meet the standard. Activated carbon quality that has not met national standards is made possible due to the non-optimal carbonation process. Consequently, many non-carbon elements are still bound in it. During carbonization process, most non-carbon elements such as oxygen, hydrogen, and nitrogen are eliminated as volatile gas species [13].

The activated carbon of sea pandanus leaves is also characterized by using Surface Area Analyzer (SAA). It has surface area of 19.667 m^2/g. It is better than surface area of activated carbon of sea pandanus leaves by [9], i.e. 3.575 m^2/g.

3.2 Identification of capacity adsorption of activated NaOH carbon of sea pandanus leaves

Identification of capacity adsorption in this study was obtained by using Atomic Absorption Spectrocopy (AAS). It is based on concentration before and after activation process. Data of adsorption efficiency and capacity adsorption activated NaOH carbon to Fe^{3+} ions can be observed in the table 2 and figure 1.

Table 2. Data of adsorption efficiency and adsorption capacity of activated NaOH carbon to Fe^{3+} ions.

Ion Fe^{3+}	Initial Concentration (mg/L)	Final Concentration (mg/L)	Absorbed Concentration (mg/L)	Adsorption Efficiency (%)	Adsorption Capacity (mg/g)
1	18	0.098	17.902	99.453	0.895
2	15	0.119	14.881	99.205	0.744
3	12	0.129	11.871	98.920	0.593
4	9	0.134	8.866	98.503	0.443
5	6	0.140	5.860	97.668	0.293
6	3	0.140	2.860	95.336	0.143

The greater the concentration of Fe^{3+} ions, the adsorption capacity of adsorbent increased. At a dilute concentration of the solution, the less number of Fe^{3+} ions causes the chance of analytes interacting with the adsorbent is also smaller. Conversely, at more concentrated concentrations, the greater number of Fe^{3+} ions causes greater opportunities for analytes to interact with the adsorbent. Therefore, the greater the concentration of Fe^{3+} ions solution, the adsorption capacity tends to increase. High adsorption efficiency values (>99%) show that analytes are almost completely absorbed. But it never reached 100%. It is possible that contact between analytes and adsorbents in batch system cannot be perfect.

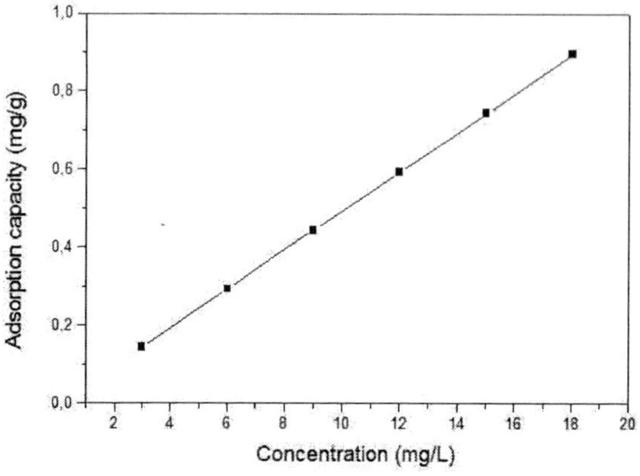

Figure 1. Curve relation between concentration vs capacity adsorption of Fe^{3+} ion.

Table 3. Data of adsorption efficiency and adsorption capacity of activated NaOH carbon to Cu^{2+} ions.

Ion Cu^{2+}	Initial Concentration (mg/L)	Final Concentration (mg/L)	Absorbed Concentration (mg/L)	Adsorption Efficiency (%)	Adsorption Capacity (mg/g)
1	24	0.156	23.844	99.350	0.192
2	20	0.077	19.923	99.616	0.996
3	16	0.084	15.916	99.475	0.796
4	12	0.052	11.948	99.570	0.597
5	8	0.048	7.952	99.400	0.398
6	4	0.084	3.916	97.900	0.196

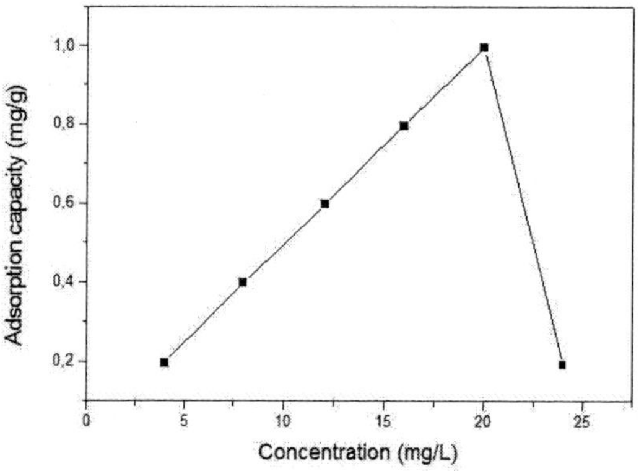

Figure 2. Curve relation between capacity adsorption vs concentration of Cu^{2+} ions.

Adsorption capacity of activated NaOH carbon to Cu^{2+} ions can be observed in the table 3 and figure 2. Based the data, can be stated that the adsorbent adsorption efficiency value of Cu^{2+} ions greater, especially on the concentration of 4, 8, 12, 16 and 20 mg/L Cu^{2+} ions. However, the decrease in adsorption capacity occurs at concentrations of 24 mg/L Cu^{2+} ions. This is possible due to the desorption process after optimal adsorption capacity is achieved.

The adsorption capacity was then determined at the largest adsorption capacity. By this study adsorbent capacity of activated NaOH carbon from sea pandanus leaves to Fe^{3+} and Cu^{2+} ions was 0,895 mg/g and 0,996 mg/g respectively. Adsorption capacity of this adsorbent to Cu^{2+} ions is slightly greater than that of Fe^{3+} ions. This is due to Cu^{2+} ion particle size slightly smaller than Fe^{3+} ion particle size. Actually, adsorption capacity of activated NaOH carbon from sea pandanus leaves is still possible greater than the value considering the efficiency that has not reached 100%. Therefore, in this study the adsorption test with a greater concentration should be carried out.

3.3 Identification of isotherm adsorption type of activated NaOH carbon of sea pandanus leaves

The identification of isotherm adsorption type of activated NaOH carbon of sea pandanus leaves is determined, i.e Langmuir and Freundlich adsorption isotherms [7].

The Langmuir equation used is as follows:

$$\frac{C_e}{\frac{X_m}{m}} = \frac{1}{ab} + \frac{1}{a} C_e$$

The Freundlich equation used is as follows:

$$log\left(\frac{X_m}{m}\right) = \log k + \frac{1}{n} logC_e$$

Information C_e : concentration of metal ions in solution after adsorption
X_m/m : the mass of metal ions adsorbed per mass of activated carbon
b : Langmuir's affinity or constant parameters
a and k : maximum capacity of adsorption (mg/gram)

Table 4. X_m/m, $C_e/(X_m/m)$, log C_e, and log X_m/m value of Fe^{3+} ion adsorption.

C_o (ppm)	C_e (ppm)	X_m/m (mg/g)	$C_e/(X_m/m)$	log C_e	log (X_m/m)
3	0.1399	0.1430	0.9783	-0.8542	-0.8447
6	0.1399	0.2931	0.4773	-0.8542	-0.5330
9	0.1347	0.4432	0.3039	-0.8706	-0.3534
12	0.1295	0.5935	0.2182	-0.8877	-0.2266
15	0.1192	0.7440	0.1602	-0.9237	-0.1284
18	0.0984	0.8951	0.1099	-1.0070	-0.0481

Table 4 and 5 show concentration of Fe^{3+} and Cu^{2+} ions before and after adsorption process. Further, the data is processed into the Langmuir and Freundlich equation. Graph of isotherm type of Fe^{3+} ion can be seen at figure 3 and 4, while graph of isotherm type of Cu^{2+} ion can be seen at figure 5 and 6. Based on the graph is obtained, the R^2 Langmuir and Freundlich equation values are different. For Fe^{3+} ion, R^2 Langmuir value of 0.4565 and Freundlich of 0.6042. For Cu^{2+} ion, the R^2 Langmuir value is 0.0077 and Freundlich is 0.184. However, R^2 values of the Langmuir and Freundlich equation for Fe^{3+} and Cu^{2+} ions are $\leq$ 95% [14]. So, isotherm adsorption of activated carbon does not follow the equation of both, Langmuir and Freundlich.

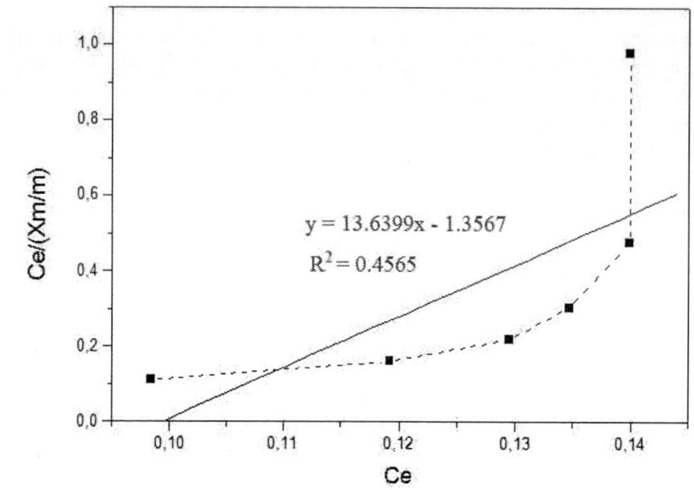

Figure 3. Graph of Langmuir equation of Fe^{3+} ion, C_e vs $C_e/(X_m/m)$.

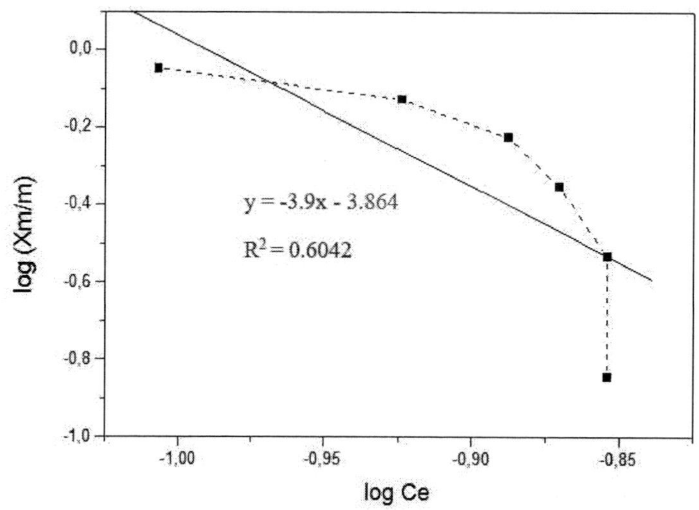

Figure 4. Graph of Freundlich equation of Fe^{3+} ion, log C_e vs log X_m/m.

Table 5. X_m/m, $C_e/(X_m/m)$, log Ce, and log X_m/m value of Cu^{2+} ion adsorption.

C_o	C_e	X_m/m	$C_e/(X_m/m)$	log C_e	log X_m/m
4	0.0840	0.1958	0.4290	-1.0757	-0.7082
8	0.0480	0.3976	0.1207	-1.3188	-0.4006
12	0.0516	0.5974	0.0863	-1.2874	-0.2237
16	0.0840	0.7958	0.1055	-1.0757	-0.0992
20	0.0768	0.9962	0.0771	-1.1146	-0.0017
24	0.1560	1.1922	0.1309	-0.8069	0.0763

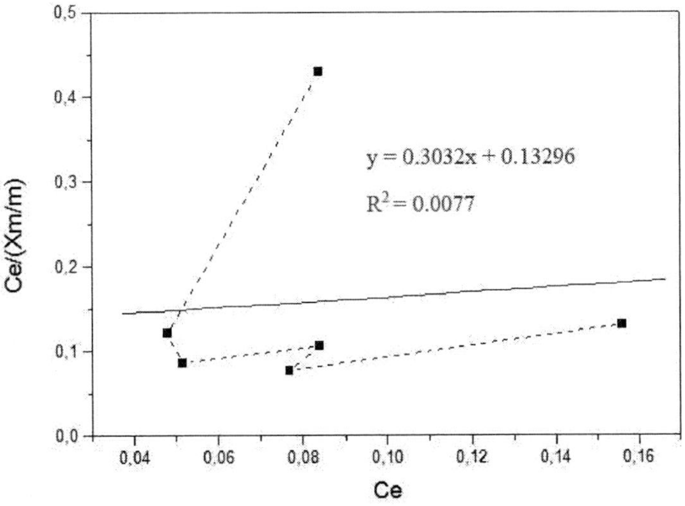

Figure 5. Graph of Langmuir equation of Cu^{2+} ion, C_e vs $C_e/(X_m/m)$.

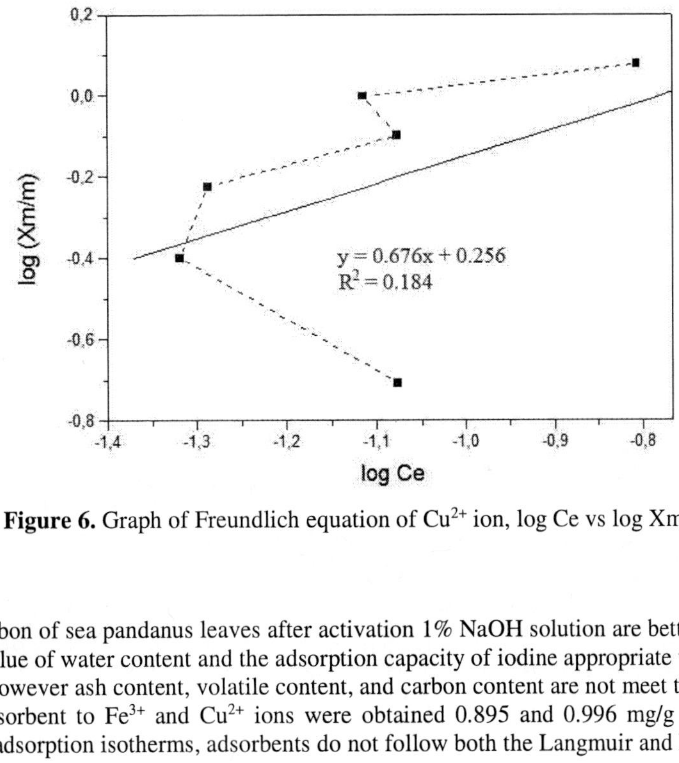

Figure 6. Graph of Freundlich equation of Cu^{2+} ion, log Ce vs log Xm/m.

4. Conclusion

Characters of carbon of sea pandanus leaves after activation 1% NaOH solution are better than carbon before activation.The value of water content and the adsorption capacity of iodine appropriate with the standard SNI 06-3730-1995. However ash content, volatile content, and carbon content are not meet the standard. Capacity adsorption of adsorbent to Fe^{3+} and Cu^{2+} ions were obtained 0.895 and 0.996 mg/g respectively. For the identification of adsorption isotherms, adsorbents do not follow both the Langmuir and Freundlich types.

References

[1] S Marwati, R T Padmaningrum and Marfuatun 2009 *Jurnal Penelitian Saintek* 14(**1**) *17-40*
[2] Fu F and Qi W 2011 *J Environ Manag* **92** 407-418

[3] Naim R, Kisay L, Park J, Qaisar, Zulfikar, Noshin M and Janil K 2010 *Int. J. Environ. Res* 4(**4**) 735-40
[4] Mazumer D, Ghosh D and Paddhyay P B 2011 *IJCEE* 3(**2**) 101-10
[5] Auliah A 2009 *Jurnal Chemica* 10(**2**) 14-23
[6] Danarto Y C and Samun T 2008 *Ekuilibrium* 7(**1**) 13-16
[7] Handayani M dan Eko S 2009 *Proc Semnas Sains dan Teknologi Nuklir* PTNBR–BATAN Bandung 130-6
[8] Kristianingrum S, Endang D S dan Annisa F 2014 *Jurnal Sains Dasar* 3(**1**) 45-54
[9] Setiyowati L 2014 *Sintesis adsorben daun pandan laut (Pandanus tectorius) dan uji sifat adsorptifitasnya terhadap ion logam Pb dan Cr dalam pelumas bekas* Skripsi (UNY: Program Studi Kimia)
[10] Maslahat M Agus T and Prima W S 2015 *Jurnal Sains Natural Universitas Nusa Bangsa* 5(**1**) 92-100
[11] Yacob A R, Zaiton A M, Ratna S D D and Vicinicvarri 2008 *The Malaysian Journal of Analytical Sciences* 12(**1**) 264-71
[12] Wulandari F, Umiatin, dan Esmar B 2015 *Spektra: Jurnal Fisika dan Aplikasinya* 16(**2**) 60-64
[13] Bansal, R and Goyal M 2005 *Activated carbon adsorption* (Boca Raton: CRC Press)
[14] Ngapa Y D 2017 *Jurnal Kimia dan Pendidikan Kimia* 2(**2**) 90-96

Characterization of yeast hydrolysate enzymatic (yhe) from yeast fermented in the variation of rice flour

R Agustini *, I G M Sanjaya and Lupita
Department of Chemistry, Faculty of Mathematic and science, Surabaya State University, Indonesia

*Email: rudianaagustini@unesa.ac.id

Abstract. Research has been conducted to determine the chemical and physical characteristics of Yeast Hydrolysis Enzymatic (YHE) based on its fermentation media. The fermentation medium used is white, red, and black rice. The procedures carried out include the sample preparation, the fermentation, making pineapple crude and the enzymatic hydrolysis stage. In this study measured Cr^{3+} content using voltammetry method, and measurement of protein content using Kjehdahl method. There are three kind of YHE: YHE from yeast which is grown in white rice (YHE-Er), red rice (YHE-Rr) and black rice (YHE-Br). YHE from yeast grown in various growth media has different Cr^{3+} content. The lowest yield of Cr^{3+} was YHE-Wr which was 32 x 10^{-5}% while YHE-Br was 37 x 10^{-5}% and YHE-Br was 40 x 10^{-5} %. The protein content of YHE-Wr, YHE-Rr and YHE-Br were respectively 38.45; 38.48; and 39.23%.

1. Introduction

Yeast Hydrolysate Enzymatic (YHE) is a yeast extract that is processed through an enzymatic hydrolysis process. Yeast is a unicellular fungus, it is known to have several types and one of them is *Saccharomyces cerevisiae*. Saccharomyces cerevisiae is one of the yeasts that has been widely known and widely used, especially in the fermentation process. Saccharomyces cerevisiae is a microorganism contained in baker's yeast and there are no other microbes in it [1]. Based on market research, it is known that the world yeast extract needs approximately $ 1.000 million in 2009 and this need continues to increase until now. Over the next five years, projects that Yeast Extract will register a 4.3% CAGR in terms of revenue, reach US$ 1920 million by 2023, from US$ 1490 million in 2017 [2]. On the other hand, the price of this product is relatively expensive. Yeast extract can be used for various needs, among others: as a medium for microbial growth, as a basic ingredient of pesticides and fertilizers or bio fertilizer which is very useful in agriculture, and as a flavour-enhancing ingredient that can be used by the food industry and for home needs, preparing family food menu every day [3], it can be used as a supplement to meet the nutrition of people who experience poor nutrition, Food & Beverages (soups, processed foods, dairy products, bakery, sauces & savory flavours), animal feed, and pharmaceuticals). The yeast extract is known to contain peptone and acid various amino acids have been used for centuries as a major nutritional component in culture media. This compound is a source of carbon, minerals, vitamins and growth factors. The results showed that YHE from fermented yeast using tapioca waste and sucrose then hydrolyzed enzymatically as follows: total nitrogen (NT) 25.97%, α-amino nitrogen (N-α) 20.07%, price (N -α) / (NT) 0.77%, fat content 1.18%, pH (2% in solution) is 5, total carbohydrate is 58.99%, moisture content is 7.19%, vitamin: thiamine, riboflavin, pyridoxine, and niacin) is 10.307% proven to be used as microbiological media material [4].

One of the efforts to meet the needs of the global market for yeast extract is one of the alternatives is the production itself using materials available in the surrounding environment, including rice flour. Rice flour can be used as a fermentation medium. One of the good fermentation media for fermentation with *Saccharomyces cerevisiae* is flour whose main content is starch [5]. In addition to meeting needs, it can also increase the economic value of rice. These foodstuffs are available abundantly in Indonesia, Thailand, Korea, Japan, and other countries. This ingredient is known to be rich in carbohydrates, namely starch. There are several kinds of rice, including white rice, brown rice, and black rice. Several studies reveal that these three types of rice have different chemical content, including protein, starch, minerals, and antioxidants [6-7]. Black rice shows higher amounts of minerals, higher than anthocyanins, faster hydrolysis levels, higher amino than white rice acids. Black rice also has the highest mineral content when compared to brown rice or white rice, such as iron, zinc, manganese, and phosphorus [6]. Mashed brown rice contains 7.30% protein, 4.20% iron, and 0.34% vitamin B1 [8]. Amylum is a complex compound that cannot be directly used by yeast as a source of energy and growth. This compound must be degraded into a simple compound, glucose. This degradation process can be done in many ways, one of which is enzymatical. Enzymatic hydrolysis is known to be most effective than others. The enzymes used in the process are α-amylase and glucoamylase. Amylase starch will be degraded into simpler constituent parts such as dextrin, isomaltose, maltose, and glucose, and then degraded by glucoamylase to produce glucose. Glucose is used for yeast growth.

Yeast which is successfully grown can be further processed, which is extracted through an enzymatic hydrolysis process. Enzymatic hydrolysis is the most effective process compared to others [3]. Yeast is known to be rich in protein. Protein is an essential molecule for the structure and function of all living things. This compound is composed of various amino acids which are linked to peptide bonds and form various complex structures [9]. This protein can be degraded into a simpler component, namely peptone and various amino acids enzymatically, using bromelain. Amino acids are needed for cell growth.

Yeast besides being known to contain iron, zinc, manganese, and phosphorus also contains chromium. The content of chromium in yeast every 1 oz (28.35 grams) contains 3.3 microgram [10]. This mineral is needed by the body, even though it has to be fulfilled a little. Chromium has several types, including chromium (III) or Cr^{3+}. Valence mineral (III) is the most stable and is known to be the safest to consume. Chromium serves to help enter glucose in body cells. Glucose can enter the body's cells with the help of the hormone insulin. Regulation of blood glucose levels by chromium is not widely understood. It is recently known that chromium interacts with the low-molecular-weight chromium (LMWCr) [11]. The interaction of chromium with LMWCr will activate the role of insulin. If the amount of hormone insulin in a person's body is insufficient or if the body's cells do not respond to insulin, there will be a build up of glucose in the blood (hyperglycemia) and the person suffering from DM. Therefore, the content of chromium in food is very important to know.

2. Experimental Section

2.1 The Process of Making Bromelin

Pineapple fruit is washed, cut, and mashed using a blender. The crushed pineapple is filtered with a cloth and then centrifuged at 1500 rpm for 10 minutes. The resulting filtrate was added with 35% ammonium sulfate and stirred slowly using a magnetic stirrer in cold conditions for 45 minutes and allowed to stand for 24 hours in the refrigerator. The mixture was centrifuged at a speed of 6000 rpm for 15 minutes; the resulting residue was a crude bromelain extract. The residue is ready to be used for yeast hydrolysis [12].

2.2. The Process of making YHE

The initial stage in making YHE is to prepare flour as fermentation medium, namely red, white, and black rice flour which passes the 100 mesh sieve. Each of flour was dissolved in distilled water in a ratio of 1: 5, then gelatinized at 100°C and cooled. The formed gel was hydrolyzed using α-amylase and glucoamylase at 50°C for 24 hours. Hydrolysate is added with yeast bakery's with a ratio of 1: 5, then fermented at 37°C for 36 hours. After fermentation, plasmolysis was done by adding 35% NaCl, incubated at 60°C for 48 hours, then centrifuged at 1500 rpm for 10 minutes. The residue obtained is fermented Yeast from rice flour, namely yeast-white rice, yeast-brown rice, and yeast-black rice added bromelain crude and incubated for 48 hours at 37 ° C [12]. Hydrolysates obtained were YHE-white rice, YHE-brown rice, and YHE-black rice which were ready to be tested for water content, Cr^{3+} content, and protein.

2.3. Determination of trivalent chromium levels (Cr^{3+}).

The first step is to make a standard Cr^{3+} solution. The standard solution used is $CrCl_3.6H_2O$, with variations in concentrations of 10, 20, 40, and 100 ppm. The next stage, preparing samples of YHE-white, YHE-red, and YHE-black rice flour. Each sample was cooled in a furnace at 500 °C for 4 hours cooled and each dissolved using 1 mL concentrated HCl and 1 mL concentrated HNO_3. The mixture was allowed to stand for 24 hours, then each one was dissolved with 25 mL of aqua-demineral then taken 10 mL of sample and added 5 mL of citrate buffer pH 3, and also added 10 mL of KCl 0.1N. Each standard solution and samples were tested for Cr^{3+} using voltammeter. The obtained voltammogram was analyzed using the originPro7.0 application so that the current values of the standard solution and the sample were known [13].

2.4. Determination of protein content using the Kjeldahl method

The method of determining the N-total method of Kjeldahl was adopted from AOAC [14]. YHE-black, white and red rice samples were weighed about 5 grams each and put into the Kjeldahl flask. A sample of 1.4 grams of catalyst and boiling stone was added. 5 ml of concentrated H_2SO_4 is added (in a fume hood). The sample is reconstructed until the color becomes green then left to cool. The sample was added 60 ml of aquadest, shaken and put into Erlenmeyer 300 ml. In the distillation stage, the sample solution was diluted by adding 40% NaOH, the flask was placed on the distillation apparatus. Distillation is stopped if all N has been caught by sulfuric acid in the Erlenmeyer flask. In the Titration process, the excess H_2SO_4 is used to capture N titrated with Sodium Hydroxide. Titration is stopped if the solution changes from purple to turquoise. Blank solutions are also made, namely replacing the sample with distilled water. The final step is to calculate the nitrogen-based content on the result of the titration. The results of this calculation will get a percent value of TKN (Total Kjeldahl Nitrogen). Here is the calculation formula:

$$\%TKN = \frac{(\text{Tr. Sample} - \text{Tr. Blank}) \times 14.007 \times N}{\text{Mg. Sample}} \times 100\%$$

Note:
- Tr. Sample = number of titrant for sample (ml)
- Tr. Blank = number of titrant for blanks without sample (ml)
- N = titrant normality
- mg. Sample = sample weight (mg)

3. Discussion

The initial stage carried out in this study is menumbuhkan yeast dalam media tepung beras. Yeast yang digunakan adalah *bakery yeast*, yang diketahui mengandung *Saccharomyces cerevisiae*.

Fermentation is a way to convert a substrate into a certain desired product by using microbial assistance [15]. Before stepping into the fermentation process, first, do the gelatinization process. Gelatinization is a process when starch granules are heated with enough water to develop starch granules and produce a gel form [16]. This process occurs the breakdown of intermolecular bonds from starch in the presence of heat and water given [17]. The heat and water used in the gelatinization process cause high granule freezing and amylose is able to diffuse out of the granule [18]. Each gel-shaped rice is easily accessible and degraded by α-amylase and glucoamylase so that glucose is produced which is used by yeast for cell growth. Degradation is characterized by decreasing the level of viscosity. The sample which was originally in the form of a gel turned into liquid after the process of degradation or enzymatic hydrolysis. This is due to the breaking of the glycosidic bond from starch. Then, the hydrolyzate obtained is used for yeast fermentation. Yeast hasil fermentasi selanjutnya didegradasi secara enzimatis menggunakan bromelin nanas sehingga diperoleh YHE dan dikarakterisasi. Characterization includes: moisture content, protein content and Cr^{3+} content.

Determination of water content was carried out by the gravimetric method using a Moisture Analyzer. Moisture Analyzer utilizes infrared or halogen lamps as heat sources. The results of flour-Wr, flour-Rr, and flour-Br water content are found in table 1. In this research also determined the water content of rice flour and yeast as raw material of YHE.

Table 1. Water content of rice flour, yeast, and YHE

Sample	Water content (%)
White rice flour (flour-Wr)	14.04
Red rice flour (flour-Rr)	11.67
Black rice flour (flour-Br)	12.86
Yeast-White rice (yeast-Wr)	77.00
Yeast-Red rice (yeast-Rr)	74.06
Yeast-Black rice (yeast-Br)	72.21
YHE-White rice (YHE-Wr)	52.95
YHE-Red rice (YHE-Rr)	58.29
YHE-Black rice (YHE-Br)	62.67

The data in table 1 shows that white rice flour has the highest water content when compared to brown and black rice, which is 14.04%. Water content of YHE is smaller when compared to yeast both YHE-Wr, YHE-Rr and YHE-Br. YHE-Br has higher water content than YHE-Wr and YHE-Rr. Water content of material affect the shelf life. High moisture content causes susceptibility to microbial activity. The water content in the ingredients is the growth medium of mold and fungus.

Cr^{3+} levels on YHE were determined using the voltammetry method, cyclic voltammetry. The voltammetry provides analysis of various metals and provides both qualitative and quantitative information [19]. The first step taken to determine Cr^{3+} is to dispose of Cr^{3+} standard solution with variations in concentrations of 5, 10, 20, 40, 80, and 100 ppm. The next step of each standard solution is measured by its current strength with a voltameter and will produce a voltammogram. The data obtained were analyzed using originPro 7.0 and the current value of the current will be known. Figure 1 shows the voltammogram of the Cr^{3+} standard solution in various concentrations.

According to the Nerst law, strong currents are directly proportional to concentration [19], the greater the concentration of the solution the greater the current strength. Based on the voltammogram results in figure 1, the current strength of each solution can be determined. Figure 2 shows the relationship between current strength and the concentration of Cr^{3+} standard solution.

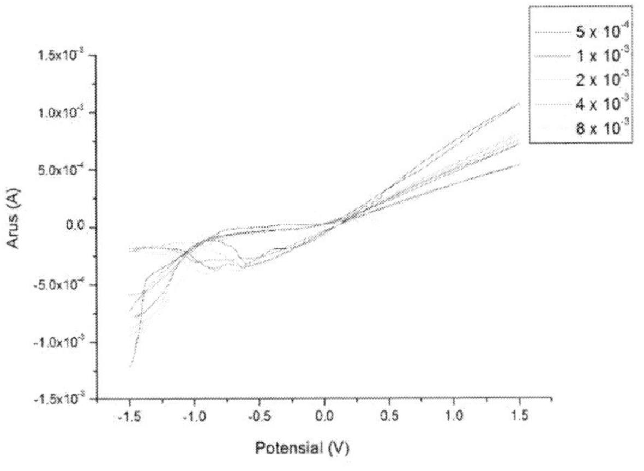

Figure 1. Voltammogram of Cr^{3+} standard solution in various concentrations

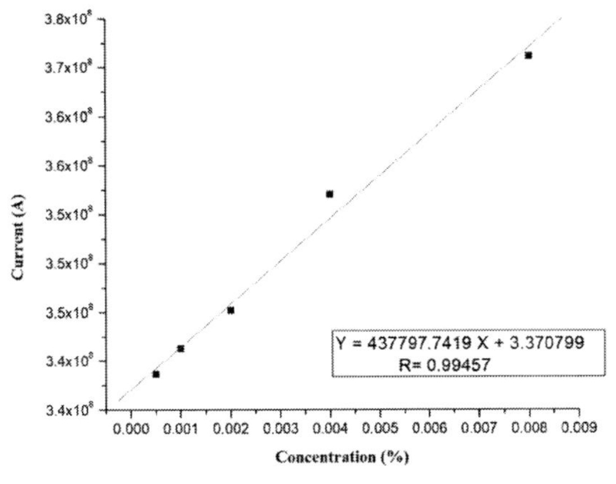

Figure 2. The relationship between current strength and the concentration of Cr^{3+} standard solution

The mapping between concentration and current strength of each standard solution can be obtained by the equation Y = 437797,7 X + 3.370799. Furthermore, measurements of YHE-Wr, YHE-Rr, and YHE-Br sample current strength were measured. The results are entered into the equation Y = 437797,7 X + 3.370799 and the Cr^{3+} concentration of each sample will be known. Table 2 shows the results of measuring Cr^{3+} on YHE from yeast grown in various types of rice flour.

Table 2. Results of Cr^{3+} measurement in YHE

Sample	Cr^{3+} content (%)	Difference of Cr^{3+} between flour and yeast	Difference of Cr^{3+} between yeast and YHE
White rice flour (flour-Wr)	20 x 10^{-4}		
Red rice flour (flour-Rr)	19 x 10^{-4}		
Black rice flour (flour-Br)	20 x 10^{-4}		
Yeast-White rice (yeast-Wr)	101 x 10^{-4}	81	
Yeast-Red rice (yeast-Rr)	103 x 10^{-4}	84	

Yeast-Black rice (yeast-Br)	112 x 10^{-4}	92	
YHE-White rice (YHE-Wr)	32 x 10^{-4}	12	69
YHE-Red rice (YHE-Rr)	37 x 10^{-4}	18	66
YHE-Black rice (YHE-Br)	40 x 10^{-4}	20	72

Table 2 shows that kandungan Cr^{3+} in rice flour, yeast or YHE differ in all measurements. The highest Cr^{3+} levels in YHE from yeast grown in flour-Br are 40x10^{-5} %, while Cr^{3+} in YHE from yeast grown in red rice flour has almost the same value and even close to the yield YHE –Br is 37 x 10^{-5} %. YHE from yeast grown in a type of flour-Wr has the smallest level of 32 x 10^{-5} %. The research results also showed that there were also differences between yeast and YHE, when compared with yeast, the Cr^{3+} content was smaller both in YHE-Br, YHE-Re, and YHE-Br, respectively 69, 66, and 72%. This is probably due to Cr^{3+} being released during the centrifuge process. The following table 3 shows YHE protein levels of sample.

Table 3. YHE protein levels from yeast grown in variations of rice flour

Sample	Protein content (%)	Difference of protein between flour and yeast	Difference of protein between flour and YHE	Difference of protein between yeast and YHE
White rice flour (flour-Wr)	7.45			
Red rice flour (flour-Rr)	9.06			
Black rice flour (flour-Br)	10.60			
Yeast-White rice (yeast-Wr)	40.00	32.55		
Yeast-Red rice (yeast-Rr)	38.66	29.60		
Yeast-Black rice (yeast-Br)	40.19	29.59		
YHE-White rice (YHE-Wr)	38.45		31.00	1.55
YHE-Red rice (YHE-Rr)	38.48		29.42	1.71
YHE-Black rice (YHE-Br)	39.23		28.63	0.96

Table 3 show that the highest protein content in YHE is YHE-black rice which is 39.23%. When compared with the fermentation media, namely flour-Wr, flour-Rr, and flour-Br, the YHE-Br has increased from 10.60% to 39.23%. YHE-Wr also experienced an increase compared to its fermentation media, from 9.06% to 38.45%. Whereas YHE-Br also experienced an increase compared to the fermentation media, from 7.65% to 38.48%. Increased protein levels in YHE-red, white and black rice occur due to the addition and growth of *Saccharomyces serevisiae* during the fermentation process [20]. Cr^{3+} is one of the minerals that is needed for cell metabolism, for growth. Growth is strongly related to protein metabolism. Chromium increases the rate of protein synthesis [21].

YHE black rice has a greater binding ability to Cr^{3+} than YHE-Wr and YHE-Rr. Variations in the type of rice flour give a difference in the results of chromium content because it has a different composition. Crude fiber is known to have a good mineral binding ability, so it can be said indirectly that black rice has the ability to scrape Cr^{3+} larger than others. Thus the impact on Cr^{3+} black rice content is higher than the others. The highest crude fiber content is black rice [20], this supports the research findings which state that the ability to bind the highest minerals, Cr^{3+} is YHE black rice (YHE-Br).

Based on Table 3, it is known that the protein content of each YHE is different. The highest protein level is YHE-Br. The difference in levels of this protein is likely to have something to do with the Cr^{3+} content contained in rice flour. Cr^{3+} is a mineral that is needed by yeast for growth. The highest content of Cr^{3+} black rice compared to the others. This will affect yeast growth in black rice. If yeast growth is high at the time of fermentation, it will affect the high levels of yeast protein produced. Similarly, when the yeast was processed into YHE-Br, the protein content was also the highest, which was 39.23%.

4. Conclusion

Based on the research that has been carried out, the conclusions of this study are as follows: 1) the content of Cr^{3+} from YHE-Wr by $32x10^{-5}$ %, YHE-Rr by $37x10^{-5}$ %, and YHE-Br by $40x10^{-5}$ % ; and 2) YHE-Wr, YHE-Rr, and YHE-Br content respectively 38.45; 38.48; and 39.23%.

References

[1] Risha T J 2011 Effect of pH, Temperature of α-amylase enzyme hydrolysis and bread yeast concentration to produce ethanol using rice bran *Thesis paper* Surakarta: Eleven March University
[2] *https://thefairreporter.com/global-yeast-extract-market-2018-industry-growths-with-cagr-4-3-in-forecast-to-2023/*
[3] Godfrey T and Reichet J 1986 *Industrial enzymology, the application of enzyme in industry* (Great Britain: Stockton Press)
[4] Agustini R and Prahara Mirwa 2015 Yeast hydrolysate enzymatic (YHE) degradation results using pineapple bromelin as preparation material for microbiology and biofertilizer culture media *Competitive grant research report* Not published (Surabaya: Research Institute and Community Service, Surabaya State University)
[5] Purwitasari A and Pangastuti 2004 *Bioteknologi* **1** 37-42
[6] Kang H J, Hwang L K, Kim K S and Choi H C 2006 *J. Agric. Food. Chem.* **54** 4833-38
[7] Kuswaha 2016 *Black rice researce, histori, and Development* (Swizerland: Springer Nature)
[8] Directorate of Community Health Development 1995 *List of Indonesian food nutrition composition* (Jakarta: Ministry of Health) p **77**
[9] Lehninger A L 1990 *Basics of Biochemistry* ed Maggy Thenawidjaja (Jakarta: Erlangga Publisher)
[10] National Center of Biotechnology Information 2001 *Dietary reference intakes for vitamin a, vitamin k, arsenic, boron, chromium, copper, iodine, iron, manganese, molybdenum, nickel, silicon, vanadium, and zinc* USA www.ncbi.mlm.nih.gov
[11] Cefalu W T, Wang Z Q, Zhang X H, Baldor L C and Russell J C 2002 *Journal of Nutrition* **132**(6) 1107-14
[12] Agustini R and Sanjaya IGM 2017 Utilization of yeast hydrolysis enzymatic (YHE) which is produced in various growth media as a drug type 2 by assessing chromium content (III*). Report of the superior university research grant* Not published (Surabaya: Research Institute and Community Service, Surabaya State University)
[13] Yulianto E and Setiarso P 2014 *Making of chitosan modified carbon paste electrode for Cr (VI) metallic analysis with ion interfering Fe (II) and Zn (II) by cyclic stripping voltammetry* (Surabaya: Chemistry, Surabaya State University)
[14] AOAC 2000 *Official Analytical methods of Association of Official Analytical Chemists* Washington D.C
[15] Waites MJ, Morgan NL 2001 *Industrial microbiology: An introduction* (USA: Blackwell Science)
[16] Rohaya MS, Maskat MY, and Ma'aruf AG 2013 *Sains Malaysiana* **42** 1707-14
[17] Daomukda N, Moongngarm A, Payakapol L, and Noisuwan A 2011 Effect of cooking methods on physicochemical properties of brown rice. 2nd *International Conference on Environmental Science and Technology IPCBEE* Vol. 6, IACSIT Press, Singapore.
[18] Harper JM 1981 *Extrusion of food* vol *I* (Boca Raton-Florida CRC Press, Inc)
[19] Wang J 2001 Analitical electrochemistry (New york: Wiley-VHC)
[20] R Agustini, I G M Sanjaya and Widodo A 2018 The use of yeast hydrolysate enzymatic (YHE)

produced in various growth media as anti-dm type 2 by assessing chromium content (III) *Higher Education Research Report* (Surabaya: Institute of Research and Community Service in Surabaya State University)
[21] Burton J L 1995 *Anim. Feed Sci. Tech.* **53** 117-133

IOP Conf. Series: Journal of Physics: Conf. Series **1156** (2019) 012007 doi:10.1088/1742-6596/1156/1/012007

Spin state transition in iron(II): a review on bis-[(2,6-bis(pyrazol-3-yl)pyridine]iron(II) complex

K H Sugiyarto*

Department of Chemistry Education, Faculty of Mathematics and Natural Sciences, Yogyakarta State University, Indonesia

*E-mail: sugiyarto@uny.ac.id

Abstract. Iron(II) has been known to exist as high-spin ($^5T_2$) and low-spin ($^1A_1$) states. The first and the second states are due to the weak and strong ligand fields of the corresponding complexes, respectively. The magnetic moments of the two states are then immediately distinguished, being about 4.9-5.4 BM due to four unpaired electrons, but 0-0.6 BM due to no unpaired electrons in the corresponding electronic configuration of $3d^6$ system for the former and the latter, respectively. It could be reasonably understood that in the "middle" ligand field strength the magnetic moment may no longer reflect neither fully high-spin nor fully low-spin but to be in between the two, and thus in this situation the concentration of the two fractions may be induced by temperature or pressure. Thus, the term spin-state transition or spin crossover in iron(II) or spin equilibium of $^1A_1 \leftrightharpoons {}^5T_2$ are then introduced. Indeed, many complexes of $[FeN6]^{2+}$ were found to be temperature dependence associated with the spin-state transition. In this paper, the iron(II) complexes of six N-coordinating agent of five-six membered rings and its character of magnetism, Mössbauer and UV-Vis electronic spectra as well as bond lengths are discussed.

1. Introduction

Iron(II) is the $3d^6$ system, and in free ion gas, it leads to ^5D ground term (state) with other excited terms, ^3H, ^3G, ^3F (2), ^3D, ^3P (2), ^1I, ^1G (2), ^1F, ^1D (2) and ^1S (2). In [Fe(II)N6]-octahedral configuration, the interaction of those terms results in $^5T_2$ ground state for the weak ligand field, but $^1A_1$ for the strong one. Figure. 1 shows a simplified of the change of the ground state of $^5T_2$ (high-spin) and $^1A_1$ (low-spin) and the *cross-over* region. The typical parameters to observe of the two states are the magnetic properties, Mössbauer and electronic spectral properties and Fe-N bond length. The magnetism is an average of bulk of solid sample, high-spin ~ 4.9-5.4 BM (4 unpaired electrons) and low-spin 0-0.6 BM (no unpaired electron) [1]. In general, the types of spin-state transitions in iron(II) readily observed through magnetism of the solid are abrupt-discontinuous, gradual-continuous, incomplete-continuous, and thermal-hysteresis transitions (figure 2) [2]. Two steps transitions may be also observed, though very rarely.

Unlike magnetism which represents a bulk property of the two spin states, Mössbauer spectra of Fe(II) show each characteristic parameters of both spin-states individually (ΔE_Q and $\delta_{i.s}$), and thus the presence of the two spin states should be readily distinguished. The parameters of -0.2 – 0.5 mm s^{-1} ($\delta_{i.s}$) and 0-2.0 mm s^{-1} (ΔE_Q) with respect to iron foil is typical for the singlet state, $^1A_1$, while 0.5-1.7 mm s^{-1} ($\delta_{i.s}$) and 1.0-4.0 mm s^{-1} (ΔE_Q) is typical for the $^5T_2$ [3] . The electronic spectral property for the high-spin iron(II) appears in about 13000 cm^{-1} while the low-spin one usually appears at 17000cm^{-1} [4]. While the bond length of the high-spin Fe(II)-N is usually observed at about 2.2Å, it is about 1.9Å for the low-spin Fe(II)-N, being about 0.2Å in diffference [5].

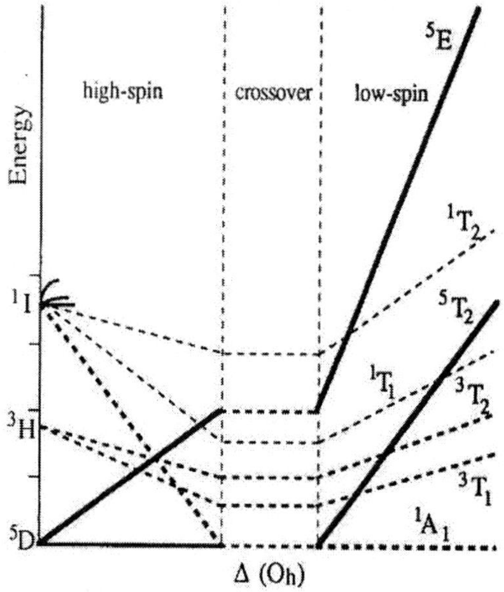

Figure 1. Simplified Tanabe-Sugano, diagram for d^6-iron(ii)

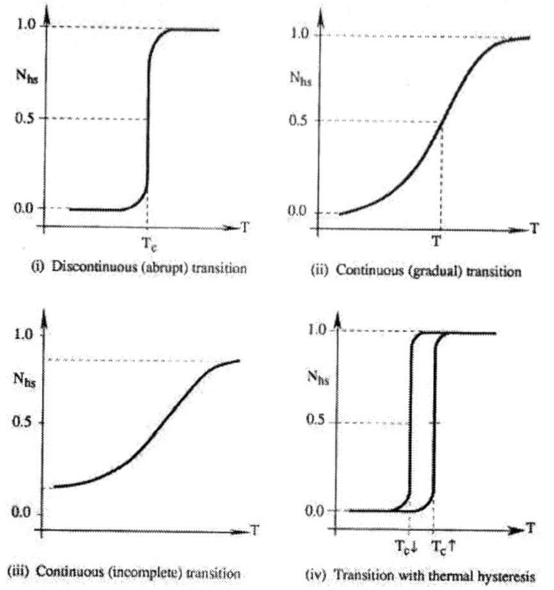

Figure 2. Types of spin state transition in solid

2. Spin transition in iron(II) complexes of 2,6-bis(pyrazol-3-yl)pyridine

2.1. Complex of $[Fe(bpp)_2](BF_4)_2.nH_2O$ (n = 0-2); *bpp = 2,6-bis(pyrazol-3yl)pyridine* [6]

The synthesis of *bpp* was adapted from Lin and Lang [7]. The single step direct preparation of the complex from the corresponding stoichiometric precursors, $FeCl_2.4H_2O$, 3*bpp*, and $2NaBF_4$ in aqueous-alcoholic solution resulted in the desired hydrated complex. On heating, the corresponing anhydrous complex were obtained, with the change in color from brown for the hydrated to bright-yellow for the anhydrous complex. Thus, while at room temperature, the dihydrated complex was composed of the mixture about 50% low-/high-spin ($^1A_1/^5T_2$) with the moment of 3.5 BM. The corresponding anhydrous complex was in fact stabilized by the fully high-spin $^5T_2$ state with the moment of 5.4 BM. The magnetic moment of both complexes were found to be temperature dependence, associated with the spin state transition in iron(II) as shown in figure. 3. The magnetism of the dihydrated complex shows gradual-continuous type within the experimental temperatures, being about 3.5 BM at 300K down to 0.9 BM at 100K which is fully low spin in A (figure 3). However, the anhydrous complex exhibits abrupt-discontinuous-complete spin crossover of $^1A_1 \rightleftarrows {}^5T_2$ in iron(II) on slow cooling, and it is associated with an hysteresis loop of width ΔTc 10K (Tc↓: 170-177K, and Tc↑: 180-187K) as shown in figure 3B. Thus, the anhydrous complex was stabilized by fully quintet state at room temperature with the moment of about 5.40 BM.

The occurence of the transition was confirmed by Mössbauer effects (figure 4). For the dihydrated complex, the Mössbauer spectral property recorded at room temperature shows three main lines, indicating the mixture of the two spin states. It could be well resolved to two sets of parameters, first with $\delta_{i.s}$ (0.27 mm s^{-1}) and ΔE_Q (0.75 mm s^{-1}) which is typical for singlet, $^1A_1$ state, and second with $\delta_{i.s}$ (0.90 mm s^{-1}) and ΔE_Q (2.14 mm s^{-1}) which is typical for quintet, $^5T_2$ state. At low temperature (77K) only a doublet-set of parameters was observed, it is $\delta_{i.s}$ (0.39 mm s^{-1}) and ΔE_Q (0.71 mm s^{-1}) being typical for singlet, $^1A_1$ state. This is consistent with the observed magnetic data. For the anhydrous complex, a single-doublet Mössbauer parameters was typically resolved for fully quintet state only at 295K; $\delta_{i.s}$ (1.01 mm s^{-1}) and ΔE_Q (2.40 mm s^{-1}). At low temperature (77K) by slow cooling, again a single-doublet Mössbauer parameters was typically resolved for singlet state only; $\delta_{i.s}$ (0.37 mm s^{-1}) and (ΔE_Q) (0.68 mm s^{-1}). Detailed parameters were presented in Table 1[6].

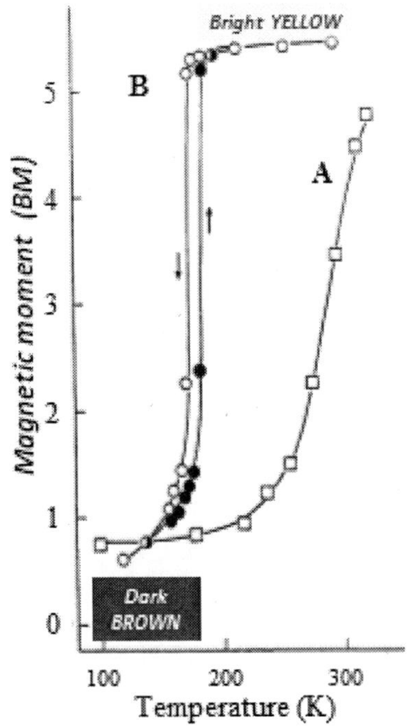

Figure 3. Temperature dependence of magnetic moment of:

A. $[Fe(bpp)_2](BF_4)_2$. $2H_2O$
B. $[Fe(bpp)_2](BF_4)_2$

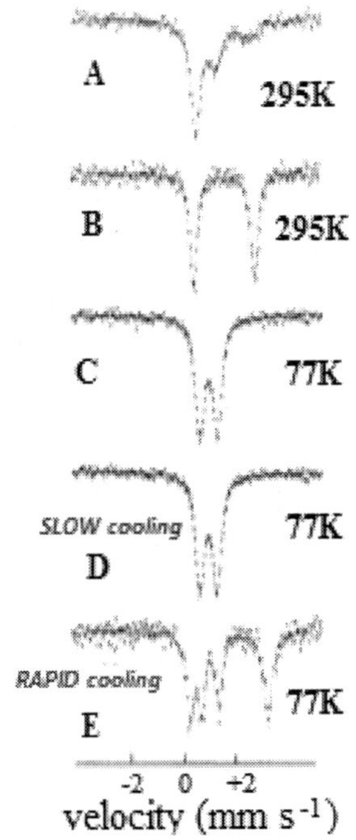

Figure 4. Mössbauer spectra of:
A. $[Fe(bpp)_2](BF_4)_2.2H_2O$ at 295K
B. $[Fe(bpp)_2](BF_4)_2$ at 295K
C. $[Fe(bpp)_2](BF_4)_2$. $2H_2O$ at 77K
D. $[Fe(bpp)_2](BF_4)_2$ at 77K (*slow* cooling)
E. $[Fe(bpp)_2](BF_4)_2$ at 77K (*rapid* cooling)

Table 1. Mössbauer spectral parameters (mm s^{-1})
Estimated error ±0.03 mm s^{-1}

Complex	T (K)	Spin state	ΔE_Q	$\delta_{i.s}$
$[Fe(bpp)_2](BF_4)_2$. $2H_2O$	295	$^5T_2$	2.14	0.90
$[Fe(bpp)_2](BF_4)_2$. $2H_2O$	295	$^1A_1$	0.75	0.27
$[Fe(bpp)_2](BF_4)_2$. $2H_2O$	77	$^1A_1$	0.71	0.39
$[Fe(bpp)_2](BF_4)_2$	295	$^5T_2$	2.40	1.01
$[Fe(bpp)_2](BF_4)_2$	77^A	$^1A_1$	0.68	0.37
$[Fe(bpp)_2](BF_4)_2$	77^B	$^5T_2$	3.01	1.13
$[Fe(bpp)_2](BF_4)_2$	77^B	$^1A_1$	0.67	0.37

^{A} Slow cooling ^{B} Rapid cooling

One interesting phenomena is that, for the anhydrous complex, about 50% of the high spin fraction was trapped on rapid cooling to low temperature (~77K), and then on the elevated temperatures but in the temperature of slow cooling low spin region, the trapped residual high spin fraction was time dependence, and thus the kinetics of high spin to low spin transformation can be observed. After rapid cooling and then bringing to 99K, the magnetic moment due to conversion to fully low spin was recorded with time. Similar treatment was continued but it was then brought to 104K, to 109K, and to 114K. Following the relevant equations, the graphs as shown in figure 5-6 suggested that the $^5T_2 \rightarrow {}^1A_1$ transformation in this residual high spin followed the first order kinetics with first-order constant as shown in table 2 [8]. Similar $^5T_2 \rightarrow {}^1A_1$ transformation has also been studied in [Fe(*ptz*)$_6$](BF$_4$)$_2$, but it was by the different method, i.e. LIESTT-Light Induced Excited Spin State Trapping [4].

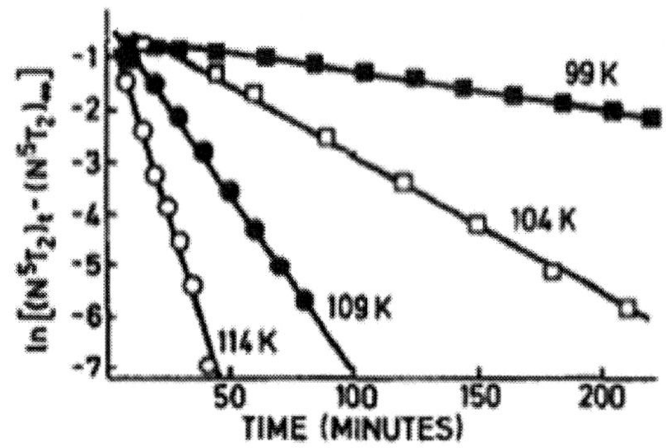

Figure 5. Plot of ln [N (^{5}T$_2$)$_t$ - N (^{5}T$_2$)$_\infty$ *vs* time for the $^5T_2 \rightarrow {}^1A_1$ transformation in [Fe(*bpp*)$_2$](BF$_4$)$_2$ at 99, 104, 109, and 114K

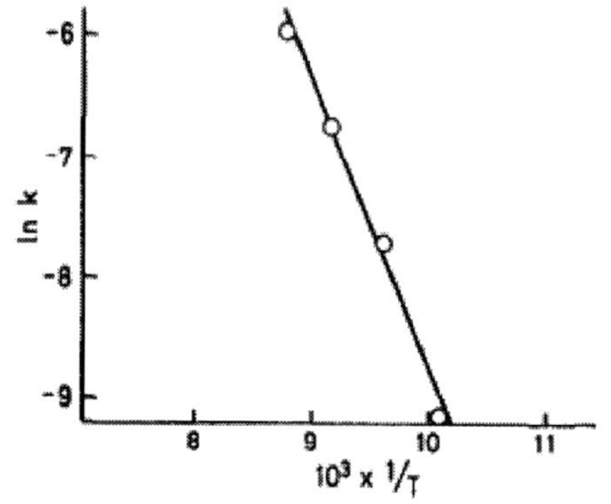

Figure 6. Plot of ln k *vs* 1/T for the $^5T_2 \rightarrow {}^1A_1$ transformation in [Fe(*bpp*)$_2$](BF$_4$)$_2$.

Table 2. First-order rate constant for the $^5T_2 \rightarrow {}^1A_1$ transformation in [Fe(*bpp*)$_2$](BF$_4$)$_2$

T (K)	$10^5\, k$ (s^{-1})
99	10.4
104	43.5
109	113
114	239

2.2. *Complexes of* [Fe(bpp)$_2$](X)$_2$.nH$_2$O (X = ClO$_4$, PF$_6$, NO$_3$, I, Br, NCS, NCSe, CF$_3$SO$_3$ (n = 0-5) [6,9-11]

Various aninonic counterparts were prepared for this cationic complex. All solid salts exhibited the spin state transition $^1A_1 \rightleftarrows {}^5T_2$ in iron(II), with various degree and types of the transitions, being continuos-discontinuous, complete-incomplete, two-steps, and hysteresis. The plot between the magnetic moment and temperature of the perchlorate and hexafluorophosphate salts were depicted in figure 7 [6]. In general, the effect of the lattice water is to stabilize the singlet state for the metal atom, *i.e.* to increase the transition temperature. Thus, complete and continuous type of transition was observed on the former, but incomplete transition and also two-steps on the latter. The Mössbauer spectrum of anhydrous [Fe(bpp)$_2$](PF$_6$)$_2$ was recorded at different temperatures, one well above, one well below, and two in the neighbourhood of the transition region [9]. At room temperature, the spectrum shows the lines that could be resolved into three doublets, both the two outer doublets are normal for high spin iron(II), and the inner doublet is normal for low spin iron(II). The parameters of the outermost doublet-1 are 1.03 mm s^{-1} ($\delta_{i.s}$) and 2.35 mm s^{-1} (ΔE_Q), those of doublet-2 are 1.01 mm s^{-1} ($\delta_{i.s}$) and 1.94 mm s^{-1} (ΔE_Q), and those of the inner doublet-3 are 0.33 mm s^{-1} ($\delta_{i.s}$) and 0.69 mm s^{-1} (ΔE_Q). From the spectra at lower temperatures, it suggests that the outermost doublet-1 does not take part in the transition to the low spin state and thus results in a residual high spin fraction at low temperature as indicated by incomplete transition deduced from the magnetic momentf figure 7b [9].

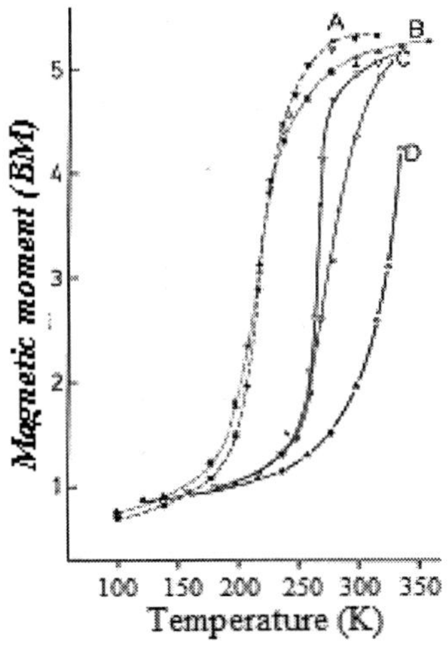

a.

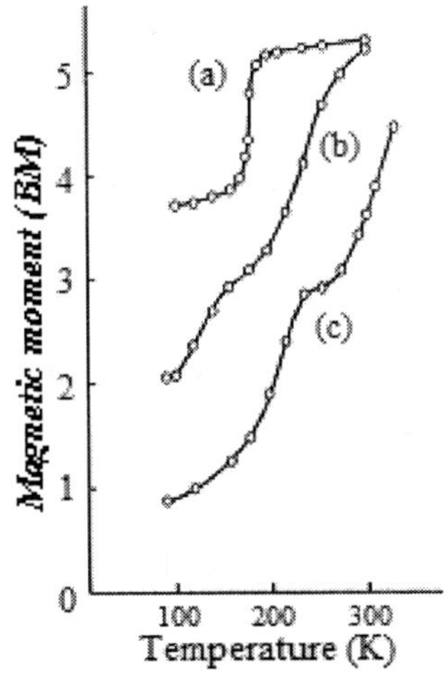

b.

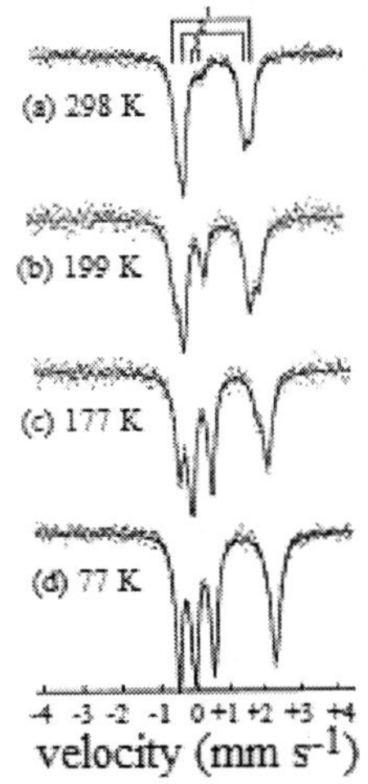

c.

Figure 7.a. The magnetic moment of anhydrous $[Fe(bpp)_2](ClO_4)_2$ (sample A and B) and the monohydrate (C) and the dihydrate (D); **b.** The magnetic moment of anhydrous $[Fe(bpp)_2](PF_6)_2$ (a), the monohydrate (b) and the dihydrate (c); **c.** Mössbauer spectra of $[Fe(bpp)_2](PF_6)_2$ at 298K (a), 199K (b), 177K (c), and 77K (d).

The iodide and bromide complexes were found to be tetra- and penta-hydrate respectively, and they were essentially low spin. The corresponding anhydrous complexes were found to be hygroscopic, and the plot between the magnetic moment and temperature were shown in figure 8 together with the nitrate trihydrate. For both anhydrous iodide (figure 8a.) and anhydrous bromide (figure 8b.), the spin state transition were abrupt and associated with hysteresis for the former only with the hysteresis loop to be very narrow (ΔTc=2K), Tc↓: 203K and Tc↑: 205K, while for the bromide, the spin state transition was centred at Tc = 250K [9].

The thiocyanate complex was isolated as dihydrate, while the selenocyanate was an anhydrous complex [10]. Both complexes undergo abrupt spin state transition to low spin below room temperature (figure 8b,c). For thiocyanate, the transition exhibited two–step and accompanied by hysteresis in both steps with difference in the width of the loop. The second step was 24K (ΔTc), Tc↓: 193K and Tc↑: 219K, but the data was unmeasurable for the first step. The spin state transition in the selenocyanate was centred at Tc: 231K, and no measureable hysteresis could be detected (if any, the width of the loop ΔTc < 1K) [10].

IOP Conf. Series: Journal of Physics: Conf. Series **1156** (2019) 012007 doi:10.1088/1742-6596/1156/1/012007

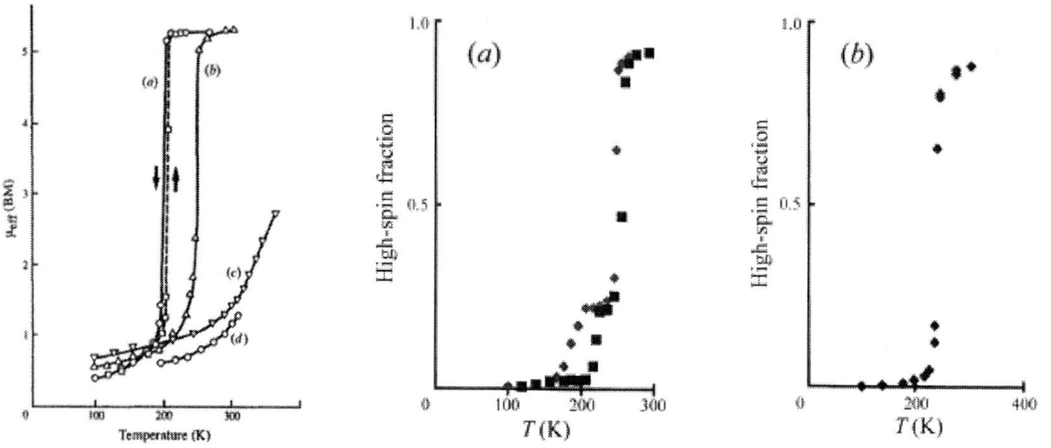

a. b. c.

Figure 8. a. The magnetic moment of anhydrous $[Fe(bpp)_2](I)_2$ (a) $[Fe(bpp)_2](Br)_2$ (b), $[Fe(bpp)_2](NO_3)_2.2H_2O$ (c); and $[Fe(bpp)_2](I)_2.4H_2O$ (d); **b**. The magnetic moment of $[Fe(bpp)_2](NCS)_2.2H_2O$ (a); **c.** The magnetic moment of anhydrous $[Fe(bpp)_2](NCSe)_2$.

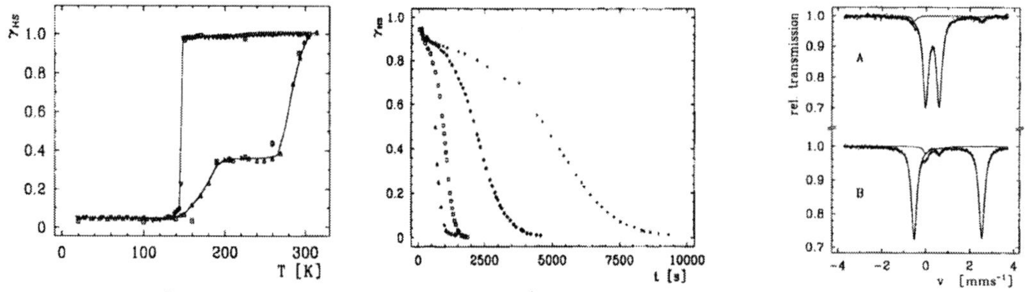

a. b. c.

Figure 9. a. The magnetic moment of $[Fe(bpp)_2](CF_3SO_3)_2$. H_2O with time; **b.** with time after rapid cooling from left-right at 118K, 113K, 108K and 104K; **c.** the Mössbauer spectra of $[Fe(bpp)_2](CF_3SO_3)_2$. H_2O recorded at low temperature on slow cooling (A) and at the same low temperature on rapid cooling (B).

The triflate (CF_3SO_3) complex was similarly isolated as trihydrated which was red-brown and low spin. However, upon heating for dehydration, the bright yellow monohydrate complex was obtained to be predominantly stable as high spin at room temperature [11]. This monohydrate, however, undergoes a discontinuous transition to the low spin state (Tc↓: 147K). This transition is in fact accompanied by unusual broad hysteresis loop which is two-step pattern on low spin → high spin heating direction, giving rise to Tc↑: 285K (ΔTc ≈ 238K) as shown in figure 9a [11]. Unlike the anhydrous tetrafluoroborate, of which the high spin fraction could be only about 50% trapped on rapid cooling, more than 90% high spin fraction of this triflate monohydrate, could be trapped, and thus the kinetics of high spin to low spin relaxation could be studied and the result was shown in figure 9b. The evidence of this trapping was demonstrated by the completely different Mössbauer spectra, recorded at low temperature (20K) by slow cooling (figure 9a, 95% low spin) and by rapid cooling (figure 9b, 95% high spin).

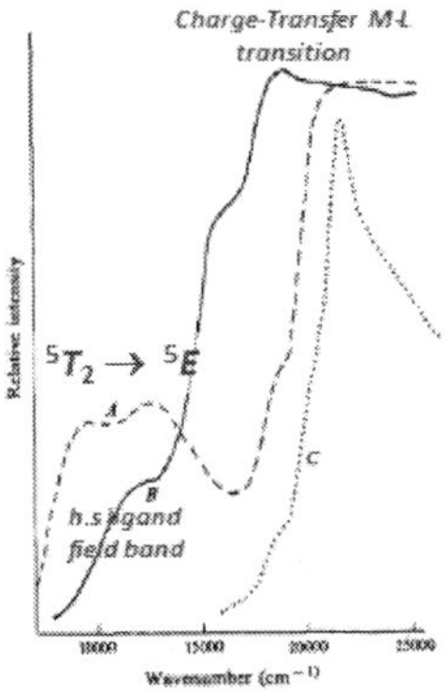

a.

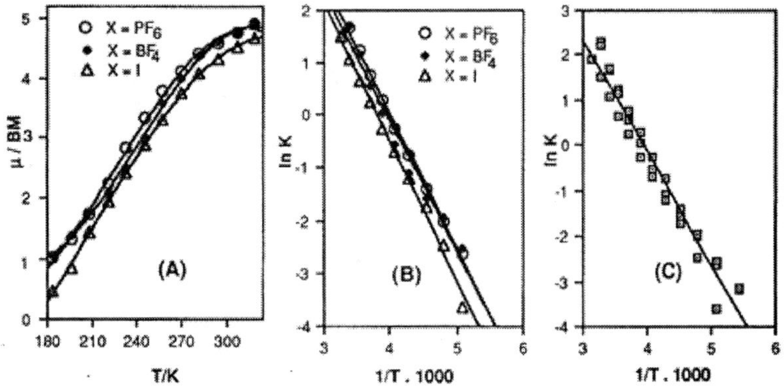

b.

Figure 10. a. Diffuse reflectance spectra of $[Fe(bpp)_2](BF_4)_2$ at room temperature (A), at 100K (B) and in solution (C); **b.** magnetic moment with temperature of $[Fe(bpp)_2](X)_2$ in aceton (A), *ln K vs* 1/T for each salt (B) and the composite plot (C).

In fact, the electronic spectral properties only revealed clearly the high spin field band, since the corresponding low spin ligand field band was masked by the strong and low-energy of the charge-transfer as indicated by its red-brown color. Thus, as shown in figure 10, the room-temperature prominent high spin ligand field band was split to a doublet band due to Jahn-Teller and/or low-symmetry effects centred at 9500 and 12500 cm^{-1}. This band is associated with $^5T_{2g} \rightarrow {}^5E_g$. At about 100K this band disappeared and a strong-low spin charge transfer then moved to lower energy, accounting for visible darkening to brown color with a prominent shoulder appearing at 16800 cm^{-1}, perhaps due to the $^1A_{1g} \rightarrow {}^1T_{1g}$ transition in singlet state iron(II) [6].

From various solid salts, different types of transitions could occur though it could not be predicted. To study if any significant anion effect is manifested in solution, three different salts,tetrafluoroborate, hexafluorophosphate and iodide were dissolved in acetone. The magnetic data were collected according to Evan's method [12]. As shown in A in figure 10.b. all salts showed essentially the same continuous type of spin state transition in iron(II). Thus, considering the actual equilibrium of $Fe^{2+}_{(low\text{-}spin)} \rightleftarrows Fe^{2+}_{(high\text{-}spin)}$, the equilibrium constant, *K*, for each temperature measurement could be calculated. The plot of *lnK vs* 1/T for each solution and for the composite of the three were depicted in B and C in figure 10b. and the thermodynamic parameters, ΔH and ΔS could then be extracted and found to be 20-22 kJ mol^{-1} and 81-84 J K^{-1}mol^{-1}, respectively [6].

3. Structural Studies of the Complex [9,10]

All salts in these instances showed a strong effect of hydrate in stabilizing the low spin state in iron(II) as evidenced by by the structure of $[Fe(bpp)_2](I)_2.4H_2O$ (figure 11a [9]. The hydrogen bonded linking iodide ions, the water molecules and >NH(pyrazolyl) of *bpp* seems very effective in the stabilization of the singlet state. In this instance the average bond length of Fe-N is 1.95Å, being typical for low spin iron(II) and thus, consistent with its magnetic moment at room temperature. In the crystal structure of $[Fe(bpp)_2](CNSe)_2$ (figure 11b), the average bond length of Fe-N is 2.17Å [10], being typical for high spin iron(II) and consistent with its magnetism at room temperature. The difference of the bond length between the two spin states is 0.22 Å, which is normally observed in the spin transition system.

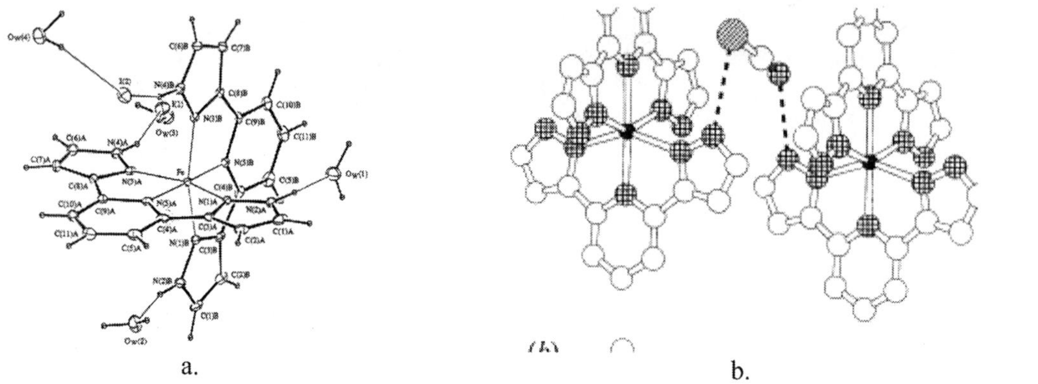

a. b.

Figure 11. a. Crystal structure of $[Fe(bpp)_2](I)_2.4H_2O$ showing effective H_2O–I–$NH_{pyrazol}$ bonding and $[Fe(bpp)_2](CNSe)_2$; **b.** showing –(NCS)–bonded to $NH_{pyrazol}$ of complex-1 and complex-2.

4. Concluding Remark

$[Fe(bpp)_2](X)_2.nH_2O$ (n =0-5) demonstrated various types of spin state transitions in iron(II). The continuous-discontinuous, complete-incomplete, thermal hysteresis, and two-step type could be generated from these samples. The interesting phenomena was the availability of the samples to perform rapid cooling to have freezing-in of the high spin fraction at low temperature so that the kinetics of high-spin → low spin relaxation could be studied.

References

[1] Sinn E 1970 *Coord. Chem. Review.* **3** 313-347.
[2] Gütlich P 1981 *Advanced Chem. Series* **194** 405-452.
[3] Goodwin H A 1976 *Coord. Chem. Reviews* **18**(3) 293–325.
[4] Decurtins S, Giltlich P, Kohler C P, Spiering H and Hauser A 1984 *Chem. Phys. Lett.* **105** 1-4.
[5] Goodwin H A, Kurcharski E S and White A H 1983 *Aust. J. Chem.* **36** 1115-24.
[6] Sugiyarto K H and Goodwin H A 1988 *Aust. J. Chem.* **41** 1645-63.
[7] Lin Y I and Lang S A 1977 *J. Heterocycl. Chem.* **14** 345-347.
[8] Goodwin H A and K H Sugiyarto 1987 *Chem. Phys. Letters.* **139** 470-474.
[9] Sugiyarto K H, Craig D C, Rae D and Goodwin H A 1994 *Aust. J. Chem.* **47** 869-890.
[10] Sugiyarto K H, Scudder M L, Craig D C and Goodwin H A 2000 *Aust. J. Chem.* **53** 755-765.
[11] Sugiyarto K H, Weitzner K, Craig D C and Goodwin H A *Aust. J. Chem.* **50** 869-873.
[12] Evans D F 1959 *J. Chem. Soc.* 2003-2005.

Silica-nanoparticles in slow release supplement: preparation and characterization

K S Budiasih [1*], Z Ikawati [2], Z Marsha[1] A Aris[1] and R Chrismara[1]

[1] Department of Chemistry Education, Faculty of Mathematics and Natural Sciences, Universitas Negeri Yogyakarta, Yogyakarta 55281
[2] Faculty of Pharmacy, Universitas Gadjah Mada, Yogyakarta 55281

*E-mail: kunsb@uny.ac.id

Abstract. The nanosilica was prepared from bamboo leaves ash by calcination method in 800°C for an hour. The result was 350 g silica from 500 g of its ash. The SEM shows a homogeneous surface of SiO_2 and the EDX consists of 56,73% Si. The FTIR test indicates the Si-O-Si functional groups. The XRD shows $2\theta = 21,6$ which corresponds to JCPDS No. 39-1425 and 18-1170. The supplement was vitamin C tablet which prepared into slow release supplement model, by mixing of three component : generic ascorbic acid, magnesium stearate, and nanosilica in certain composition. Some assessments were carried out to determine the tablet's characteristic: the size, homogeneity, and hardness. Physical testing of the slow-release vitamin C tablet was found to meet the requirements of tablet by Indonesian Farmakope. In vitro assessment was represented by dissolution test in simulation solutions of stomach by 0,1M HCl and the intestine by citric acid. The result showed that the dissolution of the samples were slower than the control tablets. This indicates that the addition of silica-nanoparticlesaffect the characteristics of slow release supplement.

1. Introduction

Silica was found in several biomass, including the side product of agriculture. The silica in the form of nanoparticles form biomass were developed in recent years, ie from rice husk ash [1] and coconut shell [2]. Application of biomass based silica-nanoparticles were reported in the preparation of several industrial product, i.e. in food industry [3] and drug delivery [4]. The size of nanoparticles give an advantage in mobilization as nanocarrier. It was conducted in the delivery of curcumin [5]. Kumar et al, (2016) developed the silica-nanoparticles from bamboo leaf. Bamboo leaves contain more than 70% of silica [6]. This research on the preparation of oral supplement was supported by silica-nanoparticles (SNs). This paper reported the result of preliminary experiment with vitamin C as the supplement. The utilization of SNs is the part of the medical application of silica-nanoparticles as drug carrier. Slow release supplement is a model of SNs supported drug.

Vitamin C, in this case, was selected as the sample of slow release supplement. People use vitamin C as the supplement which is usually needed in increasing the body resistance. The development of slow release drug / supplement is needed to decrease the frequency of the drug/supplement consumption. The addition of supporting material as drug carrier will accommodate this purpose. One of the recent research of drug carrier in nano medicine is silica-nanoparticles (SNs) [7]. In vivo study is needed to find the bioavailability and biocompatibility in the human body, which is begin with the application in animal model [8-9].

2. Material and Method

2.1. Materials
The bamboo leaves were collected from Sleman Yogyakarta, Indonesia. The main chemicals (HCl, Sodium citrate, Magnesium Stearate, and buffer) were from Merck, vitamin C from IPI, aquadest from The Chemistry Laboratory, UNY.

2.2. Preparation of silica from bamboo ash
The bamboo ash was prepared by direct drying the leaves under the sun, followed by combustion in traditional stove, and calcined by muffle furnace at 800 ^{0}C in 6 hours. The product was then washed in 0,1 M HCl (by ratio 1:6) and well stirred in 60 minutes, and filtered by Whatman filter paper after 24 hours and then washed in water until neutral. The residue was then heated in the oven at 110 °C in 2 hours to remove the retained water [6].

2.3. Characterization of nanosilica
The morphology of the silica was determined by Scanning Electron Microscopy-Electron Dispersive X-Ray Analyser (SEM -EDX) of JEOL JED-2300 model, and then continued to determine the crystalinity by X Ray diffraction (XRD Lab-X Type 6000 Shimadzu Japan). The infrared spectrum was recorded by Fourier Transform InfraRed (FTIR) Horizon MB3000.

2.4. Preparation of BNs-supported supplement
Preparation of SNs supported supplement was conducted by mixing the materials: vitamin C, magnesium stearate and SNs in certain composition and pressed to round shaped tablets. The control sample was the mixture without SNs. The characteristic of tablets include the homogeneity, size, and hardness of the tablets. The main characteristic is the dissolution process which was determined by dissolution test in variation of media vs. time of dissolution. Dissolutions test was conducted in both HCl and sodium stearate media, as the simulation media of stomach and intestine.

3. Result and Discussion

3.1 Preparation of nanosilica from bamboo leaves
The yield from 700 grams bamboo leaves was 502 grams charcoal which was then calcined at 800°C in 6 hours to get 350 grams of greyish ash of silica.

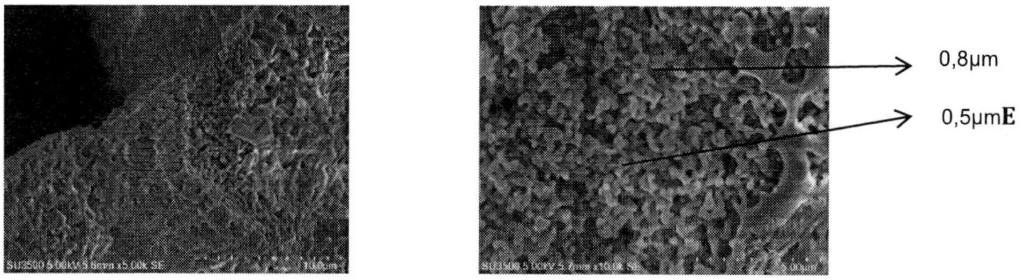

A B

Figure 1. The SEM image of silica [(A= 5000x, B = 10.000x]

3.2 The Sem-EDX result of the nanosilica
The morphology of the silica was characterized by Scanning Electron Microscope *-Electron Dispersive X-Ray analyser).* The SEM image of the silica from bambo leaves was shown in figure1. Figure 1 shows the soft and homogenous surface of silica. The particles have ellipsoid and sphere shape. Stanley et al., (2014) also reported the *plate and spheres*shape with the size of the aggregate wasin range of 0.5-1.0 μm [10].

The EDX result is shown in figure 2 indicating that the composition of the ash are 56.73% Si and 43.26% oxygen. The pure SiO_2 contained 58.63% Si, so this result was comparable to the previous publication [11]].

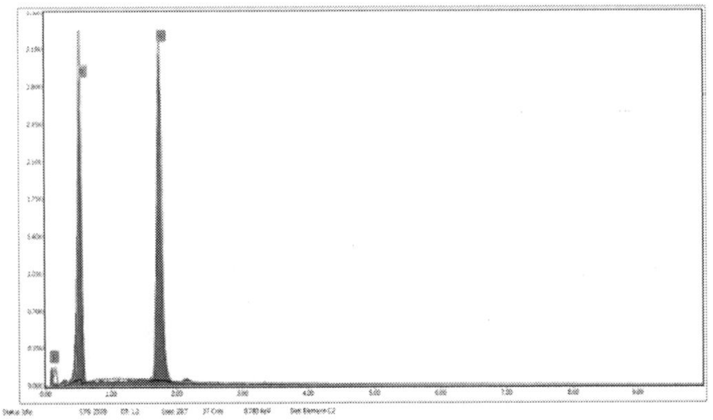

Figure. 2. The EDX result of bamboo leaf ash.

3.3. The FTIR Spectra of the nanosilica

The infrared spectrum of bamboo leaf silica after calcined at 600°C was shown in figure 3 and the vibration characteristic was listed in table 1. This spectrum showed the vibration of –OH, C-H, Si-O-Si, and Si-OH, which are associated with silica

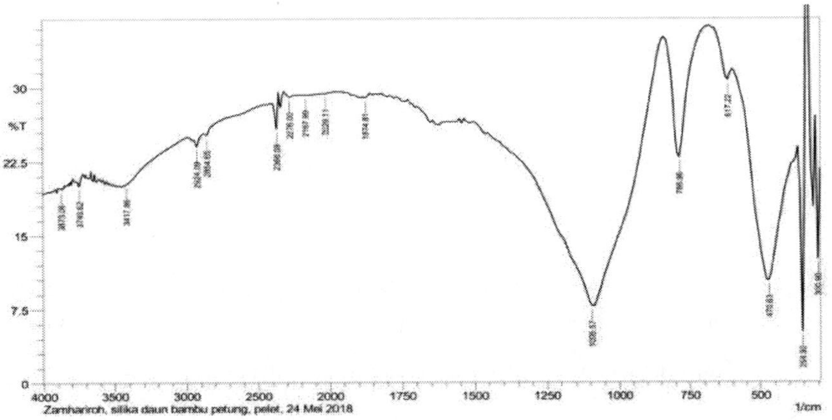

Figure 3. The Infrared spectrum of bamboo leaf silica.

Table 1. The characteristic vibrations of the silica.

Wavenumber (cm^{-1})	Vibration	References
3749.62	Stretching Vibration of –OH form Si-OH	Stanley and Nesaraj, 2014 [12] Priyanto, 2015 [13]
2368.59	Stretching Vibration of C-H	Wang et al, 2000 [14]
1095.67	Asymmetric stretching of Si-O of Si-O-Si	Stanley and Nesaraj, 2014 [12];
786.96	Stretching Vibration of Si-O in Si-OH	Priyanto, 2015 [13]

IOP Conf. Series: Journal of Physics: Conf. Series **1156** (2019) 012008 doi:10.1088/1742-6596/1156/1/012008

3.4 The X Ray Diffaction analysis of the nanosilica
The X- Ray diffraction analysis on the silica shows the pattern on the range of 2θ of 2° to 80° . The XRD pattern shows a sharp peak on 2θ=21.6° associated to Si. According to PDF No. 39-1425 and 18-1170 from JCPDS (figure 4b), Si has characteristic peaks on 2θ=20.5 °, 21.6 °, and 23.3°. The formation of crystalline silica started on 650° C and completed in 800° C. The XRD pattern of the sample was shown in Figure 4 (a) and compared with the pattern in JPPDS (b).

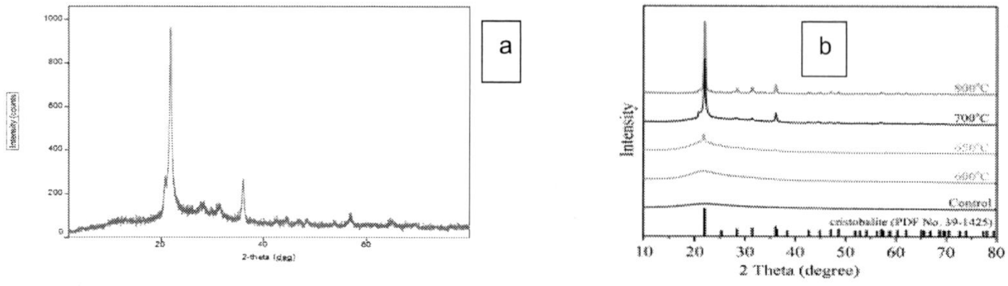

Figure 4. (a) The XRD pattern of bamboo leaf silica (b)The JCPDS data (reference).

The particle size was 7.057 nm, following Scherrer formula. It matches with the requirements of nanoparticles size, in range 5-100 nm. This product was labeledas Bamboo Nanosilica BNs). This product was then applied in the preparation of BNs suppported supplement.

3.5 Preparation and it's in vitro study of slow release supplement.
Preparation of BNs supported supplement was conducted by mixing the composition as shown in table 2. The mixture was then pressed to round shaped tablets with diameter of 0.72 cm. The weight of each tablet was about 160 mg. The control sample was the mixture without BNs.

Table 2. Formulation of BNs supported Vitamin C.

No	Sample code	composition		
		Vit.C	Magnesium stearate	BNs
1	A	140 mg	20 mg	-
2	B	100 mg	50mg	10 mg
3	C	100 mg	45 mg	15 mg
4	D	100 mg	40 mg	20 mg

The physical characteristic of tablets is determined in terms of size, weight, hardness, fragility and dissolved time. The result was listed in table 3.

Table 3. The physical characteristic of BNs - Vitamin C.

Physical characteristic	**Formula**			
	A	**B**	**C**	**D**
Mass (mg)	170.07	170.10	170.17	170.28
Diameter (cm)	0.71	0.72	0.72	0.72
Hardness (kg/cm2)	0.396	0.674	0.676	1.188
Fragility (%)	0.2	0.7	0.8	1.4
Dissolved time (minute)	8.5	5	6	6

All the result have fulfiled the requirement of oral supplement according to the 4th Indonesian Farmakope, that the standard deviation of the value is not more than 10%. The BNs addition was also found to increase the hardness of the tablets. The higher the hardness, was found the slower the supplement released. According The 4th Indonesian Farmakope, the maximum hardness of the

IOP Conf. Series: Journal of Physics: Conf. Series **1156** (2019) 012008 doi:10.1088/1742-6596/1156/1/012008

supplement tablets is 1.5 kg/cm^{-2}. It means, all the samplesmeet the requirement of the hardness. The fragility of BNs -vitamin C also meet the requirement of the standard supplement tablet, namely maximum 1%. All samples ranged in 0.4 – 0.8 %, so that the parameter of the formulation of a good tablet also has been fulfilled. The standard of dissolved time of this supplement tablet is 15 minutes (max). All samples also meet the requirements of Indonesian Farmakope [14]

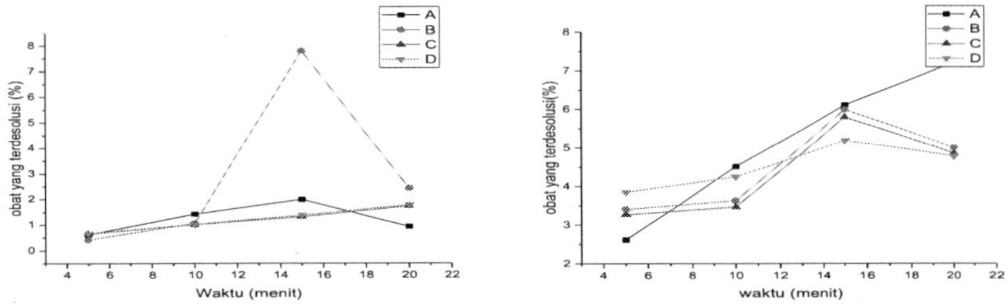

HCl medium Sodium citrate medium

Figure 5. The dissolution test in 2 simulation media

The main characteristic is the dissolution process which is determined by dissolution test by variation of media vs time of dissolution. This test was associated with the ability of the drug in the interaction with the human body [15]. dissolution test was conducted in both HCl and sodium stearate media, as the simulation media of stomach and intestine. The result of the dissolution test of the sample was shown in figure 5.

Interpretation of the dissolution data was carried out by observation of the curve pattern of the formula in two media. The HCl medium is the simulation of stomach and the sodium citrate is the simulation medium of intestine. The profile of dissolution was formed by plotting of the amount of the released particles of the supplement vs time (in minute) (figure 5). The observation was carried out in the sample of 10% weight of the tablets. The rate of the release of the supplement was proportional to the BNs contained in the tablets. The higher the silica content, results in the slower the release. The standard of dissolution time of slow release drug by Indonesian National Agency of Drug and Food Control is 60 minutes. It was provided that the addition of silica-nanoparticles was useful in preparation of slow release supplement.

4. Conclusion

The preparation of bamboo leaf silica-nanoparticles (BNs) was successfully conducted by calcination of bamboo leaves ash at 800 °C in 6 hours. The particles were crystalline and the size is 7.057 in average. The utilization of bamboo leaf silica-nanoparticles in supplement preparation meet to the requirement of the slow releass drug / supplement, referring to Indonesian National Agency of Drug and Food Control.

References

[1] Pukird, S, Chamninok P, Samran ,S, Kasian, P, Noipa, K, and Chow, L 2009 *JMMM* **19** 33-37

[2] Madakson, PB, Yawas DY, and Apasi, A 2012 *Int. J. Eng. Sci. Tech (IJEST)* **4** 03 1190-1198

[3] Chaudhry Q, Scotter M, Blackburn J, Ross B, Boxall A, Castle L, Aitken R and Watkins 2008, *Food Additives. Contaminants*, **25** 3 241–258

[4] De Jong, W H and Borm P J A 2008, *Int. J. Nanomed.* **3**(2) 133-149.

[5] Bolouki A, Rashidi L, Vasheghani-Farahani L and Piravi-Vanak Z 2015 *Int. J. Nanosci. Nanotechnol.* **11** 3, 139-146.

[6] Kumar, V, Tiwari P, Krishnia L, Kumari R, Sing, A, Ghosh A and Tyagi P K 2016 *Adv.Mater. Lett.* **74** 271-276

[7] Tang L and Cheng J 2013 *Nano Today* **8** 290-312

[8] Asefa T and Tao Z 2012 *Chem. Res. Toxicol.* **25** 2265–84.

[9] K S Budiasih, C Anwar, S J Santosa, H Ismail and I P Sari 2013 *J. Chem. Pharm. Res.* **5**(9) 34-39
[10] Tang L and Cheng J 2013 *Nano Today* **8** 290-312
[11] A K Prodjosantosa, K S Budiasih, Utomo P, Indah D and Prasiwi H 2018 *Orient. J. Chem.* **34**
[12] Stanley R and Nesaraj A S 2014 *Int. J. App. Sci. Eng.* **12**(1) 9-21
[13] Priyanto A, Ilmu F, Dan T, Islam U and Walisongo N 2015. Skripsi. Semarang: Universitas Islam Negeri Walisongo.
[14] Ditjen Pengawas Obat dan Makanan 1995 *Farmakope Indonesia. Edisi IV* (Jakarta: Departemen Kesehatan RI)
[15] Vijayalakshmi U, Balamurugan A and Rajeswari S 2005 *J. Trends. Biomater. Artif* Organs **18**(2) 101-105

Ackowledgment
Thank to The Ministry of Research, Technology and Higher Education for the Research Funding.

IOP Conf. Series: Journal of Physics: Conf. Series **1156** (2019) 012009 doi:10.1088/1742-6596/1156/1/012009

Determination of heavy metals concentration in produced water of oil field exploration in siak regency

M Hardi*, Y I Siregar, S Anita, M Ilza
Student of Environmental Science Program, University of Riau, Patimura, Kota Pekanbaru, Riau, 28127 Indonesia

Email: maul_94@yahoo.com

Abstract. Oil and gas industry exploration activities produce a lot of formation water called produced water with a total of up to millions of barrels per day. Produced water from oil fields has different chemical compositions depending on geographic location, formation depth, oil production methods, and age of oil production wells. Currently, produced water is still considered as industrial waste water containing heavy metals which are harmful to humans and the environment, and require special treatment to be able to dispose of it into the environment. To determine heavy metals content in produced water of oilfield exploration, laboratory tests were performed for Siak Regency of Riau Province. Laboratory testing parameters include arsen, barium, boron, cadmium, cobalt, chromium hexavalent, iron, copper, lead, manganese, selenium, mercury, and zinc using instrument inductively couple plasma (ICP-OES) Perkin Elmer 8300DV, spectrophotometer HACH DR2800, and LUMEX RA-915. The results show the heavy metal content of <0,01 – 1,0 mg/L, and only barium and boron have a concentration of more than 1.0 mg/l. Without advance treatment especially to eliminate heavy metal content, the produced water in Siak Regency is harmful as source of drinking water and other daily activity including for fisheri and crop plantation.

1. Introduction

Produced water is water trapped in subsurface formations that is brought to the surface during oil and gas explorations. This water has been in contact with the hydrocarbon-bearing formation for centuries, and as a result, it contains some of the chemical characteristics of the formation and the hydrocarbon itself. Produced water may include water from the reservoir, water injected into the formation, and any chemicals or surfactant added during the production and treatment processes. Produced water is sometimes referred to as brine or formation water [1].

The chemical composition of produced water vary depending on the geographic location of production area, subsurface formation with which the produced water has been in contact for thousands of years, and the type of hydrocarbon product being produced. Produced water volume also vary throughout exploration time of a reservoir. If water and chemical or surfactant injections are conducted to enhance oil production, the chemical composition in produced water may vary even more dramatically.

The metals in produced water have concentration depend on the age and geology of the reservoir from which the oil and gas are produced. The common metals found in produced waters include zinc, lead, manganese, iron, and barium [2]. Discarding without proper treatment of the produced water can result in severe environmental damage and even kill the surrounding environment, kill water and plant life, and damage the soil that will affect humans [3].

Many heavy metals are identified as toxic and carcinogenic substances also gave negative impact to the human population and other living creatures in environmental. Heavy metals have a tendency for bio-accumulate in the environment and cause a negative effect on human health. When heavy metals enter the food chain in large concentrations, they may also enter and accumulate in the human body. Ingested or exposed by heavy metals in concentrations beyond the regulated limit will create serious health problems in human. The serious problem in human caused by heavy metals exposure including abnormality of growth and development, functional damage of organ, cancer, nervous system damage, and in worse case is death.

Exposure to some metals, such as mercury and lead, may develop autoimmunity and in some people the immune system attacks its own cells. This can lead to joint diseases such as rheumatoid arthritis, and diseases of the kidneys, circulatory system, nervous system, and damaging of foetus's brain [4].

In Indonesia, the produced water is considered as a waste and become the biggest waste from oil and gas industry which produced more than a million barrels per day, and the industry handled this type of waste by discharging after treatment to the water bodies or with subsurface injection.

Subsurface re-injection is the preferred disposal method, although it can be infeasible as the formation may not be able to receive the produced water. Injection wells are sometimes operated in concert with adjacent extraction wells to enhance recovery. A combination of produced water and other constituents is used to flood the reservoir to increase the recovery of oil and gas. This process is referred to as enhanced oil recovery, and is a common practice in the industry [5].

2. Method of Study

The objective of this study is to determine heavy metals concentration in produced water of oil field exploration in Siak Regency and comparing with Indonesian Government Regulation No. 82 year 2001 on the Management of Water Quality and Control of Water Pollution: stating that Class 1, water that can be used for drinking water, and or other provisions which require the same water quality; Class 2, water which the designation can be used for water recreation infrastructure, freshwater fish cultivation, livestock, water to irrigate crops, and or other uses that require the same water quality; Class 3, water for which the designation can be used for aquaculture bargaining, breeding, water to irrigate crops, and or other uses requires the same water quality.

Produced water samples were taken from each Water Injection Pump (WIP) located in Oil Gathering Station (GS) from 6 production areas in Siak oilfield with 3 replicates taken from March – April 2017. Gathering Station (GS) is a facility in each area production with separation facilities to collect fluid from production wells to separate oil, water, and gas. WIP is units for injecting produced water after separation facilities in Oil Gathering Station into oil production reservoir for enhance oil recovery (EOR) system.

To ensure quality of laboratory data, samples were taken and handled as procedure and all requirement stated in Standard Method SM 1060, also sample transported to laboratory in same day, properly preserved, and analyzed within holding time.

3. Laboratory Analyses of Produced Water

Produced water samples were analyzed in Technology Support Laboratory, accredited for ISO 17025 to gain quality data of heavy metals analyses for Arsenic (As), cobalt (Co), barium (Ba), Boron (B), selenium (Se) cadmium (Cd), Chromium Hexavalent (Cr^{6+}), copper (Cu), iron (Fe), lead (Pb), manganese (Mn), mercury (Hg), and Zinc (Zn).

Produced Water samples were filtered with 0.45 mm and acidified with HNO_3 to pH < 2 then direct aspired to Inductively Couple Plasma Optical Emission Spectrometry (ICP-OES) Perkin Elmer 8300 DV as procedure in standard method of SM 3120 B. For mercury content, after 0.45 μm filtration and acidified, produced water samples were analyzed using instrument Mercury Analyzer LUMEX RA-915. For Cr^{6+}, filtered samples were analyzed using HACH DR2800 spectrophotometer. As quality assurance for heavy metals analyses, instruments were calibrated with standard reference materials (SRM) and certified reference materials (CRM) were performed to check accuracy and reliability of measurement.

4. Result and Discussion

The result obtained from laboratory analyses for concentration of heavy metals in produced water are collected in Table 1 along with the heavy metals limits for Indonesian Government Regulation No. 82 year 2001 on the Management of Water Quality and Control of Water Pollution.

Produced water from oilfield in Siak contained following range of mean concentration level from 6 GSs in oilfield: Barium (Ba) 1,02-2,35 mg/L; Boron (B) 5,67-7,15 mg/L; Selenium (Se) 0,03-0,04 mg/L; Cadmium (Cd) <0,03-0,01 mg/L; Chrom Hexavalent (Cr^{6+}) 0,01-0,03 mg/L; Copper (Cu) <0,01-0,02 mg/L; Iron (Fe) 0,10-0,20 mg/L; Manganese (Mn) 0,01-0,02 mg/L, and Zinc (Zn) <0,01-0,02. For other metals have concentration below method detection: Arsen (As) <0,01 mg/L; Cobalt (Co) <0,01 mg/L; Lead (Pb) <0,01 mg/L, and Mercury (Hg) <0,001 mg/L.

Table 1. Result of analyses for produced water samples and No. 82 of 2001 on the Management of Water Quality and Control of Water Pollution.

Metals	Unit	Class			Mean of Concentration					
		I	II	III	GS 1	GS 2	GS 3	GS 4	GS 5	GS 6
As	mg/L	0,05	1	1	<0,01	<0,01	<0,01	<0,01	<0.01	<0,01
Co	mg/L	0,2	0,2	0,2	<0,01	<0,01	<0,01	<0,01	<0.01	<0,01
Ba	mg/L	1	(-)[a]	(-)[a]	1,42	1,69	1,11	1,11	1,02	2,35
B	mg/L	1	1	1	7,15	6,19	5,67	5,89	6,94	7,08
Se	mg/L	0,01	0,05	0,05	0,04	0,03	0,04	0,03	0,04	0,04
Cd	mg/L	0,01	0,01	0,01	0,01	<0,003	0,003	0,003	0,01	<0,003
Cr 6+	mg/L	0,05	0,05	0,05	0,01	0,02	0,02	0,02	0,03	0,02
Cu	mg/L	0,02	0,02	0,02	0,02	0,01	<0,01	<0,01	<0,01	<0,01
Fe	mg/L	0,3	NA	NA	0,13	0,13	0,12	0,17	0,20	0,10
Pb	mg/L	0,03	0,03	0,03	<0,01	<0,01	<0,01	<0,01	<0,01	<0,01
Mn	mg/L	0,1	NA	NA	0,02	0,02	0,02	0,02	0,01	0,02
Hg	mg/L	0,001	0,002	0,002	<0,001	<0,001	<0,001	<0,001	<0,001	<0,001
Zn	mg/L	0,05	0,05	0,05	0,02	0,01	0,02	<0,01	0,02	<0,01

[a]Not regulated

Refer to Government of Indonesia regulation No. 82 year 2001, the heavy metals having concentration above the limit of class 1 are barium, boron, and selenium. Meanwhile for class 2 and class 3 only boron above the regulation limit.

Because of its toxic effect on plants, animals and human, heavy metals are listed in environmental pollutant category Heavy metals are persistent in nature, therefore getting accumulated in soils and plants. Dietary intake of many heavy metals through consumption of plants has long term negative effects on human health [6].

The heavy metals can have a significant impact on health also the impact on the environment. Heavy metals when entered into the body of water directly can cause special effects on living things, such as the case of minamata, damage to the nervous system, influence on fetal development, carcinogenicity and impaired immune function, so that it can be said that all heavy metals can become toxic when accumulated in the body for a long time [7].

Selenium (Se) is present in nature and often in association with sulfur containing minerals. It is normally found in concentrations of 50–90 μg/kg, but higher concentrations can be associated with some volcanic, sedimentary and carbonate rocks. Selenium is one of essential element for humans, and there are indications that selenium status may be marginal in many parts of the world. High intakes of selenium are associated with a number of specific diseases and the potential for adverse effects [8]. Even the human body actually can tolerate quite high levels of selenium without adverse effects on health. However, at high doses (>900 μg selenium/day), selenium can elicit toxic effects collectively termed

IOP Conf. Series: Journal of Physics: Conf. Series **1156** (2019) 012009 doi:10.1088/1742-6596/1156/1/012009

selenosis, with symptoms including gastrointestinal upset, hair loss, nausea, irritability, fatigue and mild nerve damage [9].

Barium is present as a trace element in sedimentary rocks. It is not found free in nature and occurs in a number of compounds, commonly in form of barium sulfate (barite) and to a lesser extent, barium carbonate (witherite). Barium is not a crucial element for human nutrition. High concentrations of barium will cause vasoconstriction by its direct stimulation of arterial muscle, peristalsis as a result of the violent stimulation of smooth muscles and convulsions and paralysis following stimulation of the central nervous system. Death may occur in a few hours or a few days depending on the dose and solubility of the barium salt. The acute toxic oral dose is between 3-4g [10].

Barium compounds that easily dissolve in water such as barium acetate, barium chloride, barium hydroxide, barium nitrate, and barium sulfide can cause serious health effects. Barium carbonate does not dissolve in water, but will dissolve in acidic condition in the stomach; it can also cause harmful health effects. Consuming very large amounts of barium compounds that dissolve in water or in the stomach can cause changes in heart rhythm or paralysis in humans. Some people who do not seek medical treatment immediately after eating or drinking a very large amount of barium have died. Some people who eat or drink somewhat smaller amounts of barium for a short period may experience vomiting, abdominal cramps, diarrhea, difficulties in breathing, abnormal blood pressure, numbness around the face, and muscle weakness [11].

Boron is a non-metal element in Group 13 of the Periodic Table. Properties of boron are very close to carbon and silicon. Boron is a trace element which have an important role in mineral and hormonal metabolisms, cell membrane functions, and enzyme reactions. Boron is almost completely absorbed through gastrointestinal and respiratory systems and mostly present in body tissues and fluids as boric acid $B(OH)_3$ and in lesser amounts as $B(OH)^{4-}$ anion [12]. Boron concentration in produced water is significantly high and makes it unusable if not properly remove. The health implication of boron consumption is enormous because according to the medico-biological investigation, boron compounds belong to second class of toxicological danger [13]. Exposure to large amounts of boron (about 30 g of boric acid) in short periods of time may affect the stomach, intestines, liver, kidney, and brain and the worse is lead to death.

The lethal dose of boron (as boric acid) through gastrointestinal was reported to be 2–3 g in infants, 5–6 g in children, and 15–20 g in adults. Liver, kidney, central nervous system, and gastrointestinal effects and skin lesions have been found in lethal cases following ingestion of boron, but death has been attributed to respiratory failure [14].

Boron enters aquatic environments in many ways such a result from human activities and natural sources. Boron is soluble in water and may cause toxic effects in aquatic animals and fish are more sensitive than aquatic invertebrates to boron exposure during chronic exposures [15]. The concentration-response curve for boron is likely to be U-shaped for many species, with harm effects observed at very high and very low concentrations, while no harm effects are observed at the intermediate concentrations. The most sensitive tests report that acute effects on fish are in the range concentration of 10-20 mg/L [16].

5. Conclusion

This study provided information that produced water in Siak Regency contains some heavy metals with concentration above regulation limit. Without proper treatment especially to eliminate heavy metals content, produced water in Siak Regency is harmful as source of drinking water and other daily activity including for fishery and plantation based on Indonesian Government Regulation No. 82 year 2001 on the Management of Water Quality and Control of Water Pollution. Produced water injection to oil production reservoir to support enhance oil recovery (EOR) in Siak oilfield is for now the best practice of produced water management for protecting people and environmental.

References

[1] Hagstrom, E. L. Lyles C, Pattanayek M., DeShields B and Berkman M P, 2016. *Environ. Claim J.* **28** 2 122-139.

[2] Veil J A, Markus P G, Elcock D, and Redweik R J, 2004. Describing Produced Water from Production of Crude oil, Natural Gas, and Coal Bed Methane. A White Paper.

[3] Tiana Afiffah Nadia, 2015. Air Terproduksi: Karakteristik dan Dampaknya Terhadap Lingkungan. Jurusan Teknik Kimia, Fakultas Teknologi Industri, Institut Teknologi Bandung.

[4] Mehta R N and Saini D, 2017. *JAPEB* **1** 1-11.

[5] Clark C E, and Veil J A, 2009. Produced Water Volumes and Management Practices in the United States. Prepared for the U.S. Department of Energy, Office of Fossil Energy, National Energy Tecz`hnology Laboratory.

[6] Rajeswari T R and Sailaja N 2014 *JCHPS* ISSN: 0974-2115.

[7] Sekarwati N and Murachman Band Sunarto 2015. *Ekosains,* Vol **7**, 164-76.

[8] World Health Organization 2011. Selenium in Drinking-water, Background document for development of WHO Guidelines for Drinking-water Quality. WHO/HSE/WSH/10.01/14.

[9] Sneddon A, 2012. Selenium nutrition and its impact on health. Rowett Institute of Nutrition and Health, University of Aberdeen.

[10] World Health Organization 2004. Barium in Drinking-water, Background document for development of WHO Guidelines for Drinking-water Quality. WHO/SDE/WSH/03.04/76.

[11] U.S. Department of Health and Human Services, 2007. Toxicological Profile for Barium and Barium Compounds. Public Health Service Agency for Toxic Substances and Disease Registry.

[12] Bakirdere S, Orenay S and Korkmaz M 2010 *TOMPJ* **3** 54-59.

[13] Ezechi H E, Isa M H and Kutty S R B M, 2012. *J, Appl. Sci.* **12** (5) 402-415.

[14] U.S. Department of Health and Human Services, 2010. Toxicological Profile for Boron. Public Health Service Agency for Toxic Substances and Disease Registry.

[15] Topala A, Orucb E, Altunb S, Ceyhunc S B and Atamanalpc M 2016 *J. Appl. Anim. Res.* **44**(1) 297–302.

[16] Gülsoy N, Yavas C and Mutlu O 2015 *Excli J.* **14** 890–899.

Study of interaction of $CoPcF_{16}$ within poly 4-vinylpyridine matrix using UV/Vis spectroscopy

N A Rahim[1,*], F Audouin[2], Johannes G V[2] and A Heise[2,3]

[1]Department of Chemistry, Faculty of Science and Mathematics, Universiti Pendidikan Sultan Idris, 35900 Tanjong Malim, Perak, Malaysia
[2]School of Chemical Sciences, Dublin City University, Dublin 9, Ireland
[3]School of Pharmacy, Royal College of Surgeons in Ireland, Dublin 2, Ireland,

*E-mail: saidah@fsmt.upsi.edu.my

Abstract. Metallophthalocyanine (MPc) complexes based on abundant transition metals has greatly interest as catalyst due to their attractive features including a rich redox behaviour, the potential to bind with metals or non-metals in the central cavity. Among the transiton metal, cobalt complexes as electrocatalyst are preferable due to their efficient catalytic activity. Immobilisation of the catalyst onto surface may improve with the poly 4-vinylpyridine (P4VP) as polymer backbone. However, aggreagation of MPc is the challenge to overcome. This study aims to investigate the interaction behaviour of cobalt(II) hexadecfluorophthalocyanine ($CoPcF_{16}$) within poly 4-inylpyridine (P4VP) using UV/Vis spectroscopy. Two solutions of $CoPcF_{16}$ and P4VP were mixed and the interaction of $CoPcF_{16}$-P4VP was observed through the time using UV/Vis spectroscopy at wavelength 500 to 750 nm. The outcome suggests that coordination of the electrocatalysts is favoured by the comportment of a high concentration of 4VP. Subsequently, this $CoPcF_{16}$-P4VP electrocatalyst has a potential for water splitting or hydrogen evolution.

1. Introduction

Metallophthalocyanine (MPc) complexes have greater attention among the researchers because of their attractive characteristics such as a rich redox behavior complexes, accessible to change the ring by substitution, possible to bind with metals or non-metals in the central cavity and chemically stable [1]. MPcs are used as dyes in the paper and textile industry, photoconductors in photocopy machines [2], heavy metals sensors [3] and catalyst for water splitting [4-5]. MPcs and other similar compounds are employed as a promising catalyst for many reactions in the electrochemical area [6]. Cobalt complexes such as cobaloximes, diimine-dioximes, porphyrins and phthalocyanines were studied as electrocatalysts and photocatalysts in water splitting or hydrogen evolution [5].

MPc coordination to ligands such as polyvinyl pyridine is forced by the presence of free radical coordination sites in nitrogen atom position. Cobalt phthalocyanine may be coordinated to either single or two pyridine units as has been recognised for other metal complexes [7]. This reaction is favoured because single molecule is predicted to be more effective as electrocatalyst than large aggregate. Many studies on aggregation and coordination of ligands at the axial position have been reported in 1980-1990s, however, the results reported are often ambiguous and questionable. In this study, the effect of incorporating $CoPcF_{16}$ with the polymer in different rations, aggregation and coordination were investigated through time by UV/Vis spectroscopy.

2. Materials and methods

All solvents were used directly as received and HPLC grade or better. Cobalt(II) hexadecafluorophthalocyanine ($CoPcF_{16}$) was purchased from Aldrich and poly 4-vinylpyridine (P4VP) 160,000 g/mol from Merck.

2.1 UV/Vis spectroscopy study

$CoPcF_{16}$ (0.03 mM) and P4VP (0.045 mM) were prepared in dimethylformamide (DMF) solution. Both solutions were mixed at different ratios with maximum volume, 5 mL. The interaction between $CoPcF_{16}$-P4VP was observed by UV/Vis spectroscopy via time at wavelength 500-750 nm. The details conditions of the experiments were demonstrated in table 1.

Table 1. Concentration of solvent mixtures $CoPcF_{16}$-P4VP matrix at different ratios with total volume, 5 mL.

0.045 mM P4VP (mL)	**0.03 mM $CoPcF_{16}$** (mL)	**Concentration P4VP (mM)**[a]	**Concentration $CoPcF_{16}$ (mM)**[a]
0	5	0	0.030
1	4	0.009	0.024
2	3	0.018	0.018
3	2	0.027	0.012
4	1	0.036	0.006

[a] Concentration after mixing the solution was determined using equation $M_1V_1 = M_2V_2$ (M_1 = the stock concentration P4VP or $CoPcF_{16}$, V_1 = volume of P4VP or $CoPcF_{16}$ taken from the stock solution, M_2 = diluted concentration, V_2 = total volume after mixed the solution.

3. Results and discussion

The UV/Vis spectra showed the π-π* band transitions within the macrocyclic phthalocyanine ring system (650-750 nm). For $CoPcF_{16}$ in DMF solution, homogeneously blue colour was discovered and two bands obtained in UV/Vis spectra of CoPcF16. The peaks are 628 nm and 448 nm indicating Q band and B band, respectively. Figure 1 and table 2 shows the UV/Vis spectra of $CoPcF_{16}$ and $CoPcF_{16}$/P4VP in DMF solution at different molecular ratios. The results showed a bathochromic shift with the low concentration of $CoPcF_{16}$ after 30 days. Therefore, the colour solution changes from blue to green. This may be a consequence of the formation of dimers and aggregation of the monomeric species [8] or coordination between P4VP and the Co centre [9]. The lowest concentration of $CoPcF_{16}$ (0.006 mM) contains an excess pyridine group (0.036 mM) availables for coordinating with MPc. Similar UV/Vis spectra results obtained from the film of $CuPcF_{16}$ in DMF solutions and aggregation formed at higher concentration [10].

In the interaction of MPc/polymer matrix, two possible chemical reactions ought to be brought into consideration which are the coordination of MPc to the polymer chains and the aggregation of MPc molecules in solution. Aggregation of MPcs is an extensively suggested characteristics of the compounds due to their significant π-π stacking. Electronic spectra in solution and films is depending on the substitution of phthalocyanine at α-, β- or hexadeca positions. Ogata *et al.* (2003) reported the MPC's interaction as follows (i) cofacial aggregation at α- position is impossible to occur, resulting the Q band appears at the same wavelength in solution or film; (ii) cofacial aggregation at β- position exists in polar solvent performing a broad Q band in the film; (iii) hexadeca substituted the compound exhibiting Q bands at the same wavelength in solution and film [11].

Table 2. Wavelength (nm) and absorbance of $CoPcF_{16}$-P4VP matrix in DMF solution after 30 days

Concentration $CoPcF_{16}$ (mM)	Q λ_{max} (nm)	Absorbance
0.030	628	1.71
0.024	705	1.37
0.018	717	1.59
0.012	727	1.35
0.006	730	0.56

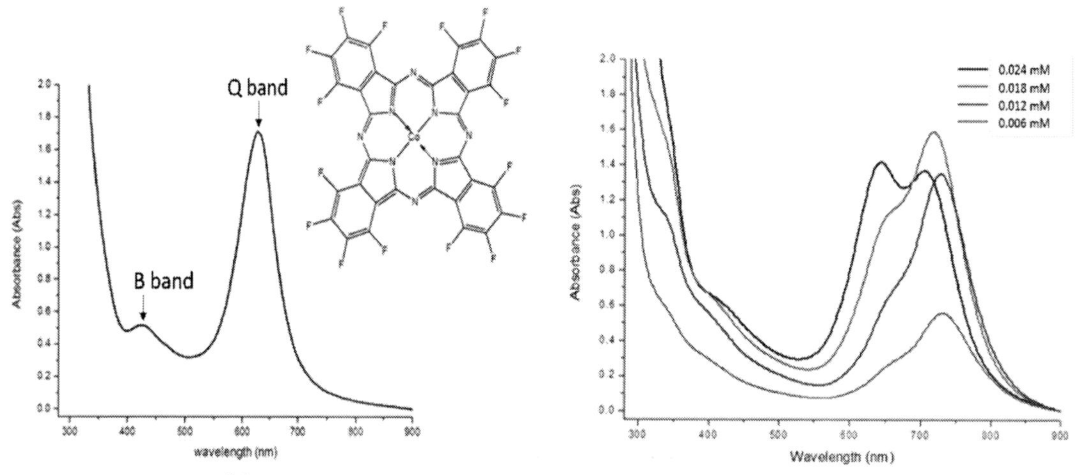

(a) (b)

Figure 1. UV/Vis spectrum (a) $CoPcF_{16}$ in DMF solution; (b) $CoPcF_{16}$-P4VP matrix in DMF solution at different concentration of $CoPcF_{16}$ after 30 days. Inset: Structure of $CoPcF_{16}$

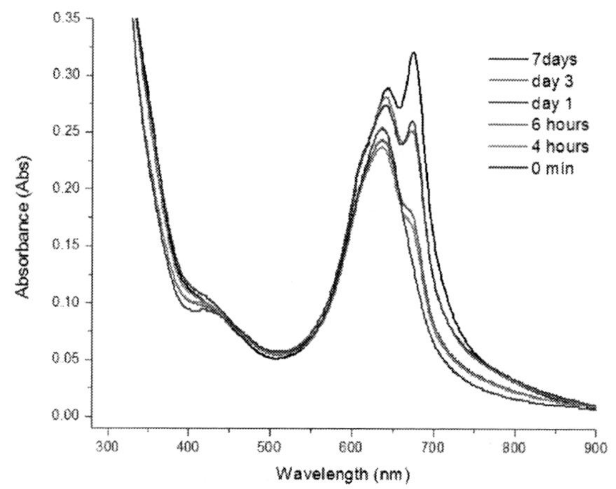

Figure 2. UV/Vis spectra of $CoPcF_{16}$-P4VP matrix *via* time at 0.006 mM of $CoPcF_{16}$.

For the coordination of $CoPcF_{16}$/P4VP matrix at low concentration, experiments were conducted for 7 days with 0.006 mM $CoPcF_{16}$. A broad Q band obtained at 636 nm in the beginning of the experiment.

Later, a small peak was observed at the lower intensity area at 673 nm. This new peak intensity increased via time till day 7 as exhibites in figure 2. The results suggest that dimeric structures formed in solution led to the aggregation and axial coordination between pyridine from P4VP chain and the Cobalt centre [12-13]. Nevin reported that the formation of dimeric species and aggregation resulting a shifted peak at 30-50 nm from the monomeric peak [14]. Ghani claimed that the unsubstituted CoPc is unstable and formed aggregates in solution with the value of maximum solubility in DMF solution is 0.174 mM [15]. However, the solubility of CoPc in polar or organic solvents increased with the substitution of electron-withdrawing group such as fluorine atom [16]. These observations indicated that $CoPcF_{16}$ is dispersed homogeneously in DMF solution before forming aggregates through time and at high concentration. There are three types of binding that probably occur in the solution (figure 3): (i) pyridine bind with the Co centre; (ii) the coordination formed between pyridine from the same polymer chain (intrachain); (iii) the coordination formed between different polymer chains (interchain).

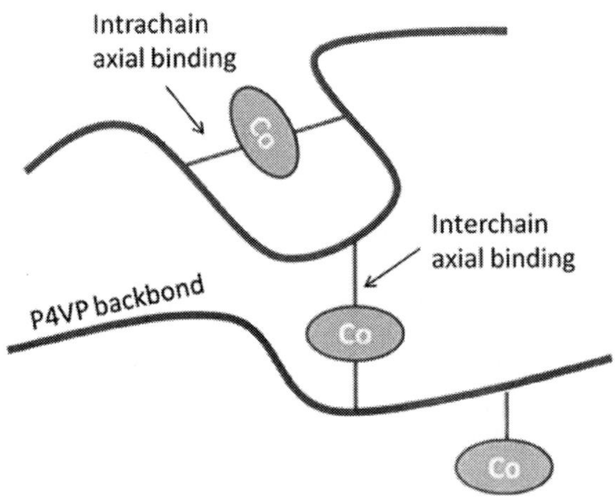

Figure 3. Cartoon illustration of possible binding sites between $CoPcF_{16}$ and a P4VP chains in DMF solution.

The effect of axial coordination of $CoPcF_{16}$/P4VP matrix was examined by interaction of $CoPcF_{16}$ and pyridine solution. Two conditions of experiment were selected which were an addition of 0.1% (v/v) pyridine solution to the 0.030 mM $CoPcF_{16}$ in DMF solution and 0.1% (v/v) 0.030 mM $CoPcF_{16}$ in DMF to pyridine solution. The UV/Vis spectrum was observed over 7 days and the results present in Figure 4. For 0.1% (v/v) pyridine in $CoPcF_{16}$, the Q band (650 nm) increase in intensity while the B band reduce via time. Whilst, the Q band obtained at 665 nm with a shoulder at 600 nm for 0.1% (v/v) $CoPcF_{16}$ in pure pyridine solution. After 300 minutes, both spectroscopic features have achieved their maximum intensity showing the coordination of pyridine with the Co complexes. The shoulders obtained suggest the dimeric species which was formed in the solution as demonstrated for other phthalocyanine complexes [7].

In contrast, $CoPcF_{16}$/P4VP with the massive amount 4-vinylpyridine showed a changeable UV/Vis spectrum via time, while $CoPcF_{16}$/pyridine spectrum is more stable. Obviously, the Q band shifted from 628 nm in DMF solution to 650 nm in diluted pyridine to 665 nm in pure pyridine. These outcomes are congruent with the model suggested above. At low pyridine concentration, one pyridine is bound to the metal centre, whereas two pyridine are coordinated to the metal centre at high pyridine concentration. Nevertheless, further work is needed to substantiate this idea because less aggregation is found in pure pyridine solution as compares to the P4VP containing solution. For P4VP solution, the broad Q bands (630 nm) is reduced with time due to the forming of new band at 670 nm. The following bands that

IOP Conf. Series: Journal of Physics: Conf. Series **1156** (2019) 012010 doi:10.1088/1742-6596/1156/1/012010

correspondg to the increment of coordination of $CoPcF_{16}$ to the polymer chains. The coordination P4VP to $CoPcF_{16}$ by one or two unit pyridine remains unclear. The small variations observed in UV / Vis spectra of "pyridine coordinated compounds" is elucidated by solvent effects.

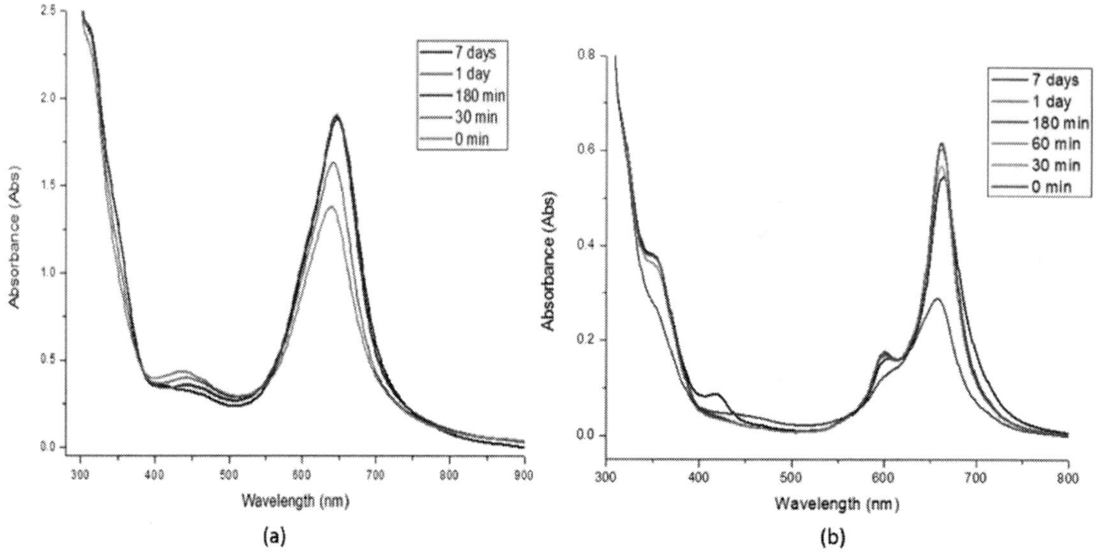

Figure 4. UV/Vis spectra of $CoPcF_{16}$-P4VP matrix *via* time (a) 0.3 mL pyridine (0.1% v/v) added into 3 mL $CoPcF_{16}$ 0.03 mM ; (b) 0.3 mL $CoPcF_{16}$ 0.03 mM (0.1% v/v) added into 3 mL pyridine.

4. Conclusion

The spectroscopic studies demonstrated that the solution that contains P4VP and $CoPcF_{16}$ change via time. The Q bands showed bathochromic shift indicating aggregation form in solution. The amount of coordinated $CoPcF_{16}$ increases with the amount of pyridine in solution especially P4VP. Based on the results, the coordination sites is suggested either one pyridine bind to the metal centre or two pyridine bind to the metal centre as interchain or intrachain conditions.

References

[1] Claessens C G, Hahn U and Torres T 2008 *The Chem. Records* **8** 75-97
[2] Gregory P 2000 *J. Porphyr. Phthalocya.* **4** 432-7
[3] O'Malley S, Schazmann B, Diamond D and Nolan K 2007 *Tetrahedron Lett.* **48** 9003-7
[4] Inglis J L, MacLean B J, Pryce M T, Vos J G 2012 *Coord. Chem. Rev.* **256** 2571-2600
[5] Artero V, Kerlidou M C and Fontecave M 2011 *Angew. Chem. Int. Ed.* **50** 7238-66.
[6] Zagal J H, Griveau S, Silva J F, Nyokong T and Bedioui F 2010 *Coord. Chem. Rev.* **254** 2755-91.
[7] Snow A W and Jarvis N L 1984 *J. Am. Chem. Soc.* **106** 4706-11
[8] Suchan A, Nackiewicz J, Hnatejko Z, Waclaweka W and Lis S 2009 *Dyes Pigments* **80** 239-44
[9] Yoshida T, Kamato K, Tsukamoto M, Iida T, Schlettwein D, Whohrle D and Kaneko M 1995 *J. Electroanal. Chem.* **385** 209-25
[10] Man F, Wang S and Li X 2012 *J. Phys. Chem. Solids.***73** 589-92
[11] Ogata H, Higashi R and Kobayashi N 2003 *J. Porphyr. Phthalocya.* **7** 551-7
[12] Zheng Z Y, Gupta S L, Huang H and Yeager E B 1991 *J. Appl. Electrochem.* **21** 973-81
[13] Nyokong T 1995 *Polyhedron* **14** 2325-9
[14] Nevin W A, Liu W and Lever A B P 1987 *Can J Chem.* **65** 855-8
[15] Ghani F, Kristen J and Riegler H *J. Chem. Eng. Data.* **57** 439-49
[16] Selçukoğlu M and Hamuryudan E 2007 *Dyes Pigments* **74** 17-20

Acknowledgment
The authors recognised the sponsorship given by Ministry of Higher Education (MOHE), Malaysia and Universiti Pendidikan Sultan Idris (UPSI).

Phytochemical and antioxidant evaluation of ethanol extract leaves of *dendrophthoe falcata* (loranthaceae) hemiparasitic on *melia azedarach* host tree

S Atun*, Z Q A'yun, N Lutfia and S Handayani
Department of Chemistry Education, Faculty of Mathematics and Natural Sciences, Universitas Negeri Yogyakarta, Indonesia, Jl. Colombo 1, Yogyakarta 55281, Indonesia

*E-mail: sriatun@uny.ac.id

Abstract: The aim of the study was to identify the content of total phenolic and flavonoids, and to test the antioxidant activity of leaves extract of *Dendrophthoe falcata* hemiparasite on *Melia azedarach* plants. The research method was carried out by extracting dry powder from the leaves of *D. falcata* with ethanol. Ethanol extract was partitioned using consecutive *n*-hexane, chloroform, and ethyl acetate solvents. Each fraction was then analyzed for the content of total phenolic and total flavonoids by spectroscopic methods. The method for testing antioxidant activity using 2,2-diphenyl-1-picrylhydrazyl (DPPH) by in vitro. The results showed that leaves of *D. falcata* were rich in phenolic compounds and flavonoid. Furthermore, in the antioxidant test showed ethyl acetate fraction had the highest activity, with IC_{50} 6.66 ± 0.11 μg /ml. This study shows that *D. falcata* leaves have potential as natural antioxidants.

1. Introduction

Dendrophthoe falcata is one of the hemiparasitic plant species found in tropical and sub-tropical regions. This plant is usually attached to the host *Mangifera indica* (Anacardiaceae), *Melia azedarach* (Meliaceae) and *Psidium guajava* (Myrtaceae). The *D. falcata* plant is used in the original treatment system as cooling, bitter, substance, aphrodisiac, narcotics and diuretics and is also useful for lung disease, asthma, menstrual disorders, and antifertilitas [1-2]. Methanolic extract and water of leaf *D. falcata* in animal models has a potent anti-inflammatory effect and antioxidant [3]. Several studies have shown that loranthus plants are also known to be a polio virus prevention [4], antimicrobial and anti-inflammatory [5], neuroprotective [6], antioxidant [7], and antidiabetic [8].

D. falcata is one of the parasitic plant species (Loranthaceae) which is widely found and used for traditional medicine in some countries such as China and India. The content of chemical compounds from parasites is generally influenced by the host plants [9]. A study comparing the chemical compounds found in the parasitic *Loranthus longiflorus* plant parasite *Casuarina equisetifolia* and *Ficus religiosa* showed different compound components. *L longiflorus* found in *C. equisetifolia* shows the main compound of myristic acid, vinyl ester (82.81%) which also has antimicrobial and anticancer activity. While *L. longiflorus* contained in *F. religiosa* plant is a major component of vinyl dekanoate (71.7%) and showed no antimicrobial and anticancer activity [10]. Similarly, the parasitic *Loranthus micranthus* in various plant stems such as *Persian americana*, *Baphia nitida*, *Kola acuminata*, *Penthaclethra macrophylla*, and *Azadiracta indica* show variation in various major component types depending on the host plant [11]. Previous research the leaves of *D. falcata* reported to contain compounds identified as lupeol, 3-β-acetoxy-12-ene-11-one and β-sitosterol [12]. In this article we will report the phytochemical and antioxidant evaluation of ethanol extrat from leaves of *D. falcata* plant that hemiparasite on *Melia azedarach*.

2. Experimental Section

2.1. Apparatus and reagent
The tools used include the evaporator Buchi Rotavapor R-114, spectrophotometer, vortex mixers, and glassware. The materials used include organic solvents such as ethanol, methanol, *n*-hexane, ethyl acetate, and acetone, 2,2-diphenyl-1-picrilhidrazil (DPPH, Aldrich), ascorbic acid (Aldrich), folin-ciocalteu reagent, sodium carbonate, gallic acid (Sigma), rutine (Sigma), aluminum nitrate, potassium acetate, acetic acid, ammonia, sulfuric acid, wagner reagent, iron (III) chloride, acetic acid, leaf of *D. falcata* hemiparasite on *Melia azedarach*, and distilled water.

2.2. Preparation of extracts from leaves of D. falcata
The leaves of *D. falcata* hemiparasite on *Melia azedarach* were collected in October 2017 and determined by botanists. Voucher specimens (DF-01-2017) are stored in the Organic Laboratory. The dried powder leaves of *D. falcata* (2.5 kg) were extracted using ethanol for 24 hours. The extract was then evaporated so that the concentrated extract was obtained and produced brown residues for about 140 g. Ethanol extract of *D. falcata* (120 g) was partitioned using *n*-hexane, chloroform, and ethyl acetate, respectively. Each separated fraction was then concentrated and dried, so the concentrated fraction was obtained of *n*-hexane (11 g), chloroform fraction (19.8 g), ethyl acetate fraction (65.1 g), and residual ethanol fraction (10 g).

2.3. Phytochemical screening
The qualitative phytochemical screening was performed using reagent for terpenoid test, alkaloid test, phenolic test, and saponin test. Terpenoid and steroid test with Salkowski reagent, alkaloid test with Wagner reagents, phenolic test with with ferrous (III) chloride test, and foam test for saponin [13].

2.3.1. Terpenoid and steroid test. 2 ml of each *D. falcata* fractions (1%) were added with 10 drops of CH_3COOH and H_2SO_4 concentrated in 2 drops. The solution is shaken gently and left for several minutes. Positive test of steroids shows producing blue or green colour, while triterpenoid produces red or purple colour.

2.3.2. Alkaloid test. The alkaloid test was carried out by adding 2 mL of each sample to the test tube (1%) plus 2 mL of chloroform, and 2 mL of ammonia. The mixture is stirred and then filtered. The filtrate is then added to 3-5 drops of concentrated H_2SO_4 and then shaken to form two layers. The top layer is added with Wagner reagent 4-5 drops. Orange red sediment formation indicates that the sample contains alkaloids.

2.3.3. Phenolic test. The sample was dissolved in a suitable solvent, as much as 2 ml was introduced into the test tube and added a few drops of $FeCl_3$ solution, when a purple color was shown to be positive against phenolic.

2.3.4. Saponin test. Saponin test was done by foam formation method. In the test tube, 0.5 mL sample and 5 mL of distilled water were added. The mixture was shaken vigorously and the foam formed was observed. The foam then mixed with 3 drops of olive oil and shaken vigorously after being observed for emulsion formation.

2.4. Determination of total phenolic
Analysis of total phenolic content from samples was carried out using folin-ciocalteu reagents [14]. 200 μL of sample were mixed with 0.4 mL of folin-ciocalteu reagent, 0.4 mL of aquadest, and 2 mL of sodium carbonate (15% b/v). The mixture was incubated at 50°C for 10 minutes. Then the absorbance was measured at 760 nm. The total phenolic content was calculated according to the standard curve

determined for gallic acid as a reference. Total phenolic content is expressed as mg Gallic Acid Equivalent (GAE) per g of plant extract.

2.5. Determination of total flavonoids

Total flavonoid content was determined by the spectrophotometric method, with slight modifications [15]. In the test tube, 0.5 mL sample was diluted in ethanol (80%), then add 0.1 mL of aluminum nitrate (10%), 0.1 mL of potassium acetate (1M), and 4.3 mL of ethanol (80%). The mixture was incubated at room temperature for 40 minutes. The absorbance of the solution was then measured by a spectrophotometer at 415 nm. The total flavonoid content was calculated based on the standard curve determined by rutine as a reference. The total flavonoid content is expressed as mg Equivalent Rutine (RE) per g of plant extract.

2.6. Determination Antioxidant Activity Test by DPPH Method

The antioxidant activity of each sample was tested using the DPPH (2,2-diphenyl-1-pikrilhidrazil) method, with a slight modification [16]. About 5 mL of the sample in various concentrations was mixed with 5 mL of DPPH solution in methanol (0.12 mM) and incubated at room temperature for 30 minutes. Absorption of each sample was determined using spectrophotometer at 516 nm. As a positive control, ascorbic acid was used. Each sample was tested with three repetitions. The inhibitory activity was calculated as the percentage of DPPH decrease compared to the control. Furthermore, from the calculation of inhibitory activity can be used to calculate antioxidant activity expressed as IC_{50} (concentration of samples that have a 50% inhibitory activity).

3. Results and discussion

Phytochemical studies of plant leaves of *D. falcata* hemiparasites on *Melia azedarach* plant (figure 1). The results of qualitative analysis which include the content of terpenoids, alkaloids, phenolics, and saponins are listed in table 1.

Figure 1. Leaves of *D. falcata*

Table 1. Qualitative phytochemical screening of the extracts and fractions from leaves of *D. Falcata*

Qualitative phytochemical screening	Extract ethanol total	*n*-hexane fraction	Chloroform fraction	Ethyl acetate fraction	Residual fraction of ethanol extract
Terpenoid & Steroid (Salkowski Test)	+++	+++	++	-	-
Alkaloid (Wagner Test)	+	-	+	-	-
Phenolic ($FeCl_3$ test)	+++	-	+++	+++	+++
Saponin (Foam test)	+++	-	-	-	+++

Note: Note: - = absent, + = present, ++ = present a lot; +++ = present very large number

In this study, quantitative analysis of total phenolic levels was used by spectrophotometric methods. Each sample was added with Folin-Ciocalteu reagent. Absorption of each mixture is measured at a wavelength of 760 nm. As standard phenolic compounds are used gallic acid at various concentrations (50-200 μg /mL) and absorbance is measured at the same wavelength. The measurement results obtained a calibration curve with a linear regression equation $y = 0.004x - 0.032$ ($R^2 = 0.985$). The total phenolic content is expressed as mg of Gallic Acid Equivalents (GAE) per g of plant extract. Meanwhile, to determine the total flavonoid content, rutin standard compounds were used at various concentrations (50-200 μg / ml) and absorbance was measured at a wavelength of 415 nm. The measurement results obtained by the regression equation as a calibration curve with a linear regression equation $y = 0.003x + 0.099$ ($R^2 = 0.963$). The total flavonoid content is expressed as mg Rutin Equivalents (RE) per g of plant extract. The results of the phenolic content and total flavonoids from the extract and each fraction were collected in table 2.

Table 2. Results of analysis of total phenolic content and flavonoids from leaves of *D. falcata*

Sample	Total phenolic content (TPC) (mg GAE/g)	Total flavonoid content (TFC) (mg RE/g)
Crude ethanol extract of *D.falcata*	670.0± 2.83	386.4± 1.1
n-hexane fraction	0	0
Chloroform fraction	1363.3 ±3.21	347.5± 1.9
Ethyl acetate fraction	1675.0± 2.70	731.4± 2.8
Residual of ethanol extract fraction	1072.5±5.40	354.4± 3.8

This study indicate the suitability between qualitative and quantitative data. Qualitative data showed phenolic compound content of total ethanol extract, chloroform fraction, ethyl acetate fraction, and ethanol extract residue, while n-hexane fraction showed negative test. The phenolic compound is a relatively polar compound, so it will dissolve in a relatively polar solvent as well. Terpenoid compound is a compound that is relatively less polar, so there are many in the fraction of *n*-hexane and a little on chloroform. The crude ethanol extract, chloroform fraction, etyl acetate fraction, and residual ethanol fraction leaves of *D. falcata* showed a total phenolic content as much as 670.0± 2.83; 1363.3 ±3.21; 1675.0± 2.70; 1072.5±5.40 mg Gallic Acid Equivalen (GAE)/g) respectively, while flavonoid total content 386.4± 1.1; 347.5± 1.9; 731.4± 2.8; and 354.4± 3.8 mg Rutin Equivalent (RE)/g) respectively. This data is consistent with the results of previous studies showing the presence of phenolic compounds from leaves *Loranthus micranthus* Linn, such as epicatechin, epicatechin-gallate, rutin, and peltatoside [17]. While in *Scurrula ferruginea* Danser (Loranthaceae) plants found phenol compounds, among others quercetin, quercitrin, and flavonol glycoside 4-O-acetylquercitrin [18].

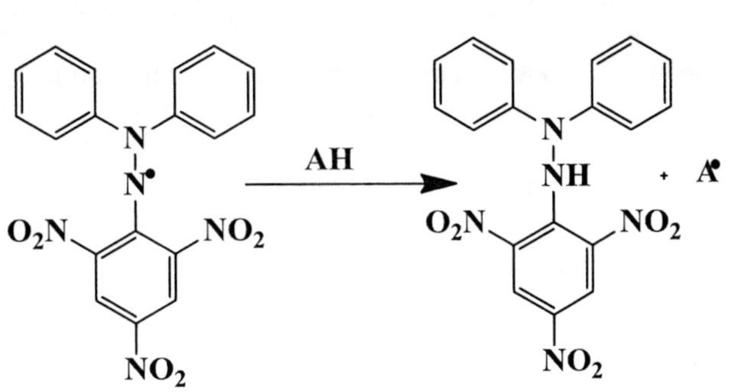

Figure 2. DPPH radical reaction with antioxidant (AH)

Test of antioxidant activity from plant samples can be done by various methods. One of the easiest methods to do is the DPPH method. The antioxidant activity test with the DPPH method is based on radical capture reactions by antioxidant compounds with the mechanism of transfer of oxygen atoms, which will produce a stable DPPH-H molecule [19]. Reduction of DPPH concentration is indicated by a decrease in the intensity of purple solution compared to blank. Absorbance of each solution was measured using spectrophotometry at a wavelength of 516 nm. Antioxidant capacity is expressed as IC_{50} (μg/mL) from DPPH capture activity at various concentrations. The absorbance data of each sample was then made into a linear regression equation. From the regression equation used to calculate the IC_{50}. The reaction captures DPPH radicals by the antioxidants shown in the figure 2. The inhibitory activity of each sample and calculation of antioxidant activity data expressed as IC_{50} are shown in table 3.

Table 3. Antioxidant activity (IC_{50}) test leaves of *D. Falcata*

Sample	Data of antioxidant test			
	Concentration (μg/mL)	Inhibition activity (%)	The regression equation	Antioxidant activity (IC_{50} μg/mL)
Crude ethanol extract of *D.falcata*	25	86.63	y = 3.230x + 9.239	12.62
	12.5	58.25	R^2 = 0.967	
	6.25	27.11		
	3.125	16.39		
n-Hexane fraction of leaves *D. falcata*	100	19.25	y = 0.107x + 8.692	392.07
	50	13.74	R^2 = 0.925	
	25	12.98		
	12.5	8.82		
Chloroform fraction of leaves *D. falcata*	25	81.21	y = 2.531x + 18.91	12.28
	12.5	52.83	R^2 = 0.995	
	6.25	34.80		
	3.125	25.47		
Ethyl acetate fraction of leaves *D. falcata*	25	91.04	y = 2.315x + 34.58	6.66
	12.5	65.69	R^2 = 0.975	
	6.25	52.58		
	3.125	3.125		
Residual ethanol extract of leaves *D. falcata*	25	92.3	y = 3.198x + 19.13	9.65
	12.5	75.28	R^2 = 0.883	
	6.25	38.21		
	3.125	20.68		
Ascorbic acid (positive control)	5.00	95.59	y = 13.76x + 32.51	1.27
	2.50	79.82	R^2 = 0.883	
	1.25	51.45		
	0.625	32.28		

The high content of phenol and flavonoid compounds from each extract and fraction is associated with high antioxidant activity as well. The IC_{50} less than 10 μg/mL indicate very active antioxidant activity, such as positive controls used ascorbic acid. Ascorbic acid is a widely known antioxidant compound, exhibiting very high antioxidant activity and is commonly used as a positive control. The results showed that IC_{50} of ethyl acetate and chloroform fraction from leaves *D. falcata* were below than 10 μg/mL so that antioxidants were very active. The previous studies also showed the stem of *D. falcata* plant contained high phenolic and flavonoid compounds and showed very high antioxidant activity and medium antimicrobial activity [20].

4. Conclusion

From this study it can be concluded that leaves from *D. falcata* extract contains high phenolic and flavonoid. The most phenolic and flavonoids content is found in ethyl acetate fractions. The antioxidant test results showed total ethanol extract, *n*-hexane fraction; chloroform fraction, ethyl acetate fraction, and residues of ethanol fraction leaves of *D. falcata* with IC_{50} values 12.62 ± 0.15, 392.07 ± 4.22, 12.28 ± 0.42, 6.66 ± 0.11, and 9.65 ± 0.52 μg/mL respectively. The leaves of *D. falcata* can be developed as natural antioxidants.

References

[1] Moghadamtousi S Z, Noor M, Kamarudin A, Chan C K, Goh B H and Kadir H A 2014 *Am. J. Chin. Med.* **42** (1) 23-35.

[2] Downey P O 1998 *Cunninghamia* **5** (3) 685-720.

[3] Patil S, Anarthe R, Jadhav R and Surana S 2011 *Iran J. Pharm. Res.* **10** (2) 253-9.

[4] Wong D Z H, Kadir H A, Lee C L and Goh B H 2012 *J. Ethnopharmacol* **139** 256-264.

[5] Govindappa M, Channabasava R, Sowmya D V, Meenakshi J, Shreevidya M R, Lavanya A, Gustavo S and Sadananda T S 2011 *Pharmacogn J.* **3** (25) 82-90.

[6] Wong D Z H, Kadir H A, Lee C L and Goh B H 2012 *J. Nat. Med.* **66** 544–551

[7] Chandrakasan L and Neelamegam R 2012 *Plant Archives* **12** (1) 31-40.

[8] Ukpanukpong R U, Adekunle I A, Uyabeme R N and Adeleke C O 2017 *JAMPS* **13** (1) 1-12.

[9] Chandrakasan L and Neelamegan R 2011 *J. Pharm. Res.* **4**(9) 3072-74

[10] Osadebe P O and Ukwueze S E 2004 *Bio-Research* **2**(1) 18-23.

[11] Chandrakasan L and Neelamegam R 2017 *J. Med. Plant. Stud.* **5**(1) 135-139.

[12] Rajdoula M R, Ahsan M, Hasan C H and Masud M M 2017 *J. Pharm. Sci.* **16**(2): 215-19.

[13] Harborne A J 1998 *Phytochemical Methods a Guide to Modern Techniques of Plant Analysis* (Netherlands-Springer) 3rd ed.

[14] Hagerman A, Harvey-Mueller I, Makkar A H, Mueller I H and Makar H P S 2000 *Quantification of Tannins in Tree Foliage–a Laboratory Manual* FAO/IAEA Vienna.

[15] Malencic D, Milan P and Jegor M 2007 *Molecules* **12**(3) 576-81.

[16] Biswas M, Haldar P K and Ghosh A K 2010 *J. Nat. Sci. Biol. Med.* **1**(1) 29-34

[17] Agbo M O, Agbo C S, Nworu F B C, Okoye and Osadebe P O 2014 *EXCLI J.* **13** 859-68.

[18] Devehat R, Tomasi S, Fontanel D and Boustie J 2002 *Z. Naturforsch.* **57**(c), 1092-95

[19] Molyneux P and Songklanakarin P 2004 *J. Sci. Technol* **26**(2) 211-219.

[20] S Atun, S Handayani, A Rakhmawati, N A Purnamaningsih, B I N Naila and A Lestari 2018 *Orient. J. Chem.* **34** (5): 2342-2349.

Acknowledgements

This research was funded by the Minister of Research and Technology of the Directorate of Higher Education, Indonesia to research research competency research funds (PBK, Ministry of Research and Technology, Number: 0045 / E3 / LL / 2018, January 16, 2018). The author would like to express his gratitude for the funding.

Structure and dynamics of Hg^{2+} in aqueous solution: an Ab Initio QM/MM molecular dynamics study

CF Partana*, Suwardi and A Salim

Department of Chemistry Education, Faculty of Mathematics and Natural Sciences, Universitas Negeri Yogyakarta, Indonesia, Jl. Colombo 1, Yogyakarta 55281, Indonesia

*Email: fajar_partana@uny.ac.id

Abstract. The human body mostly (more than 80%) consists of liquid, so that the metabolic processes in the human body involve the interaction between solvents and solutes in both ionic and compounds. Metal mercury is a toxic metal, can cause poisoning. The hydration of mercury ions is observed by performing molecular dynamics simulations of molecular mechanics. The base set used is DEF2-ECP for Hg atoms, while for water molecules used dunning DZP base collection. Simulation is done using potential partners and potential 3 bodies. The simulation is done by mixing 1 Hg^{2+} ion mixed with 499 H_2O molecules. The simulation results in the form of trajectory data are further processed to determine the structure and dynamics. Based on RDF analysis shows that the average Hg and O (water) atoms are 2.22 A. Based on CND analysis, it is known that the coordination numbers of hydration of Hg in water are as much as 6 molecules of water. Thus it is known that the hydration of Hg ions in water forms a distorted octahedral molecule structure. Ligand (water) residence time in Hg based on molecular dynamics analysis shows that hydration of Hg with water is stable.

1. Introduction

Chemistry is natural science based on experiments. This is evidenced by the history of chemistry that developed rapidly through experiments. The development of chemistry no longer depends only on experiments in the past two decades. Developments in chemistry today, theoretical predictions of chemical properties can rival the accuracy of the data obtained from experiments [1]. The rapid development of technology and information also influences the development of chemistry. One such development is the use of computers as a means or equipment in chemical laboratory work. The use of computer as a work tool has been developed into an aspect of study called computational chemistry. The rapid development of computational chemistry has changed the description of a chemical system with the inclusion of a new element between experimentation and theory, namely computer experiments. In experiments, the measurement of the system is expressed in numerical form, while in theory uses mathematical equations to model a system generally. So computational chemistry is a bridge that connects experimental results in a laboratory with a theoretical basis. Researcher conducted by previous researchers, can be explained that a metal in the human body has a function, both of which can be beneficial for example in the metabolic process or harmful to the body [2], as antibodies and others. An element that belongs to a group of heavy metals such as cadmium (Cd), arsenic (As), mercury (Hg), nickel (Ni), lead (Pb) and zinc (Zn) are present in the environment both in air, water and food [3-4-5] and can enter the human or animal body. Transition metals included in class IIB such as cadmium (Cd) and mercury (Hg) which are toxic (toxic) [4]. Cadmium and mercury are in nature there are two ways, naturally and anthropogenic. Mercury metal if it enters the human body can be dangerous and have a high risk of blood vessels and can accumulate in the liver and kidneys causing metabolic disorders [6]. In the human body most (more than 80%) consist of liquid substances, so that the metabolic processes in the human body involve the interaction between solvents and solutes. Interactions that only involve water solvents are called hydration. Because the metabolic processes in the human body are mostly water, of course it involves hydration with solutes. Hydration events involve several shells of solution called dissolution cells. The shell closest to the metal cation is the first (cell) shell, the next shell is called the second shell, and so on until the outer phase. The outer phase is a shell that is not affected by the strength of the cation charge. In this phase only hydration occurs between solvents. The dissolved substance in the first shell has sufficiently strong hydration power. This solute is a ligand that is hydrated with a central ion (cation). Ligands in the first shell can move to the second shell and vice versa. Ligand exchange can affect activity. By using the help of computers and software programs that are increasingly advanced as it is today, it can be done research that examines more deeply about the interaction of solvents with solutes [7-8]. Mercury (II) is an electron pair acceptor that forms bonds with a high degree of covalence with ligands with soft atomic donors, and its special affinity with mercaptan (RSH) is known. Distorted four-tetrahedral coordination was found for mercury ions (II) in the S N, N-dimethylthioformamide donor solvent with strong electron-pair donor properties. Likewise, nitrogen donor ligands can form strong bonds. Ammonia is a solvent with very strong electron pair donor characteristics that are expected to form strong covalent bonds with mercury (II). Recently, tetraaminmercury(II) perchlorate crystals, $[Hg(NH_3)_4]$ $(ClO_4)_2$ have been made by flowing ammonia gas into concentrated mercury (II) perchlorate solution. The crystal structure at 170 K has been reported with the distorted tetrahedral configuration around the mercury ion (II) [9]. Mercury (II) complexes with monodentate ligands often show flexible coordination geometry. In a solid state the arrangement of the ligand atoms is bound to be linear, trigonal or tetrahedral. The geometry of octahedral coordination is rare but occurs for example with O-and -N donor ligands in solid solvates eg $[Hg(H_2O)_6](ClO_4)_2$, $[Hg(OS(CH_3)_2)_6]$ $(ClO_4)_2]$, and $[Hg(NC_5H_5)_6]$ $(CF_3SO_3)_2$, which crystallizes each of the acuous solutions, dimethylsulfoxide and pyridine. Investigation of mercury(II) solvated in oxygen and nitrogen donor solvents, eg water, dimethylsulfoxide and pyridine in EXAFS and LAXS shows a labile sixth coordination with dynamic distortion of octahedral symmetry [10]. Study of the structure and dynamics of solvation of metal ions in solvents, especially water and ammonia, has an important role. Shell dynamics solvation of water molecules determines the process of chemical reactions. Some reactions only occur after the water molecules are removed from the solvation shell. Certain metal complexes with amino acids are important from biological and medical aspects, for example as antibacterial and antifungal. Metal ions in ammonia become an interesting system model because they are very relevant, because both amino acids and ammonia contain nitrogen and hydrogen atoms [11].

2. Method

2.1 Determination of Hg^{2+} H_2O coordinates in cartesian coordinates

The initial geometry of Hg^{2+} in H_2O is modeled in three-dimensional cartesian coordinates by adjusting the angle and distance between atoms in the system. With the help of Gauss View, a structural form of the Hg^{2+} -H_2O coordinate system can be produced [12].

2.2 Selection of the best basis sets

Based on several basis sets are recommended to be tested for basis set pairs that do not cause significant changes in charge to Hg^{2+} ions, and have a bond energy curve profile to the distance of Hg^{2+}-O according to Lennard-Jones potential curve profile. In addition to those used also must have a base set superposition error value (basis sets superposition error, BSSE) which is relatively small. Next the best basis sets of pairs is used as a reference for calculating the pair and 3-body functional potential.

IOP Conf. Series: Journal of Physics: Conf. Series **1156** (2019) 012012 doi:10.1088/1742-6596/1156/1/012012

2.3 QM/MM DM simulation
QM/MM simulation is done after MM simulation. The final configuration stage results from the MM simulation is used as the initial configuration of the QM/MM simulation. MM / MM simulation is carried out at least 20ps. The distance of the MK area was determined by looking at the RDF results from the MM simulation trajectory analysis. By looking at the RDF chart from the trajectory analysis of molecular mechanics (MM) simulations, the first shell of hydration / hydration can be determined. The DM MK / MM simulation requires a softener function that is useful to allow water ligands to freely move from the MM area to the MK and vice versa. The softening function is used for the MK and MM transition areas of 0.2 Å. The system style is denoted by F system with the equation as shown in the F system equation = FMM + S (FMK-FMM / MK). FMM is the force in the MK region, FMK / MM is the MM style in the MK region, and S is the softener function. The use of this softener function allows water ligands to freely move from the MK to MM region or vice versa [13].

3. Result and Discussion

3.1. The best basis sets
The activity that was carried out for the first time in DM MM simulation research on Hg^{2+} ion in water was to determine the set of base atoms. The first step is to find the type of set of atoms that are most suitable for use in DM / MM simulation. The types of support base groups can be known through taking the basis of various research journals that have been conducted. The calculation method used in the base set selection process is a method at Hartree-Fock theory level. The publication of several previous studies, shows that a specific set of support bases for the hydration system of Hg^{2+} ion in water (H_2O) is not yet available. Some previous studies that carried out research related to Mercury atoms used different sets of bases. The set of support bases that are commonly used for research in simulations for Mercury atoms include: LANL2DZ ECP, DZVP, DZVPP.

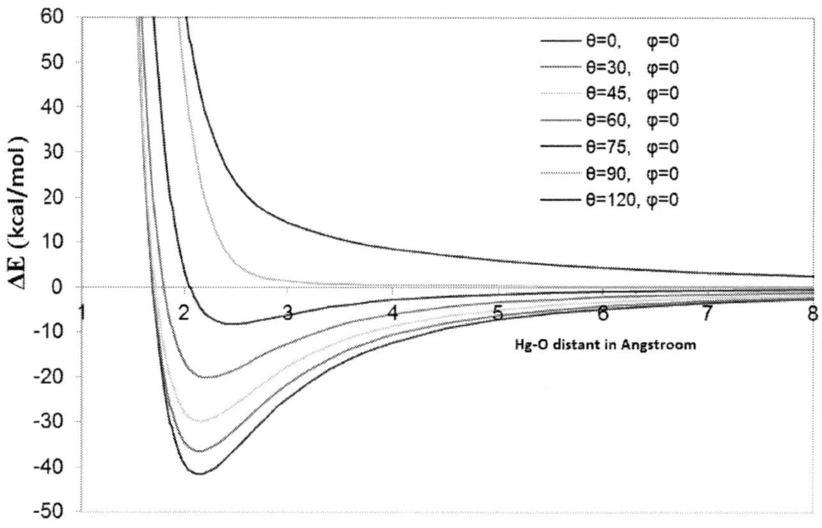

Figure 1. Energy Potential curves at various distances of the Hg^{2+} -H_2O system

Based on the set of bases that have been used by previous researchers, this study begins with a series of trials to find the basis of the most suitable for the hydration system studied. In addition to being based on the basis set that has been used by previous researchers, in order to find the appropriate base set, testing is also conducted on the base set that has an Electron core potential. This is done because Mercury atoms are located in the transition class in the periodic system (class IIB transition elements. Thus the base set analyzed consists of 5 (five) base sets, namely: three types containing ECP: LANL2DZ,

CRENBLJC, DEF2-SVP. base set for H and O atoms, used in this study is the set of DZP-Dunning bases that have been tested for reliability. Based on trials of several basis sets (LANL2DZ ECP, CRENBLJC ECP and DEF2-SVP), a base set that best matches the four hydration systems to be studied and the corresponding base set forms the Lennard-Jones potential curve, the corresponding base set is DEF2-SVP for Mercury and DZP-Dunning atoms for O and H atoms. The graph of potential profile with distance and angle variations is shown in figure 1. After obtaining a suitable base set, the base set that is in accordance with the hydration system is used in the simulation for the system studied, namely: hydration of Hg^{2+} ion in water.

3.2. The 2-body and 3-body Potential

3.2.1. *The 2-body potential equations for Hg^{2+}–H_2O.* The first step is to determine the 2-body potential equation of the Hg^{2+}-H_2O system. Determination of the 2-body potential equation of Hg^{2+}-H_2O, is carried out by collecting energy points. Energy points are collected by placing water molecules (H_2O) in the center of cartesian coordinates, (0,0,0), while Hg^{2+} ions are placed around the water coordinates (H_2O), with distances varied from 1.4 Å to with 15 Å, with varying angles θ and φ. Variation of angle θ is determined starting from 0° to 180° (0° ≤ θ ≤ 180°), while for angles φ starting from 0° to 90° (0° ≤ φ ≤ 90°). Data on energy points of potential energy at various distances and angles θ and angle φ are then processed further through a 2 body fitting program with the most suitable modification of the rank. The form of pair potential equation of Hg-H_2O is as follows:

$$E_{fit}^{2bd} = \frac{q_{Hg(II)}q_O}{r} + \frac{A_O}{r^5} + \frac{B_O}{r^9} + \frac{C_O}{r^{11}} + \frac{D_O}{r^{12}} + \sum_{i=1}^{3}\left(\frac{q_{Hg(II)}q_H}{r_i} + \frac{A_H}{r^4} + \frac{B_H}{r^5} + \frac{C_H}{r^6} + \frac{D_H}{r^{12}}\right) \quad (1)$$

a,b,c,d, Ai, Bi, Ci, dan *Di* is a fitting parameter, rMi, is the i-atom distance from the Hg^{2+} and H_2O ions, qi is the i-atomic charge of the H_2O molecule, and q_M is the total charge of Hg^{2+} ion. The results of the analysis of energy points on a variety of distances and variations in angles θ and φ of the Hg^{2+}-H_2O system are presented in table 1.

Table 1. The optimized parameters of the analytical two-body potential and three-body correction functions

2-body	A (kcal mol^{-1})	B (kcal mol^{-1})	C (kcal mol^{-1})	D (kcal mol^{-1})
Hg(II)-O	-4291.14	181896.74	−802564.93	3691064.19
Hg(II)-H	−644.85	12474.54	−2081.79	551.51

3-body	A_1 (kcal mol^{-1} Å^{-1})	A_2 (kcal mol^{-1} Å^{-1})	A_3 (kcal mol^{-1} Å^{-1})
O-Hg(II)-O	4.474	0.551	0.590

A, B, C and D are parameters of the potential pair equation (kcal/mol Å)

3.2.2. *Determination of 3-body potential equations for H_2O–Hg^{2+}–H_2O.* The preparation of 3-body potential is needed, this is done in order to minimize the influence of many body interactions between water molecules and or ammonia molecules. Collection of energy points in determining 3-body potential equations for hydration of mercury ions in water using the turbomol program (scan3bd). The steps in determining a 3-body potential equation are similar to steps when determining potential 2-body potential equations. Interaction energy is collected in various geometric configurations by varying the distances (r1 and r2) of mercury ions with water and. The interaction energy points are carried out at a distance of 1.5 Å to 7.0 Å at various variations of tau (τ) and angle φ. The angle τ used is selected from 0° to 180°, while the angle φ is determined from 90° to 180°. After obtaining energy points at various distances and various angles φ and τ, fitting is done to obtain the 3-body equation. The potential equation for 3-body hydration system is Hg^{2+} ion in water (H_2O–Hg^{2+}–H_2O). Form 3-body potential equations for hydration:

$$V_{3bd} = a1e^{-a2(r1+r2)}e^{-a3(r3)}(\mathrm{CL} - \mathrm{r1})^2\mathrm{CL} - \mathrm{r2})^2$$

information :

V_{3bd} : function of potential 3-body equation
CL : the cut-off limit which is normally set to 6.0 Å, at a distance above 6.0 Å the interaction is outside the 3-body potential count
r1 : distance of Hg with O_1 (Å)
r2 : distance of Hg with O_2 (Å)
r3 : distance of O_1 with O_2 (Å)
a1,a2,a3: parameters of potential 3-body equations

Energy points obtained from variations in distance (r1 and r2) and on variations in angles φ and τ which were collected in the determination of the 3-body equation for the Hg^{2+} ion system in water in this study were 10,560, the obtained energy interaction points then fitting is done by using the fit3bd program for water. The parameters of the potential equation for 3-hydration body Hg^{2+} in water are presented in table 1.

3.3. Structure of hydration of Hg^{2+} ions

3.3.1. Radial Distribution functions (RDF). RDF or distance distribution function in this hydration system states the function of the distance distribution of water molecules (H_2O) to $Hg2^+$ ions. The distance of the central ion with the oxygen atom from the water molecule is denoted by Hg^{2+}-O, the distance distribution function (RDF) for the hydration system of Hg^{2+} ion in water in molecular dynamics/quantum mechanics simulations is presented in figure 2.

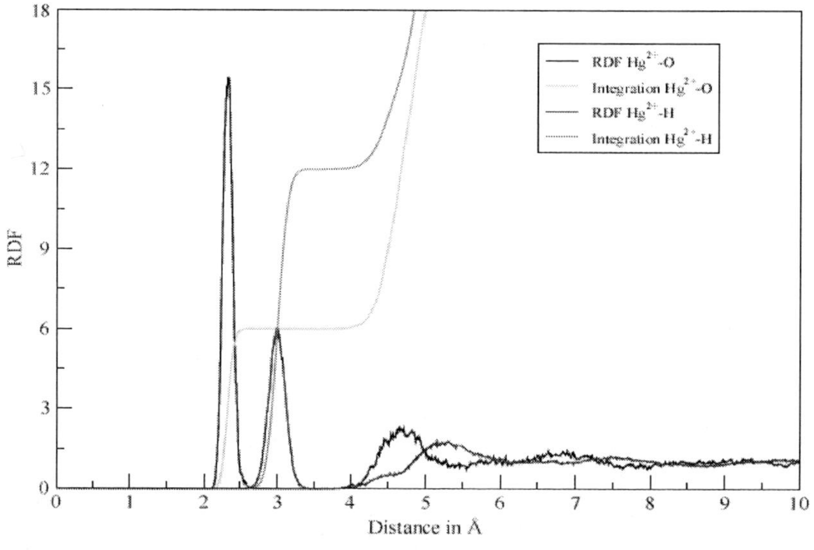

Figure 2. RDF of Hg^{2+} ion in water are obtained by the QM/MM MD simulations

IOP Conf. Series: Journal of Physics: Conf. Series **1156** (2019) 012012 doi:10.1088/1742-6596/1156/1/012012

A summary of the results of the RDF analysis of the hydration of the Hg^{2+} ion system in water is presented in table 2.

Table 2. Characterization values of RDF for Hg^{2+} ions in water are obtained by the QM/MM simulations

Simulation Method	α-β	r_{M1}	r_{m1}	n_1	r_{M2}	r_{m2}
QM/MM	Hg^{2+}-O	2.1	2.7	6.0	4.2	5.5
	Hg^{2+}-H	2.7	3.4	12.	4.2	6.9

explanation:
α-β : ligand atom(β)-center atom (α)
r_{M1} / r_{M2} : the maximum peak of the first and second solvation shell (Å)
r_{m1} / r_{m2} : the minimum peak of the first and second solvation shell (Å)
n_1 : number of ligands in the first shell

3.3.2. Coordination Number Distribution (CND). CND or distribution of coordination numbers, illustrates the distribution of coordination numbers that have central ions with related ligands. The number of coordination numbers or ligands during the hydration simulation of Hg^{2+} in water with the method of quantum mechanics/molecular mechanics is presented in figure 3.

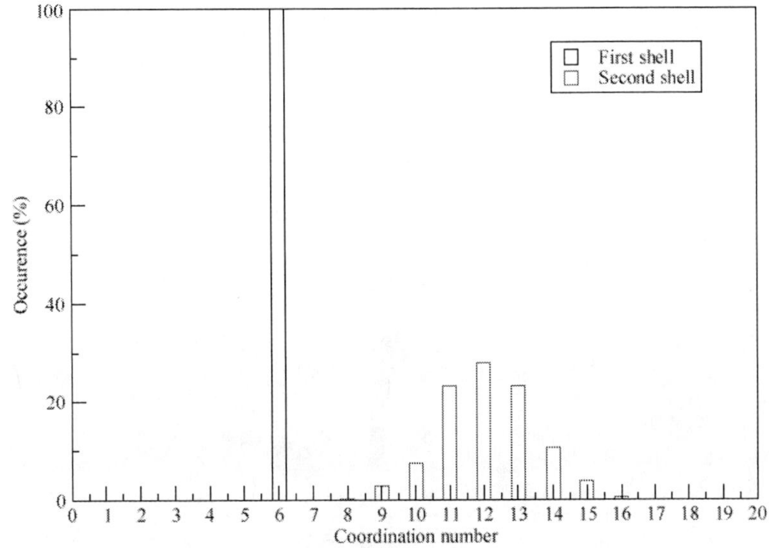

Figure 3. Distributions of the coordination numbers calculated within the first and second solvation shell of Hg^{2+} in water based on QM/MM MD simulation

Figure 3 is a CND graph from the simulation of Hg^{2+} ions in water. The figure gives information that the number of ligands surrounding the central atom for the hydration system of Hg^{2+} ion in water the number of ligands is 6. So it can be said that the complex ion $Hg(H_2O)_6^{2+}$ which is formed is an ion molecule that is less stable enough. The number of ligands in the second shell looks quite high. This fact shows that the ligand in the second shell is very dynamic and flexible.

3.3.3. Angular Distribution Function (ADF). ADF or angle distribution function is a function that describes the distribution of angles during the molecular dynamics simulation takes place. ADF provides information about the bond angle distribution formed between ligand 1 (atom O_1) - central ion (Hg) - ligand 2 (atom O_2). The ADF Hg^{2+} ion in water is presented in figure 4.

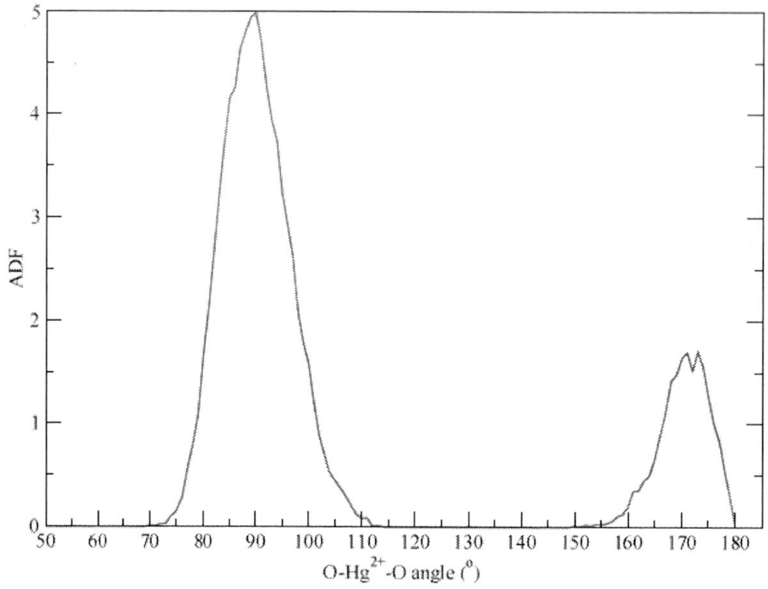

Figure 4 The O-Hg^{2+}-O angle (°) distribution of solvation structure of Cd^{2+} in water based on QM/MM MD simulation

Figure 4 shows that the angle O_1-Hg^{2+} -O_2 is between 90° and 175°. The graph shows the two highest peak angles, at 90°. While the angle of one ranges from 175°. Based on the results of the analysis of RDF, CND, and ADF above, it can be concluded that the hydration structure of Hg ions in water forms a distorted octahedral.

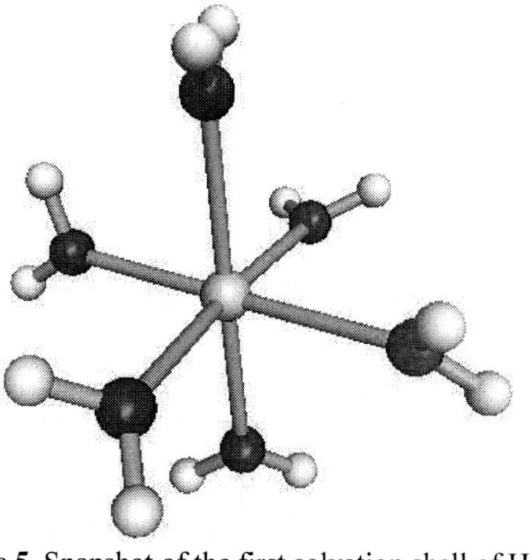

Figure 5. Snapshot of the first solvation shell of Hg^{2+} in water during the hybrid QM/MM MD simulations

Graph of ligand exchange hydration of Hg^{2+} ion in water, shows the hydration dynamics between Hg^{2+} ions and water molecules during the simulation process. In addition to information on the hydration distance of Hg^{2+} ions with water molecules, it is also seen the number of ligands that hydrate both in the first and the second shell. This ligand exchange chart can also show the presence or absence of water ligands from the first shell to the second shell and vice versa from the outer shell to the deeper shell. The presence or absence of ligand exchange during the ongoing simulation is presented in figure 6.

Figure 6 is a ligand exchange graph from a simulation and Hg^{2+} ion in water. It was seen during the simulation that it had never experienced a ligand move either towards the first shell or towards the second shell. This shows that the hydration structure of Hg^{2+} ion molecules in water is quite stable, in this case the Hg^{2+} ion binds 6 water molecules as ligands. Ligand exchange is seen in the second of shell to the bulk phase, from the graph shows the dynamics of the movement of water molecules, both the movement from the second shell to the third shell and the bulk phase, and vice versa from the bulk phase or the third shell to the second shell.

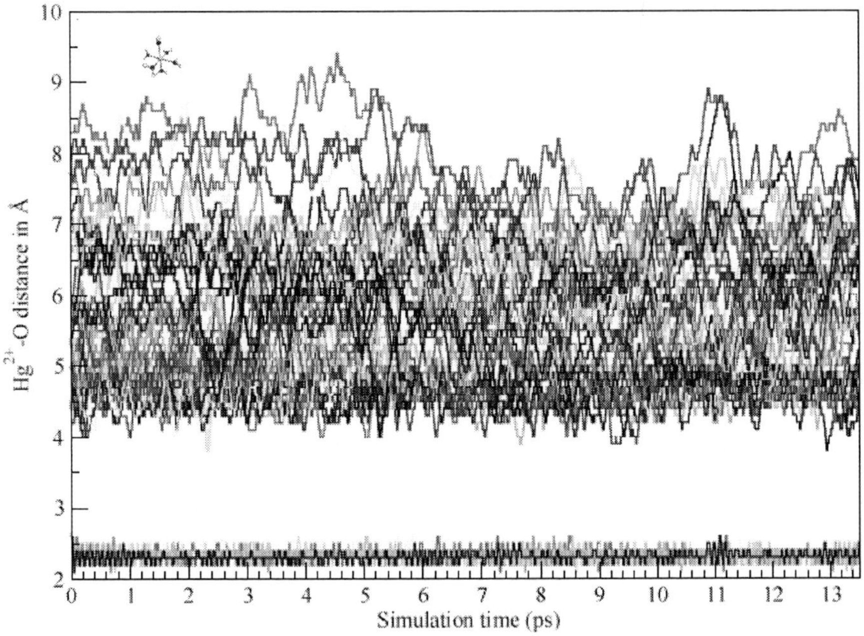

Figure 6. No water ligands migration between first and second solvation shell are observed: the ligand exchanges of H_2O are observed between the second solvation shell and bulk.

4. Conclusion

The Hg^{2+} ion in water binds 6 water molecules as ligands. Hydration of Hg^{2+} ions has a distorted octahedral structure. The residence time of ligands of water in the hydration of Hg^{2+} ions is relatively long, and thus the first shell solvation are stable.

5. References

[1] Armunanto R, Schwenk C F, Setiaji B A H, and Role B M 2003 *J.Chem. Phys.* **295** 63-73
[2] Armunanto R, Schwenk C F, Randolf B R and Rode B M 2004 *J.Chem. Phys.* **305** 135
[3] Cotruvo J Dr 2011 World Health Organization/SDE/WSH/03/04/80/Rev/1
[4] Clemens S, 2006 *J.biochi* 1707-19
[5] Darmono 2016 Suara Merdeka May 14
[6] Hirata F 2003 *Molecular Theory of Solvation* (New York: Kluwer Academic Publisher)
[7] Kritayakornupong C, Vchirawongkwin V V and Rode B M 2010 *J. Comput. Chem.* **31** 1785

[8] Lim L H V, Pribil A B, Ellmerer A E, Randolf B R and Rode B M 2010 *J. Comput. Chem.* **31** 1200
[9] Nilsson K B, Maliarik M, Persson I, Fischer A, Ullström A, Eriksson L, dan Sandström M 2008 *Inorganic Chemistry* **47** 6
[10]Nilsson K B, Eriksson L, Kessler V G, dan Persson I, 2007 *J. Mol. Liq.* 131–132, 113 – 120
[11] Rudoplh WW and Cory C Pye 2000 *J.Phys. Chem. A* **104** 1627-39
[12] Rode B M and Hofer T S 2006 *Pure Appl. Chem.* **78** 525–539.
[13] Rode B M, Schwenk C F, Hofer T S dan Randolf B R 2005 *Coord. Chem. Rev.* **249** 2993–3006

New generation biofuel from polypropylene plastic waste with co-reactant waste cooking oil and its characteristic performance

H Juwono*, K A Nugroho, R Alfian, Y L Ni'mah, D Sugiarso and Harmami

Chemistry Department, Faculty of Sciences
Institut Teknologi Sepuluh Nopember (ITS)
Jl. Arief Rahman Hakim, Surabaya 60111 Indonesia

*Email: hjachmad@gmail.com

Abstract. Liquid fuels from polypropylene plastic waste and co-reactants biodiesel from *Waste Cooking Oil* have been successfully performed by catalytic cracking method. The catalyst used is Al-MCM-41: Ceramics with ratio 7:3. The catalyst was characterized by XRD, SEM, Pyridine-FTIR, N_2-Adsorption-Desorption, and the product of catalytic cracking were investigated by GC-MS. The Liquid fuels obtained was mixed with commercial fuels and MTBE with ratio (10:87.5:2.5). The results showed liquid fuels have the largest percentage of gasoline (C_7-C_{12}) are 93.92 fraction produced. The performance of liquid fuels is enhanced with the additive MTBE investigated by generator set engine with gasoline-based fuel. The liquid fuel blends on the CB50 feedstock variation has the highest percentage of thermal efficiency of 27.42%, with a density value of 722.55 Kg/m^3, a flash point of -5.9 ° C, a viscosity of 0.238 cst, and a heating value of 15,465 , 94 Kcal/Kg. Characteristics of liquid fuel blends (CB) manufactured in accordance with SNI 06-3506-1994. Addition of the MTBE additive enhances the performance of CB100 mixture by 0.74% to the variation of TCB100 and 4.66% to Premium RON 88 at maximum loading of 2090 Watt. Characteristic of liquid fuels produced were also suitable with commercial gasoline standard.

1. Introduction

Waste plastics are still a material that widely used by industry and home. The use of very high plastic causes the building up of plastic waste. The process of polypropylene (PP) cracking is one way to handle plastic waste. Cracking is a termination reaction of long chain hydrocarbons with large molecular weights into short chain hydrocarbons of small molecular weights. Two kinds of cracking process are thermal cracking and catalytic cracking [1-2]. Based on a previous study , the catalytic cracking of polypropylene use a catalyst of Zeolite β (BEA) of 5% w/w produces 85% gasoline with Si / Al ratio of 30 at 300-350 °C [3].

In the research that has been done by Bintarasari [4], pyrolysis of polypropylene plastic waste with Al-modified MCM-41 catalyst and ceramic obtained liquid fuel product at gasoline fraction with yield of 71,89%. The resulting liquid fuels have characteristics approaching the characteristics of fuel oil of gasoline type according to the quality standard of SNI 06-3506-1994. Liquid fuels from high-calorie plastic pyrolysis results into an additional source of hydrogen and other carbon rich sources in order to obtain high aromatic yields and are on the kerosene fraction to increase the calorific value of the resulting liquid fuels [5].

One of the main important route to make renewable fuel is producing a biogasoline. Biogasoline is made fromvegetables sources consisting C_4–C_{11} hydrocarbon. The next stage of this process is converting the FAMEs to bio-gasoline by a cracking reaction. Catalytic hydrocracking is one type of cracking process that converting FAMEs compounds into hydrocarbon fractions using active acid sites available in the chemical structure of a catalyst [6]. To improve the quality of bio-oil, catalytic cracking methods of biomass or co-catalytic cracking with plastics have been recommended [7-8]. Catalytic cracking of biomass and plastics is also referred to as one method to produce aromatic hydrocarbons in the final product of oil. The synergistic effect for the formation of aromatic hydrocarbons by co-pyrolysis with plastics in the biomass cracking system has found [2][9-10].

This study the hydrocarbon fuel (C_8-C_{12}) synthesis was carried out through catalytic conversion process from waste plastic with Al-MCM-41 catalyst [11]. Furthermore, the characterizations of resulting product, such as density, viscosity, flash point, boiling point, calorific value, and analysis using Gas Chromatography-Mass Spectrometer (GC-MS) were tested. The results compared with SNI 06-3506-1994 gasoline fuel oil quality and international standard HIBER11Z Hibernia Petroleum Canada 2016.

2. Experimental

2.1 Materials

Sodium aluminate (PA), tetra ethylortho silicate (PA), and ethyl triamonium bromide (PA) have been supplied by Sigma Aldrich. Phosphoric acid (PA), methyl tertbutyl ether (MTBE), methanol (99.8%), n-pentana (95%), and pyridin (99.9%) received as it were from Merck. A sampel of waste cooking oil. Hydrocarbon liquid fuels from plastic waste from previous studies [11], creamic and N_2 gas.

2.2 Procedures

2.2.1 Synthesis of Al-MCM-41. Catalyst Al-MCM-41 was synthesized by hydrothermal technique according to the procedure of previous studies [12], using a wet impregnation method.

2.2.2 Characterization of catalysts. Samples for acidity have saturated steam of pyridine absorption and measured by FTIR, morfologies by SEM. The N2 adsorption measurements of the calcined samples were performed using a Tristar II 3020. Surface area calculations were carried out using the Brunauer-Emmett-Teller (BET) methods.

2.2.3 Catalytic activity. The samples were dried until the moisture content below 5% and were squeezed using hydraulic press to obtain raw oil of the nyamplung. The transesterification reaction was carried out by the addition of methanol. FAMEs compounds obtained by 98.15%. All products were analyzed using GC-MS Shimadzu.

2.2.4 Synthesis of biodiesel waste cooking oil. The synthesis of liquid fuels was carried out in refluks reactor at 60 °C 800 rpm by adding $NaOH_{(s)}$ and methanol PA. The result of biodiesel was accommodated on erlenmeyer.

2.2.5 Synthesis of liquid fuels. The synthesis of liquid fuels was carried out in a continous reactor. The result of hydrocarbon fuel was accommodated on erlenmeyer. Repeated use of catalysts to obtain samples PM100, PM80, PM60, PM50, and PM40. Hydrocarbon fuel was mixed with Premium RON 88 and MTBE to obtain the optimum performance in gasolin engine.

Table 1. Compotition of Liquid Fuels Blend

Sample (PM)	Liquid Fuel (v/v)	Premium (88) %(v/v)	MTBE % (v/v)	Liquid Fuel Blend (CB)
PM100	10%	90%	2.45 %	CB100
PM80	10%	90%	2.45 %	CB80
PM60	10%	90%	2.45 %	CB60
PM50	10%	90%	2.45 %	CB50
PM40	10%	90%	2.45 %	CB40
PM100	10%	90%	0 %	TCB100

2.2.6 Preparation of catalyst

Unused spark plugs were rinsed clean, crushed and sorted into ceramic parts. The Al-MCM-41 catalyst was mixed with a ceramic powder with a mass ratio of 7:3. A total of 9 grams of the pelleted catalyst blend were inserted into a catalyst tube filled with glass wool [13].

2.2.7 Characterization of hydrocarbon liquid fuels and performance test in gasolin engine

Characterization of hydrocarbon fuels includes test of density,viscosity and heating value [5]. The test methods were shown in table 2. Components of chemical compounds were analyzed using Gas Chromatograph-Mass Spectrophotometry (GC-MS). Performance of liquid fuels Blend was characterized in Gasolin Engine at Motor Fuel Laboratory Mechanical Engineering Industry Department.

Table 2. Method of Physical Properties

Parameter	**Testing Method**
Density	ASTM D.4052
Viscosity	ASTM D.445

3. Result and Discussion

3.1 Synthesis of Hydrocarbon Fuels

Hydrocarbon fuels produced from different variation volume by catalytic conversion process took place for 2 hours refer to [14-15]. The measurement of hydrocarbon fuels volume as a cracking result of each variation shown on table 3.

Table 3. Synthesis of Hydrocarbon Fuels

Feedstock Variation	Liquid Fuel (mL)	Waste (mL)	Type Liquid Fuel
100% (P):0% (J)	45	162	PM100
80% (P):20% (J)	100	235	PM80
60% (P):40% (J)	72	220	PM60
50% (P):50% (J)	30	217	PM50
40% (P):60% (J)	38	230	PM40

3.2 Characterization

3.2.1 FT-IR. The acidity of catalyst was measured by pyridin adsorption and FTIR was presented in figure 1 [16]. The spectra shown in figure 1 shows the interaction between pyridine compound and the active site of Al-MCM-41 catalyst at wave number 1537-1554 cm^{-1} showing the active side of

Bronsted Acid with an amount of 0.0670 mmol / g. While the wave number 1437-1454 cm^{-1} with the amount of 0.0386 mmol / g shows the interaction between pyridine compound with Al-MCM-41 active side Lewis acid [17].

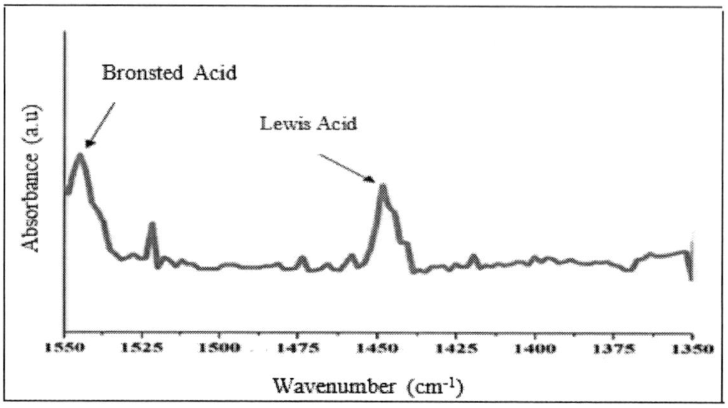

Figure 1. Spectra FTIR of pyridine adsorption

Table 4. The Result of FTIR

Type of Acid	Wavenumber (cm^{-1})	Amount Of Acid (mmol / g)
Lewis	1437-1454	0.0386
Bronsted	1537-1554	0.0670

3.2.2 Gas Chromatography-Mass Spectroscopy. The chromatogram of the product (FAMEs compounds) of the transesterification reaction can be seen in figure 2.

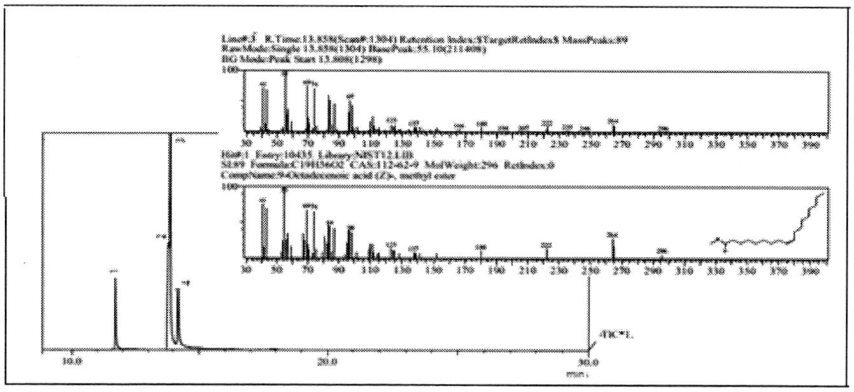

Figure 2. GC-MS chromatograms of FAMEs compounds

The chromatogram showed that the FAMEs compounds were 98.15% consisting of peak (1) 20.86% methyl palmitate, (2) methyl oleate 59.25% and (3) 18.04% methyl stearate. The GC-MS analysis showed that the catalytic conversion process produced not only a gasoline fraction (C_8-C_{12}), but also short chain hydrocarbon fractions (C_1-C_7) and long chain hydrocarbon fractions (C_{12}-C_{16}). The composition of each fraction could be seen in figure 3.

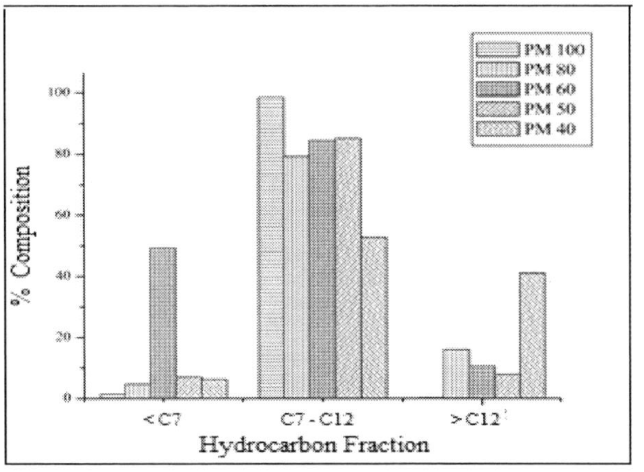

Figure 3. Composition of Hydrocarbon Fraction

3.2.3 Surface Morphology of Catalyst. The morphology SEM of the catalysts was shown in figure 4. In the SEM results, Al-MCM-41 is visually attached to the ceramic surface. The appearance of Al-MCM-41 is shown in A at figure 4 (a) and ceramic sightings are shown in B at figure 4 (b). In pure Al-MCM-41, fine surfaces are seen to agglomerate from solids with hexagonal structures, whereas in the Al-MCM-41: ceramic (7: 3) variation, there are fine particles with hexagonal structures attached to the ceramic surfaces that act as solids supporters [18].

Figure 4. (a) SEM Al-MCM-41 (b) SEM Al-MCM-41:Ceramic (7:3)

3.2.4 Adsorption-Desorption of N_2 Gas. The result of The N_2 adsorption–desorption was shown in figure 5. This graph indicates isotherm type IV which corresponds to a typical Al-MCM-41 catalyst graph [19]. This result proves the addition of ceramic with the mass ratio of Al MCM-41: ceramic (7: 3) does not change the isotherm graph type of Al-MCM-41. The surface area of the catalyst is increased from the previously synthesized Al-MCM-41 catalyst by [17]. This suggests that the presence of ceramics adds to the surface area of Al-MCM-41.

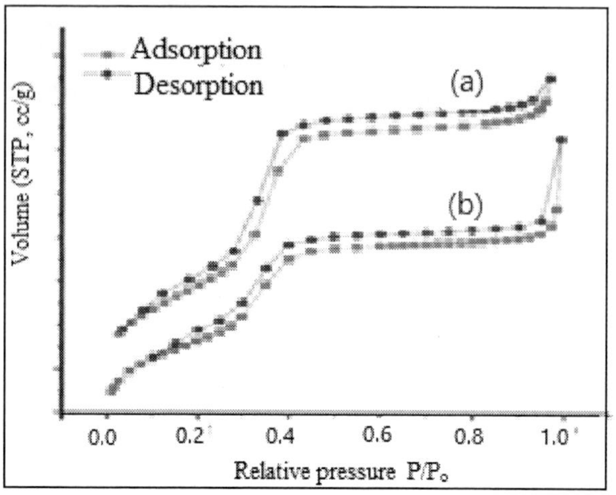

Figure 5. (a) Isotherm graph of Al-MCM-41 (b) Isotherm graph of Al-MCM-41: Cer

3.2.5 Determination of Density. Fig. 6 showed the recurrence use of catalyst affect to density of hydrocarbon fuels which weredecreasing density of the sample. The decrease of density in liquid fuel blends (CB) is caused by density possessed by premium fuel which is lower than liquid fuel (PM) that is equal to 736 Kg/m^3. CB0 type liquid fuel density of 740 Kg/m^3. The additive of methyl tertiary butyl ether has a density value of 744 Kg/m^3 with fully dissolvable properties in premium fuel so that the addition in the liquid fuel blends tends to reduce the premium fuel density value [19-21]. All liquid fuels blend of this research product compared with the criteria which based on SNI 06-3506-1994 on petroleum gasoline which states that minimum density of fuel oil is 715 kg/m^3 and maximum 770 kg/m^3 at temperature measurement 40°C.

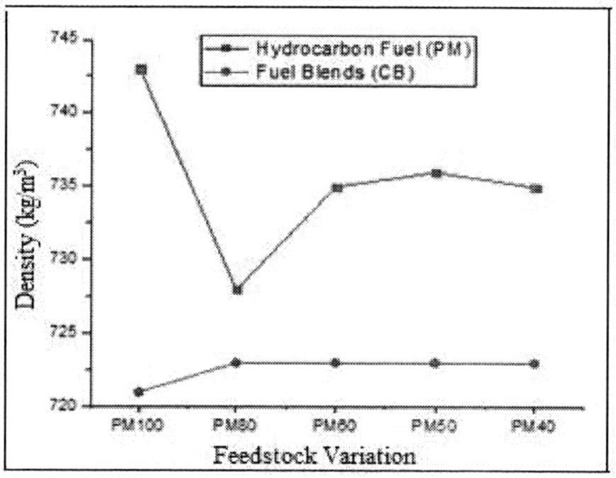

Figure 6. Density of Liquid Fuel Blends

3.2.6 Determination of Viscosity. Figure 7 showed that the viscosity of liquid fuel blends (CB) in the research tends to increase. This is due to the addition of co-reactant of Waste Oil biodiesel. The entire liquid fuel blends (CB) has a viscosity below 0.3 cSt. This is due to the viscosity of the main components of the mixture i.e. premium fuel measured having a viscosity of only 0.275 cSt and TCB100 fuel of 0.274 cSt. International standards issued by Hibernia Petroleum Canada 2016 HIBER11Z on crude oil specification testing have limits on measured kinematic viscosity values at 40°C at a maximum of 0.620 cSt [3]. All samples has lower viscosity than the standard.

IOP Conf. Series: Journal of Physics: Conf. Series **1156** (2019) 012013 doi:10.1088/1742-6596/1156/1/012013

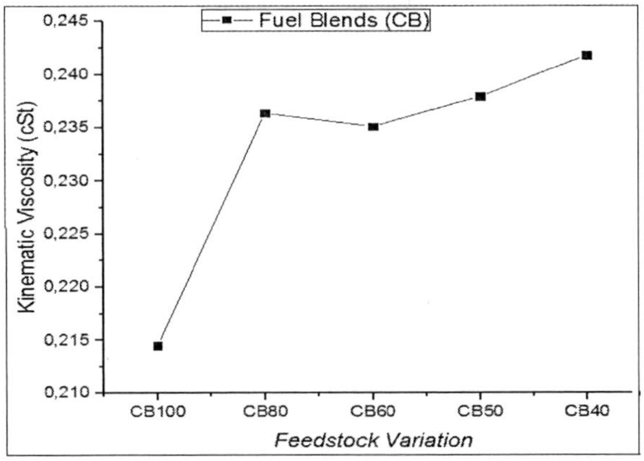

Figure 7. Viscosity of Liquid Fuels Blends

3.2.7 Calorific value. The result of the calorific value of liquid fuels blendwas shown in figure 8. The result shows MTBE has increase of sample calorific value. It was shown that CB100 has higher calorific value than TCB100 (without addition of MTBE). CB40 has higher calorific value than the other samples. The value was 19.456 Kcal/Kg. All samples has fullfilled with the HIBER11Z standard of Hibernia Petroleum which has a minimum of 10.347 Kcal/kg [19].

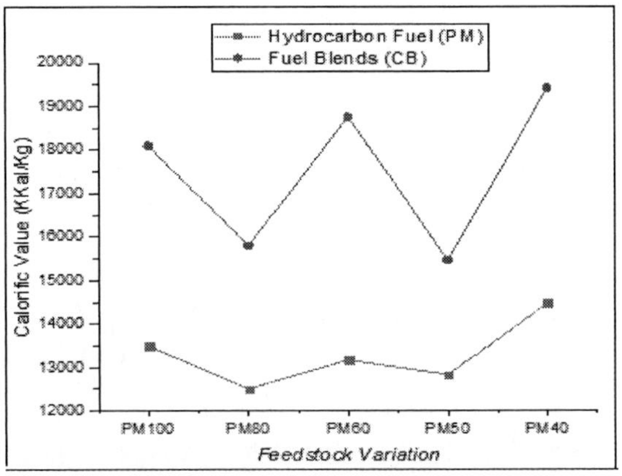

Figure 8. Calorific Value of Liquid Fuels

3.2.8 Thermal Efficiency. The result of thermal efficiency of liquid fuels were shown in figure 9. Percentage of thermal efficiency of liquid fuel blend increases with the increase of loading (watt) on gasoline engine. The greatest efficiency value was shown in liquid fuel blend CB50 ((10% (PM50): 90% (Premium), 2.45% MTBE)) obtained by 27.42% thermal efficiency at 2090 Watt. While the other variation and Premium RON 88 has lower thermal efficiency than CB50 at the same power. MTBE also increase the percent thermal efficiency of the samples. It was shown that CB100 has higher percent thermal efficiency than TCB100 (without addition of MTBE). Addition of 10% MTBE additive can increase the percent of thermal efficiency of gasoline [22].

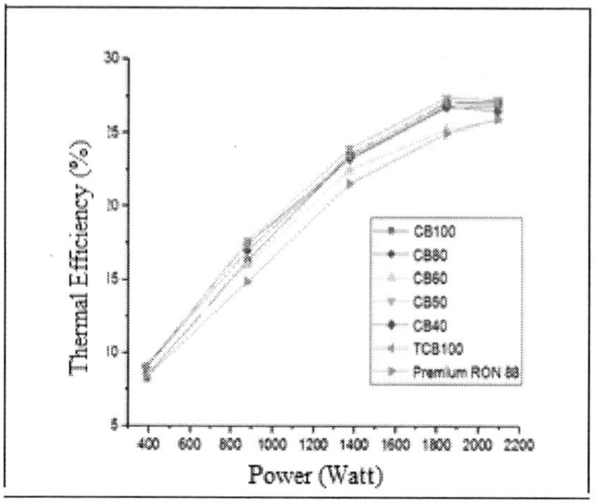

Figure 9. Thermal Efficiency of Liquid Fuels

The product of liquid fuel from catalytic cracking have probability to use as an alternative fuel, based on data result, regarding its possible as new generation fuel.

4. Conclusion

The following conclusion can be drawn from the synthesis of Al-MCM-41 catalysts and the catalytic hydrocracking of FAMEs. Hydrocarbon fuels (C_7-C_{12}) from propylene plastic waste (PP) have been synthesized through catalytic conversion method using Al-MCM-41 catalyst of 11.18 gram and also repeated use of catalyst three times. The Liquid fuels obtained was mixed with commercial fuels (Premium RON 88) and methyl tertiary butyl ether (MTBE) with ratio (10:87.5:2.5). The results showed liquid fuels have the largest percentage of gasoline (C_7-C_{12}) are 93.92 fraction produced. The performance of liquid fuels is enhanced with the additive MTBE investigated by generator set engine with gasoline-based fuel. The liquid fuel mixture on the CB50 feedstock variation has the highest percentage of thermal efficiency of 27.42%, with a density value of 722.55 Kg / m3, a flash point of -5.9 ° C, a viscosity of 0.238 cst, and a heating value of 15,465 , 94 Kcal / Kg. The addition of MTBE additive improves the performance (percent thermal efficiency) of CB100 by 0.74% to the variation of TCB100 and 4.66% to Premium RON 88 at maximum loading of 2090 Watt.

5. References

[1] Ratnasari D K, Nahil M A and Williams P T 2017 *Journal of Analytical and Applied Pyrolysis* **124** 631-7
[2] Xuesong Zhang H L, Zhu L, Qian M, Yadavalli G, Wu J and Chen S 2017 *Fuel* **88** 28–38
[3] Miandad R, Barakat M A, Rehan M, Aburiazaiza A S, Ismail I M I and Nizami A S 2016 *Waste Management* **58** 250-9
[4] Bintarasari S I 2017 *Performance Efficiency of Catalytic Cracking of Polypropylene (PP) Plastic Waste Using Al-MCM-41 Catalyst*, in *Chemistry* (Surabaya, Indonesia: Institut Teknologi Sepuluh Nopember)
[5] Kosivtsov Y, Sulman E, Lugovoy Y, Kosivtsova A, Stepacheva and Antonina 2015 *Bulletin of Chemical Reaction Engineering & Catalysis* **10**(3)
[6] Juwono H, Triyono T, Sutarno and Wahyuni E T 2013 *Indones. J. Chem.* **13 (2)** 171-8
[7] Kai F J L, Xiaoyi Yang b and Ronga L 2014 *International journal of hydrogen energy* **39** 3690-97
[8] Bulushev D A R and Julian R H 2011 *Catalysis Today* **171** 13
[9] Wang J, Bi P, Zhang Y, Xue H, Jiang P, Wu X, Liu J, Wang T and Li Q 2015 *Energy* **86** 488-99
[10] Jae J, Coolman R, Mountziaris T J, and Huber George W 2014 *Chem. Eng. Sci.* **108** 33-46

[11] Juwono H, Triyono T, Sutarno S, Wahyuni E T, Ulfin I, and Kurniawan F 2017 *Indones. J. Chem.* **17**(2) 316
[12] Roik N V and Belyakova L A 2013 *J. Solid State Chem.* **207** 194-202
[13] Aghakhania M S, Najafi A A K Sh and Mortazavi Y 2014 *J. Ind. Eng. Chem.* **20** 3037 - 45
[14] Artetxe M G L, Amutio M, Bilbao J and Olazar M 2014 *Chem. Eng. Sci.* **116** 635-44
[15] Imam Khabib S K and Wahyuni S 2014 *Indones. J. Chem* **14** (2) 192 - 8
[16] Ayodelea O B, Hamisu U, Farouk, Mohammedc J, Uemurad Y and Dauda W M A W 2014 *J. Taiwan Ins. Chem. E.* 1-11
[17] Juwono H, Triyono T, Sutarno S, Wahyuni E T, Harmami, Ulfin I, and Kurniawan F 2017 *Bull. Chem. React. Eng. Catal.* **12**(3)337-43
[18] Hagen J 2004 *Industrial Catalysis - A Practical Approach 2nd edn* (New York: Wiley-VCH)
[19] Kalargaris I, Tian G and Gu S 2018 *Fuel* **211** 797-803
[20] Dasilva R, Cataluna R, Menezes E, Samios D, and Piatnicki C, 2005 *Fuel* **84**(7-8) 951-9
[21] Topgül T, 2015 *Fuel Processing Technology* **138** 483-9
[22] Heydariaraghi M, Ghorbanian S, Hallajisani A and Salehpour A 2016 *J. Anal. Appl. Pyrol.* **121** 307-17

Acknowledgments
The authors acknowledge PDUPT Scheme of Grants as well as Higher Education Directorate of Republic Indonesia for Financial Grant in 2018, the supported of this work.

Pre-treatment of glass substrates and post treatment of the surface of single and multiple Chitosan film by heated as wettability

E Rahmawati* **and S Agustina**

Department of Physics Education STKIP Bima Yogyakarta Indonesia

*Email: rahmawatieka89@gmail.com

Abstract. In this approach, chitosan thin films have been prepared using single and multiple dip coating method. The glass substrates was heated 100 ^{0}C as a pre-treatment and the fabricated thin film were post heated in various temperature: 100 ^{0}C, 210 ^{0}C, 220 ^{0}C and the effect of pre and post heat treatment on their wetting, structural and morphological properties has been investigated. Scanning electron microscopy (SEM and contact angle measurement have been applied for the characterization and study of thin film surfaces. The results of SEM exhibit the aggregation and formation of chitosan films on the surface of substrate changed through applying heat pre-treatment substrates and increasing of heat post temperature to the chitosan surface.Wettability studies showed that the contact angle of prepared thin films was influenced by various parameters such as heat pre-treatment on glass substrates, variation of heat post-treatment andsurface morphology.

1. Introduction

Physical and chemical properties of the polymer will influence the functionality, performance, and ultimately the applicability of a polymer itself [1-3]. One of the important and interesting features of many polymers is their surface properties. There are a large number of studies have been focused on the characterization and modification of the surface properties [4-7]. Polymer surface modification can be achieved using physical and chemical methods, in which existing functional groups are altered and new physical properties are produced. The purpose of surface treatment is mainly to improve certain mechanical and performance-related properties. Some unique properties are mainly use in industrial, medical and biological application based polymers are hardness, bacterial repellence, biocompatibility, appropriate wettability, surface conductivity.With the development of coating preparation technology, a substrate surface pre-treatment, e.g. sand blasting, grinding and polishing or honing, is required for improving the performance of the coatings [8-9]. Different pre-treatment processing results in different substrates surface roughness, and its evaluation is very important for many fundamental problems such as friction, adhesion property, contact deformation, heat and electric current conduction, tightness of contact joints and positional accuracy[10]. Pre and post treatment of the thin films can be done to get the new surface properties of thin film. There are many scientific works on molecularly smooth surface [11-13], but just a little of them work on wettability and spreading phenomena of real engineering surfaces [14]. Different friction properties can be achieved by controlling surface roughness. Reduction in friction and wear usually observed for smooth but not polished surfaces [15].Wettability is usually quantified in terms of observed contact angles, so from a practical point of view, a simple methodology is needed to account for the heterogeneous rough surface influence on wetting and contact angle measurements, which indicates the degree of wetting when a solid and liquid interact. Large contact angles ($< 90°$) correspond to low wettability. Small contact angles ($< 90°$) correspond to high wettability. Surface with a water contact angle approaching zero is classified as super hydrophilic (super high wettability) and a surface with a contact angle >150 is usually categorized as super hydrophobic (super low wettability).Wetting phenomenon of composite materials had been investigated by Cassie and Baxter [16], where the rough surfaces is able to capture water by trapping process in the roughness properties. Pre-treatment glass substrate by Ar/O_2 ion beamsfor the as-sputtered rough zinc oxide thin films has been done [17]. Applying Ar and O_2 ion obtained the changing of morphology and resulted rougher surface than untreated one. While [18] irradiate the polystyrene thin film by ultra violet to enhanced the hydrophobicity of film effect to bio-immobilization. The different roughness surface of thin films polymers can be done by choosing the different solvent of polymers its self. Setyawan et al [19] using chloroform, toluene, xylene, THF as the solvent of polystyrene to get the difference surface of thin films coated on QCM.

In this paper we focus on the treatment of glass substrate to obtain the altered wettability of chitosan thin film by doing heat pre-treatment of the substrates and post heat treatment to chitosan filmsingle and multipledipcoating. Chitosan is inert, hydrophilic, biocompatible and biodegradable, it has attracted scientific and industrial interest in widely differing fields.

2. Experimental Section

2.1 Sample preparation

Glass substrate rectangular block was used as substrates, fixed size was 10 mm ×10 mm×4.76mm. All the substrates were cleaned for 10 minutes in methanol 97% and it be aerated then. To make chitosan thin film, we used whitish ivory chitosan with degree 84% of deacetylation which obtained from shrimp shells, it was acquired from chemistry laboratory of STKIP (Bima, Indonesia).

Coating was performed by using a single and multiple dip coating. Glass substrates was immersed in chitosan solution for 10 minutes and it be drawn vertically with a constant velocity. For multiples chitosan thin films, glass substrates was obtained by immersing and drying for each 10 minutes for three times repeating coating alternately.

Wettability alteration is done by doing heated pre-treatment of the glass substrate and heated post-treatment of the chitosan films for un-heatedpre-treatment substrates. Glass substrate was heated with 100 ^{0}Cfor an hour before coating chitosan and it be heated with 100 ^{0}C after coating. For heat post-treatment of un-heated substrates, chitosan films on glass substrates were heated on 100 ^{0}C, 210 ^{0}C, 220^{0}C for an hour.

2.2 Characterization of morphology

The morphology of the surfaces was characterized by scanning electron microscopy (SEM) standard on JJSM-6510LA.

2.3 Water contact angle measurement

The contact angle of chitosan thin films: heat pre-treatment and post-treatment were measured by using the sessile drop technique to measure the water contact angle (WCA) between drop water and chitosan film. Water droplets from a micropipette were placed on the surface of chitosan films. The drop water was captured by canon Eos 1200D with 18.0 MP 5184 x 3456 resolution and the image was analysed by CorelDraw X4.

3. Discussion

The coated chitosan film in two different single and multiple dip coating method were obtained. Applying heat pre-treatment substrate and heat post treatment chitosan films produced different morphology and wettability of films.

IOP Conf. Series: Journal of Physics: Conf. Series **1156** (2019) 012014 doi:10.1088/1742-6596/1156/1/012014

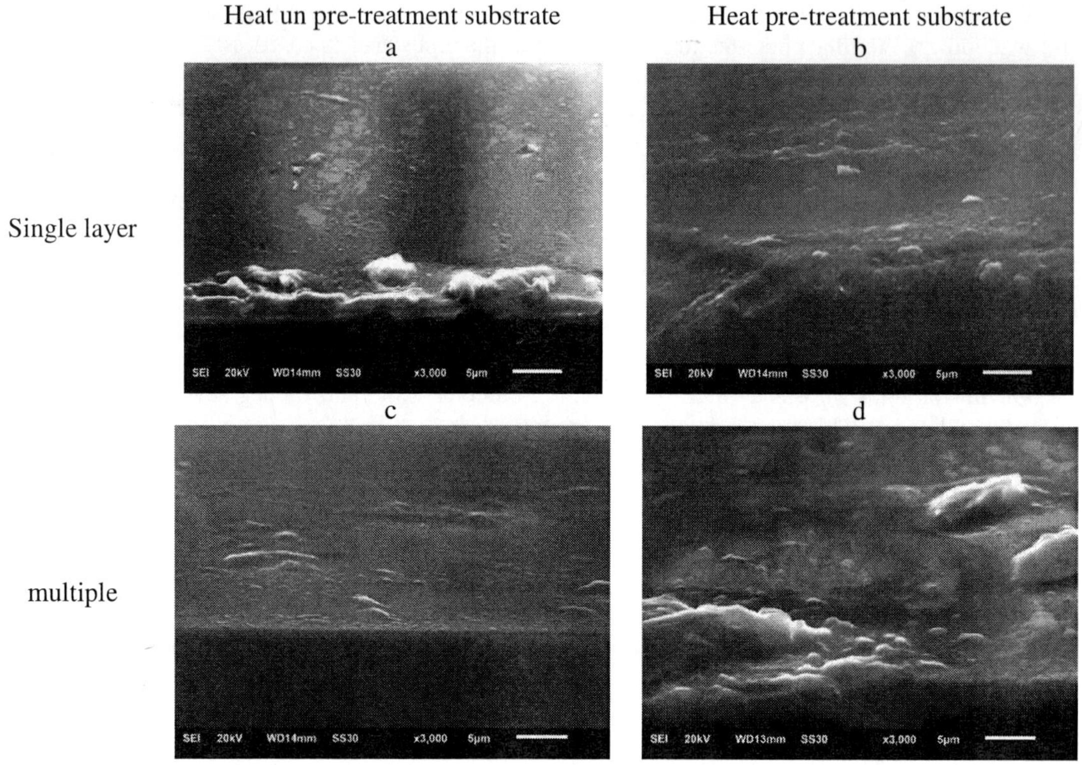

Figure 1. SEM images of chitosan thin films: a) Single layer of heat un pre-treatment substrate b) Single layer of heat pre-treatment substrate c) Multiple of heat un pre-treatment substrate d) multiple layer of heat un pre-treatment substrate.

Figure 1 shows that heat pre-treatment on glass substrate obtained the groovy and rough films for both single and multipledip coating. And it will be more groovy and rougher for multiplecoating for heat pre-treatment glass substrate. Glass particle agglomerate become a secondary particle for heat pre-treatment to glass substrate. Then all of the particle established a larger cluster in micro scale at the bottom but it still in nano scale at the top of surface. So, when chitosan was coated on the heat pre-treatment glass substrate, it will attend the groovy and rougher of surface.

Heat post treatment was one of the ways to do the surface modification get different wettability of surface. From figure 2, the SEM images indicated that the surface morphology was not significantly sensitive to the temperature post heat treatment for single dip coating films. Chitosan thin film was transparent and smooth. On the other hand, the surface morphology of the multiple dip coating film was sensitive to heat post treatment. The surface was transparent but little wavy at certain position. It indicates the increment of temperature heat post-treatment can change the distribution of grains on the surface of chitosan films. Chitosan thin film was heated at temperature 210 ^{0}C different in 7 with its glass transition temperature (Tg) 203^{0}C based on the careful DSC and DMA measurements [20]. It effected the movement of polymer chain of chitosan easily. Temperature influenced the mobility of polymer's chain as the increasing of kinetic energy; that is, the energy of objects in motion. It is actually an effect of random motion of molecules, whether they are polymer molecules or small molecules. So that, lumps that initially accumulate would move to find the best place and the surface become smoother. By increasing the heating temperature, the chitosan thin film surface become smoother all ways. Actually, to get information about chitosan grain can be obtained by AFM (Atomic Force Microscopy).

IOP Conf. Series: Journal of Physics: Conf. Series **1156** (2019) 012014 doi:10.1088/1742-6596/1156/1/012014

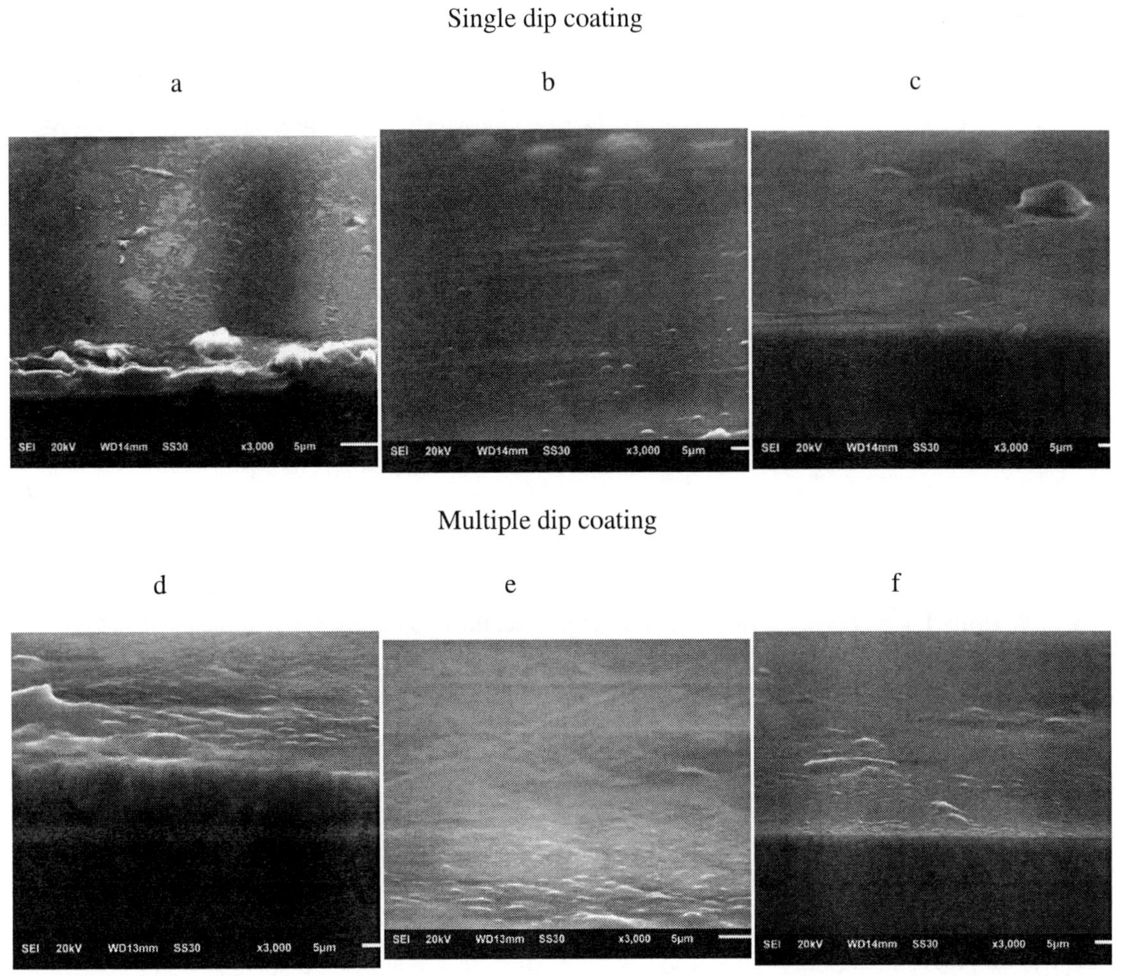

Figure 2. SEM images of chitosan thin films by post heat treatment: a) single dip coating film at temperature 100^{0}C b) single dip coating film at temperature 210^{0}C c) single dip coating film at temperature 220^{0}C d) multipledip coating film at temperature 100^{0}C e) multipledip coating film at temperature 210^{0}C f) multipledip coating film at temperature 220^{0}C.

The surface roughness will effect to wettability. Tailoring the roughness and chemical composition of the surface can be used to enhance the related hydrophilicity and hydrophobicity. In particular, nano and microtextures have been used to produce hydrophobic coating on different surfaces.

IOP Conf. Series: Journal of Physics: Conf. Series **1156** (2019) 012014 doi:10.1088/1742-6596/1156/1/012014

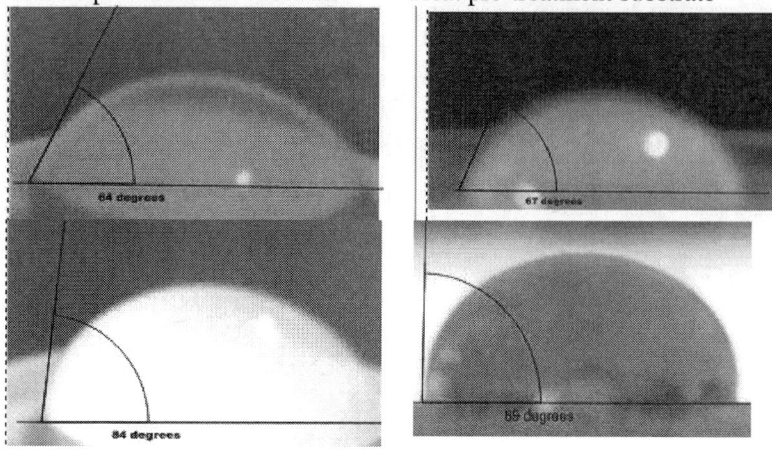

Figure 3. Image water drop on chitosan thin films between heat un pre-treatment and heat pre-treatment substrate.

Table 1. Contact angle between heat un pre-treatment and heat pre-treatment substrate.

	Contact angle (degrees)	
	Heat un pre-treatment substrate	Heat pre-treatment substrate
Single coating	64	67
Multi coating	84	89

It was explained before that glass substrate heat pre-treatment develop the rougher surface compare to heat un pretreatment. From the table 1, it can be seen that heat pre-treatment substrate obtained the higher water contact angle than untreated substrate, and it will be higher for multiple thin film. Values of measured apparent contact angle can be strongly affected by the roughness of the measured surface. When water drop onto the rough surface, the air would be trapped in the middle of roughness. Based on Cassie- Baxter, getting smaller interact surface area of particles with water, it will cause the larger contact angle. Chemically, wettability was influenced by polar molecule at surface. so that there will be an attraction between surface molecules and polar H_2O.

The contact angle value versus temperature has also been illustrated in Figure 4 and table 2 below. As can be seen, for both thicknesses, the contact angle gradually decreases with increasing in post heat temperature.

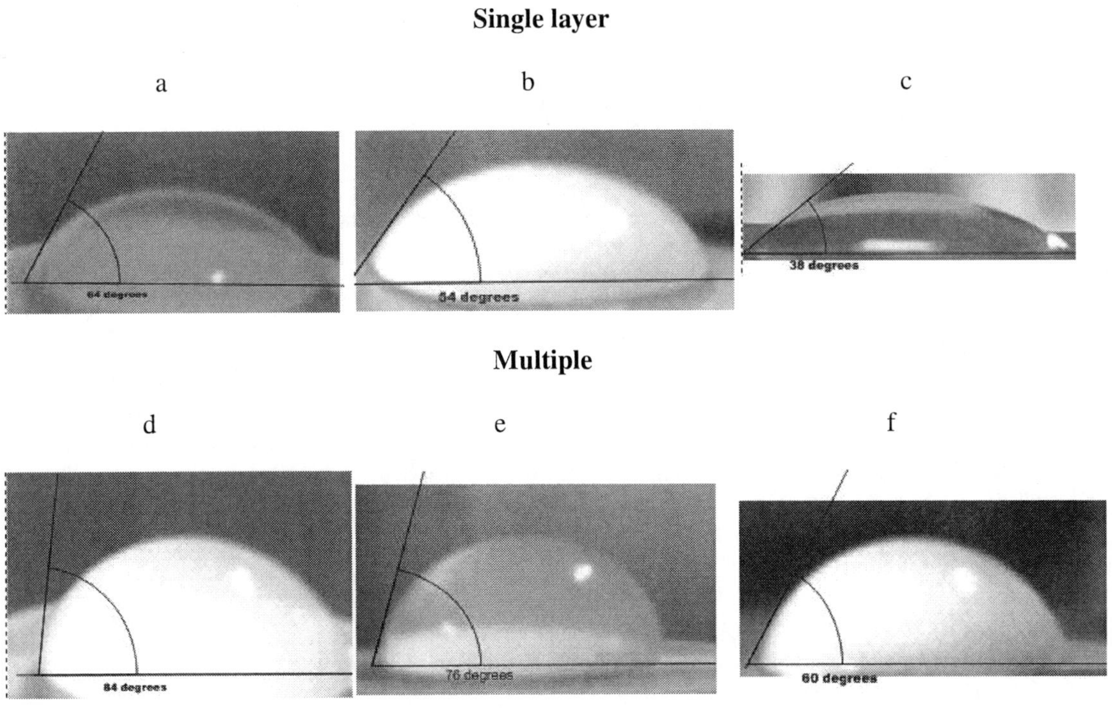

Figure 4. Image water drop on chitosan thin films was effected by temperature post heat treatment of un pre-treatment substrate : a) single layer film at temperature 100^{0}C b) single layer film at temperature 210 ^{0}C c) single layer film at temperature 220^{0}C d) multiple film at temperature 100 ^{0}C e) multiple film at temperature 210 ^{0}C f) multiple film at temperature 220 ^{0}C.

Table 2. Temperature post heat treatment of un pre-treatment substrate effect to contact angle of chitosan thin film.

	Contact angle (degrees)	
	Single coating	Multi coating
100 ^{0}C	64	84
210 ^{0}C	54	76
220 ^{0}C	38	60

It can be conclude that pre-treatment for substrates and post treatment of the surface obtained the hydrophilic surface cause the watercontact angle less than 90^0. Despite these chitosan can be modified and be used in wettability technology, pure chitosan films also show some disadvantages such as poor mechanical properties, high brittleness and high water sensitivity, thereby limitingtheir extensive applications [21-22]. Pure chitosan film displays high moisture absorption because of its hydrophilic character, thereby resulting in films with relatively poor mechanical properties at high humidity [23]. The water low resistance of chitosan thin film can be seen below. It was obtained after 20 minutes water droplet on surface.

Figure 5. chitosan low water resistance after 20 minutes water droplet.

4. Conclusion

Heat pre-treatment and post treatment of glass substrates and chitosan film result the altered wettability of chitosan thin film. Heat treatment influences the roughness of film for both single and multiple chitosan film. Glass substrate which undergo the heat treatment obtained the rougher surface and it has low wettability than other untreated surface. While the various heat post treatment surface has good wettability cause the hydrophilic properties with smoother surface. But unfortunately, the pure chitosan film has low water resistance. So it need to be modified with other material and using the other method to result the high quality thin film.

References

[1] Morent R, DeGeyter N, Desmet T, Dubruel P and Leys C 2011 *Plasma Process. Polym.* **8** 171–190
[2] De Geyter N and Morent R 2012 *Engineering and Technology* 225–246
[3] Borges A, Benetoli L, Licínio M, Zoldan V, Santos-Silva M, Assreuy J, Pasa A, Debacher N and Soldi V 2013 *Mater. Sci. Eng.* **33** 1315–24
[4] Zhao J H, Sheadel DA and Xue W 2012 *Physical* **187** 43–49
[5] Yoshida S, Hagiwara K, Hasebea T and Hotta A 2013 *Surf. Coat. Tech.* **233** 99–107
[6] Woo SK, Min H, Lee J-O and Song YH 2014 *Appl. Surf. Sci.* **295** 198–202
[7] Kolskáa Z, Řezníčkováb A and Nagyováb M 2014 *Polym. Degrad. Stab.* **101** 1–9
[8] Svahn F, Rudolphi A K and Wallén E 2003 *Wear* **254** 1092-98
[9] Harlin P, Bexell U and Olsson M 2009 *Surf. Coat. Tech.* **203** 1748-55
[10] Gadelmawla E S2002 *J. Mater. Process. Technol.* **123** 133-45
[11] Narayan Prabhu K, Fernades P and Kumar G 2009 *Mater. Des.* **30**(2) 297- 305
[12] Hay KM, Dragila MI and Liburdy J 2008 *J. Colloid Interface Sci.* **325** 472–7
[13] Extrand CW and Kumagai Y1997 *J. Colloid Interface Sci.* **191** 378–83
[14] Kubiak KJ, Mathia TG and Wilson MCT 2009 Proc. of 14th Congrès International De Métrologie Paris
[15] Kubiak KJ and Mathia T G 2009 *Wear* **267** 315-21
[16] Cassie ABD and Baxter S1944 *Trans. Faraday Soc.* **40** 546
[17] Wendi Zhang, Eerke Bunte, Martina Luysberg, Philipp Spiekermann, Florian Ruske and Jürgen Hüpkes 2011 *Surf. Coat. Tech.* **205** S223-S228
[18] Setyawan P. Sakti, LayliAmaliya, Nike F. Khusnah and Masruroh 2017 *Jurnal Teknologi* **79**(3) 61-67
[19] Setyawan PS, Rahmawati E and Robiandi F 2016 *Conf. Proc. Of ICTAP American institute of Phyisic* doi: 10.1063/1.4943712
[20] Sakurai K, Maegawa T and Takahashi T 2000 *Polymer* **41** 7051–56
[21] Thakhiew W, Devahastin S and Soponronnarit S 2013 *J. Food Eng.* **119** 140–149
[22] Liu M, Zhou Y, Zhang Y, Yu C and Cao S 2013 *Food Hydrocoll.* **33** 186–91
[23] Cervera M F, Karjalainen M, Airaksinen S, Rantanen J, Krogars K, Heinamaki J, Colarte A I and Yliruusi, J. 2004 *Eur. J. Pharm. Biopharm* **58** 69–76

Acknowledgements
The authors thank for the financial support from Directorate General of Higher Education (DIKTI), Ministry of National Education Indonesia.

IOP Conf. Series: Journal of Physics: Conf. Series **1156** (2019) 012015 doi:10.1088/1742-6596/1156/1/012015

Chemistry enrichment in tourism vocational school: The development and validation of food additives module

A Wiyarsi*, H Pratomo, E Priyambodo, Marfuatun and H Kusumaningtyas

Department of Chemistry Education, Faculty of Mathematics and Natural Sciences, Universitas Negeri Yogyakarta, Indonesia

*E-mail: antuni_w@uny.ac.id

Abstract. This study focused on the development and validation of the chemistry module in food additive topic for vocational students in the tourism field. An exploratory mixed method was conducted with four steps include qualitative, development, quantitative and interpretation. The Food Additives Module (FAM) was developed based on the basic competence of food processing and extended on chemistry content. To get the content and face validity, two experts had reviewed the FAM and gave the feedback to revise it. The analysis of FAM feasibility gained by the response of 7 chemistry teachers on four aspects includes the worthiness of chemistry concept, concept presentation, linguistics, and graphic. This study had produced the FAM module based on the contextual approach. The module had four theme namely sweeteners, preservatives, coloring and flavoring foods. A group of experts declared that the FAM was worthy as the chemistry enrichment content for vocational students of tourism field. The teacher gave a good response to the FAM quality. This study suggests that the FAM will be a useful tool for teachers to improve the chemical literacy of students in vocational tourism school.

1. Introduction

The objective of the vocational school was concerned to prepare the students in dealing with the working field. Hence, mostly students on vocational school, only focused on their specific expertise lesson instead of the other lesson. For example, in the tourism program, the students had less interest in studying the lesson outside their specific expertise such as chemistry. They feel that chemistry is not related to their specific expertise [1]. Moreover, the chemistry concepts were difficult to understand, has abstract characteristics, and not relevant to their daily life thus the students feel reluctant to learn chemistry [2]. These facts bring the decline of chemistry learning quality in the vocational tourism student. In fact, the chemistry learning is required as a prior knowledge in mastering the tourism specific expertise especially on the food processing topic. The sweeteners, preservatives, coloring, and flavoring food were the main materials in the food processing topic which were used chemistry as basic concept in understanding that topic. A learning source which provide the chemistry content with the context of tourism field is needed as a media of vocational tourism students in order to conduct a meaningful chemistry learning. The meaningful chemistry learning could be reached by correlating the chemistry concept with the daily life problem through the contextual approach.

Contextual approach may facilitate vocational tourism students in learning chemistry. Contextualizing the content of chemistry learning is potential to improve students' motivation, it is also reinforcing the meaning of learning chemistry and its perceptions [3]. The students tend to be more attracted in studying a big ideas of contextual discipline concepts than the fundamental concepts [4]. The used of contextual approach in chemistry learning facilitate the students to make a connection among the chemistry concepts with their daily life experience [5]. The contextual approach assumes that the students are naturally looking for the meaning of the learning context which appropriate with their daily life environment. Through the combination of the chemistry content with the students' daily experience could produce the meaningful understanding of the chemistry concept. The students could use their knowledge to solving some new and have never faced problems by increasing their knowledge and experience. Students are expected to be able in constructing their knowledge which be applied in the everyday life by combining the chemistry knowledge they got from the school. Thus, chemistry teacher should correlate the chemistry content with its usefulness in the students' daily life. Consequently, the students could apply their knowledge in chemistry toward the context of everyday life and bring the readiness in dealing their future working field.

The contextual approach brings the chemistry learning environment quite interesting. These learning approach lead the student understand that the tourism field is closely related to chemistry concept. They need to learn the differences of the natural and synthetic food coloring. They also learn on how and why giving natural or synthetic color to the food were safety to consume or not. The flavoring as additional materials to enhance the food taste. Which food preservatives was safety as additional materials and how much the threshold number of food preservatives to be add. These contextual problems were found in the everyday life which could be answer by the chemistry concept and bring the increasing of students' chemical literacy [6]. Hence, the used of contextual approach is needed to enhance students' chemical literacy due to the lack of Indonesian students' chemical literacy [7] especially in the vocational school.

The modification of teaching material on chemistry learning facilitate the students understand the chemistry concept easily, accurately, precisely, and clearly [8]. Consequently, a media which describe the food processing topic with the extend of chemistry content was needed in order to facilitate vocational tourism students in learning chemistry. The contextual approach in the term of daily life issues could be applied to develop several media in chemistry learning such as module. The module which used contextual approach will bring the increasing of vocational tourism students on chemical literacy. Hence, the chemistry meaningful learning will occur which implied the improving of the process and outcomes learning quality. Thus, this research was concerned to present the development and validation of the chemistry FAM for vocational students in the tourism field. These FAM was construct according to the contextual approach. Thus, the FAM was expected to improve chemical literacy of vocational tourism students.

2. Methods

2.1. Research design and procedure

This research aimed to develop and validate chemistry FAM based on contextual approach for vocational students in the tourism field. An exploratory mixed method design including the qualitative, development, quantitative, and interpretation steps were employed in this study. The research design present in the figure 1.

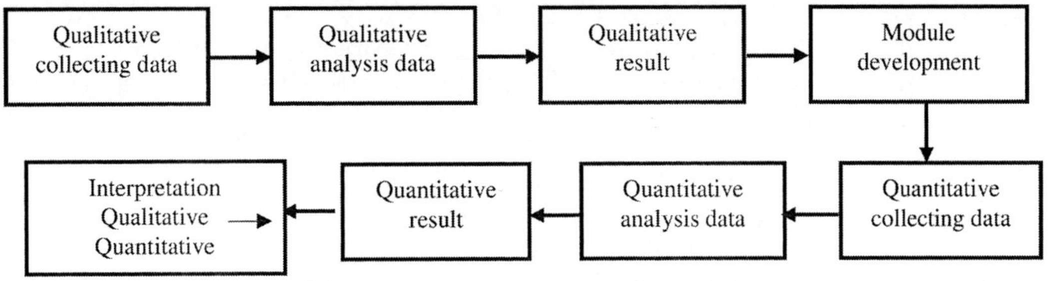

Figure 1. Exploratory mixed methods design

The qualitative step was aimed to analyze some problems on chemistry learning which occurred in the vocational tourism students. The result of this step showed that FAM was needed to enrich the knowledge of vocational tourism students. Whereas, on the development phase was aimed to provide a solution of the problems were found on the qualitative step. The determination of learning sources as reference and contextual content framework were conducted on the development step. The layout and content development were compiled to construct the FAM. The content and face validation were obtained by asking a group of experts to review the FAM. Moreover, on the quantitative step was conducted by asking the responses from a total of 7 teachers on four aspects includes the worthiness of chemistry concept, concept presentation, linguistics, and graphic toward the FAM. Finally, on the last step was interpretation of the research results according to the three steps conducted before, the qualitative, development, and quantitative steps.

2.2. Data collection and research instrument

There were three type of instruments used to collect the data in this research, there were the qualitative, development, and quantitative instrument. The qualitative instrument consists of the guide of the interview and the basic competence analysis sheet. The instrument which was used in the development step covered the validation sheets. Moreover, the instrument used on the quantitative step was the teachers' responses sheet toward the FAM covering four aspects of chemistry concept, concept presentation, linguistics, and graphic. In this case, the teachers were requested to review the FAM and gives a score from 1 (indicated very poor) until 5 (indicated very good) on each indicator of the teachers' response aspect of the FAM. A total of 25 indicators of the teachers' response toward the FAM shown in table 1.

Table 1. Indicator of the teachers' response toward FAM

Aspect	Number of Indicator	Indicator
Chemistry concept	1	The content of FAM could develop the students' chemical literacy
	2	The deepness content of FAM was appropriate with the elaboration of the tourism vocational school basic competencies
	3	The extent content of FAM was appropriate with the elaboration of the tourism vocational school basic competencies
	4	The FAM has a correct chemistry concept
	5	The FAM has contemporary, factual, and actual characteristics
	6	The content about the food additives on FAM was relevant to the needs of vocational tourism students
Concept presentation	7	The presentation of the FAM content begin with the context of vocational tourism students
	8	The presentation of the FAM content conducted logically and systematically
	9	The FAM content presentation were communicative and easily understood
	10	The FAM content presentation equipped by an illustration of the picture
	11	The method in presenting the FAM content was consistent to facilitate students
	12	The FAM content relates to the application of the students' specific expertise
	13	The presentation on FAM content encourages the students' motivation to read, collecting information, and learn the chemistry concept
Linguistics	14	The sentence on the FAM was proportional and effective
	15	The language used on the FAM was easy to understand, unambiguous, and clear
	16	The language used on the FAM was communicative, interactive, and straightforward
	17	The writing rule on the FAM was appropriate with the correct Indonesian spelling, standard rules, meaning to be conveyed, and scientific terms
	18	The used of symbol or chemistry sign on the FAM were consistence
Graphic	19	The cover of the FAM on the front and the back was harmonious, attractive, and has sufficient color contrast
	20	The chapter titles, subtitles, page numbers, illustrations, figures title, and source images were compiled in complete, clear, and precise manner
	21	The FAM cover was reflects the contents of the book and was appropriate with the content characteristics
	22	The type and thickness of the paper were appropriate and the text were printed in a good and clear manner

Aspect	Number of Indicator	Indicator
	23	The types and font sizes were arranged proportionally, neat, and attractive
	24	The paragraphs, sentences, and words were appropriate in the term of the spacing and tracking
	25	The variation of the font types were not more than 3 types and those effects was not excessive.

2.3. Data analysis
There were two kinds of data collected in this research, the qualitative and quantitative data. The qualitative data were obtained from the vocational school teachers' interview in the term that FAM was needed to enrich the vocational tourism student knowledge and a group of expert in the term of suggestion and feedback of the FAM. The experts were reviewed the module according to several aspects including the chemistry content, the contextual approach integration, the presentation of the language and Indonesian grammar rules. These qualitative data were analyzed narratively. Required revisions were made according to the experts' feedback in order to enhance the quality of the module.

Whereas, the quantitative data in the form of teachers' response toward the module were analyzed according to descriptive quantity technique. The mean score given by the teachers on each aspect of the response toward the module were classified into five categories from very poor to very good. The range interval of the five categories used in this research consisting: a very poor category if the mean score less than 1.80; within the range of 1.81–2.60 indicated poor category; 2.61-3.40 indicated sufficient category; 3.41-4.20 signified good category; and very good category if greater than 4.20. Moreover, the ideal percentage on the fourth aspects of the teachers' response were calculated.

3. Results and Discussion
The development and validation of the chemistry FAM for vocational students in the tourism field has been conducted. The FAM was developed due to the result of teachers' interviews on the qualitative phase. The result of the teachers' interviews showed a problem that a module about food additives was required to enrich the vocational tourism students' knowledge in chemistry. The module was need to be construct in contextual approach in order to present the illustration among the food processing concept in the daily life problems with the chemistry concept. The contextual approach expected to make a meaningful chemistry learning on vocational tourism students [3]. Thus, the enhancing of students' chemical literacy will occur.

Consequently, on the development phase, FAM was constructed based on the basic competence of food processing and extended on chemistry content with contextual approach. The FAM was constructed in four theme covers sweeteners, preservatives, coloring and flavoring foods. The face and content validity of FAM were conducted by asking some reviewed from a group of experts. Necessary revisions were made according to the feedback of the experts. The experts declared that the FAM was feasible and worthy enough as the chemistry enrichment media for vocational students of tourism field.

Moreover, the quantitative step was conducted to obtain the FAM feasibility through the response of 7 chemistry teachers on four aspects includes the worthiness of chemistry concept, concept presentation, linguistics, and graphic. The result of the teachers' response on the FAM feasibility presents on the table 2.

According to Table 2, it sought that the chemistry concept of the FAM had a good category. The lowest score gained according to the indicator number 1 which showed that the content of the FAM was sufficient to develop the students' chemical literacy. Even thought, the FAM was constructed according to chemistry content based on the contextual approach of the vocational tourism field which may promote students' chemical literacy. This is in line with the study conducted by [9-10] which showed that contextual approach could improve students' chemical literacy due to the real life experience which provide. Hence, a module with contextual approach on chemistry learning could enhance students' chemical literacy.

Table 2. The result of FAM feasibility

Aspect	No	Score	Category	Ideal (%)	Aspect	No	Score	Category	Ideal (%)
Chemistry concept	1	3.00			Linguistics	14	4,43		
	2	3,71				15	4,29		
	3	4,29				16	4,43		
	4	4,43				17	4,43		
	5	4,14				18	4,29		
	6	3,71							
	Mean	**3.88**	**Good**	**77.60**		**Mean**	**4.37**	**Very good**	**87.40**
Concept presentation	7	3,14			Graphic	19	4,43		
	8	3,71				20	3,71		
	9	4,00				21	4,29		
	10	4,43				22	4,43		
	11	4,14				23	4,43		
	12	3,43				24	4,29		
	13	4,00				25	4,86		
	Mean	**3.84**	**Good**	**76.80**		**Mean**	**4.35**	**Very good**	**87.00**

The same result obtained on the concept presentation aspect, the FAM gained a good category. The lowest score found on the indicator number of 7 which showed that the FAM was sufficient in presenting the chemistry content with the context at the beginning of the chapter. At the beginning of the chapter provide some issues in the daily life relating the context of vocational tourism field, such as the used of food preservatives on the tofu. The beginning of the chapter on the FAM described on why some tofu has flies and some other tofu weren't. Hence, the content on the module which begin with the context of vocational tourism students bring a good quality of chemistry learning [11-14]. However, according to the result of this study, the context presents in the beginning of the chapter need to be improved.

On the other hand, according to the linguistic aspect, the FAM reached a very good category. All of the indicators on the FAM were centered in the very good category. The FAM being developed had a communicative, interactive, and straightforward language. The term being used were easy to understand, unambiguous, and used a consistent chemistry sign and symbol. These facts lead a very good category on the linguistic aspect.

Similarly, the graphic category signified a very good category. The indicators of the graphic aspect were centered on the very good category which indicated that the FAM was attractive to read. The vocational tourism students had less interest to learn chemistry due to the chemistry content which not related to the students' specific expertise [1]. Hence, the FAM is expected to enhance students' interest in learning chemistry. The result of these study confirmed the study conducted by [15-16] which showed that modifying the learning content leads the enhancing of students' interest. Overall, the FAM which being developed showed a good category. Thus, the FAM will be a useful tool for the teachers to enhance the quality chemistry learning of students in vocational tourism school. Figure 2 showed the summarized of the four aspects of teachers' response toward the FAM.

Based on figure 1, can be concluded that the linguistic and graphic aspects had a better result compared to the concept and content presentation aspects. However, the fourth aspects had equally good category according to the teachers' response. Thus, it can be used as a reference to enrich the chemistry knowledge of vocational tourism students.

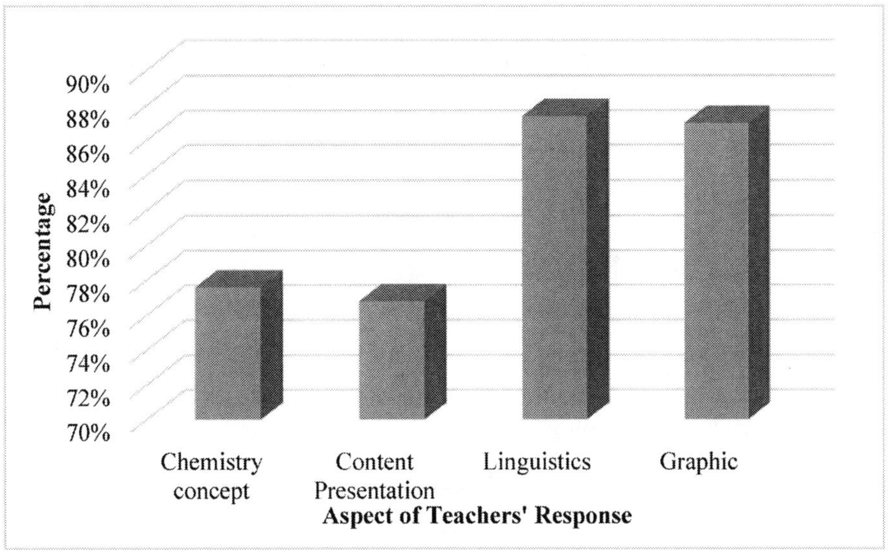

Figure 2. Ideal percentage of teachers' response toward FAM

4. Conclusion

The FAM which has been developed in this research has a good content and face validity declare by a group of expert. The correctness chemistry content, integration of contextual approach, and communicative language use in the FAM lead the experts consider that FAM is feasible to use. A good category is obtained as the result of the teachers' response toward the module. The contextual approach which use in the FAM expected to bring a meaningful chemistry learning on the vocational tourism students. Hence, this module could be used as reference for learning enrichment of tourism vocational school student to enhance chemical literacy.

5. References

[1] Wiyarsi A, Pratomo H and Priyambodo E 2017 *Proc of 3rd Int. Sem. of Sci. Educ.* (Yogyakarta) vol 3 (Yogyakarta: Yogyakarta State University) pp 359-366
[2] Ültay N and Çalık M 2012 *J. Sci. Educ. Technol.* **26** 686-701
[3] Sevian H and Bulte A M 2015 *Learning Chemistry to Enrich Students' Views on The World They Live* vol 1, ed I Eilks and A Hofstein (Rotterdam: Sense Publishers) p 7
[4] De Jong O and Talanquer V 2015 *Why Is It Relevant to Learn The Big Ideas* vol 1, ed I Eilks and A Hofstein (Rotterdam: Sense Publishers) p 11
[5] Lynch R L, Padilla M J, Harnish D and Distephano C 2001 *A Model of Excellence for Contextual Teaching and Learning in Preservice Teacher Education: Final and Summative Report* (Washington, DC: US Department of Education)
[6] Meldawati 2017 Building Students' Scientific Literacy Through Contextual Learning in The Physics Classroom *Master's thesis* University of Tampere
[7] OECD 2015 *Scientific Literacy: A Framework for PISA 2015* (OECD: Paris) pp 5-10
[8] Saglam H I 2011 *Turk. J. Educ. Technol.* **10** 36-44
[9] Anugrah I R, Mudzakir A and Sumarna O 2017 *J. Phys. Conf. Ser.* **895** 1-7
[10] Suryawati E and Osman K 2017 *Eurasia J. Math., Sci. Tech. Ed.* **14** 61-76
[11] Vaino K, Holbrook J and Rannikmae M 2012 *Chem. Educ. Res. Pract.* **13** 410-419
[12] Vos M A J, Taconis R, Jochems W M G and Pilot A 2011 *Int. J. Sci. Educ.* **33** 1407-1432
[13] Çiğdemoğlu C, Arslan H and Cam A 2016 *Chem. Educ. Res. Pract.* **18** 288-302
[14] Acar B and Yaman M 2011 *Hacettepe University Journal of Education* **40** 1-10
[15] Krapp A 2006 *Learning and Instruction* **15** 381-395
[16] Holbrook J and Rannikmae M 2007 *Int. J. Sci. Educ.* **29** 1347-1362

IOP Conf. Series: Journal of Physics: Conf. Series **1156** (2019) 012016 doi:10.1088/1742-6596/1156/1/012016

Integrated thinking ability and activities on eleventh grader students through learning cycle 7E

A Wibowo[1*] and Suyanta[2]

[1]Postgraduate of Chemistry Education, Universitas Negeri Yogyakarta, Indonesia
[2]Department of Chemistry Education, Faculty of Mathematics and Natural Sciences, Universitas Negeri Yogyakarta, Indonesia

*E-mail: aryatiwibowo.2017@student.uny.ac.id

Abstract. The study were aimed to compare the integrated thinking in the form of science process skills and analytical thinking ability; and student learning activities toward chemistry learning of eleventh grade who learned using the 7E learning cycle and expository model. The quasi experimental model with post-test only design was used as an research design. To achieve this goal, sixty students were randomly selected from one public senior high school in Bantul regency, Yogyakarta Province, Indonesia then divided into two groups- there are 28 students of the experimental group which was taught by using the 7E learning cycle and 32 students of the control group which was taught by the expository model. Data were collected using the following two instruments: 5-item integrated thinking ability test, and 29 questionnaire statements of students activity. To answer the questions, Multivariate Analysis of Variance (MANOVA) was used. The results showed that students who learned using 7E learning cycle and the expository model had statistical differences in integrated thinking ability and activities towards chemistry learning at 0.05 level of significance; the students who learned using the 7E learning cycle showed more integrated thinking ability than did the students who learned using the expository model; also the result and indicated than student activities higher than did the students who learned using the expository model. It can be concluded that 7E learning cycle model can be used to improve integrated thinking ability and students' activities. The researcher recommends that more studies should be conducted about the effect of this model on other instructional variables.

1. Introduction

We not only help students improve understanding of content but the thinking skills as well, ie. science process skills. Science process skill are needed in chemistry learning in 21st century. Thus, we need to develop students' thinking skills. Science process skills are considered to have a correlation with higher-order thinking skills [1] because they help students to develop higher-order thinking [2]. One of high-order thinking skill is analytical thinking. Science process skill and analytical thinking ability are needed in chemistry learning. Science process skills are important elements that must be possessed by students because they involve cognitive or intellectual, manual, and social skills used in problem solving [3]. Whereas, analytical thinking ability is the ability to analyze, access, evaluate, compare, and distinguish abstract concepts [4]. If science process skill is integrated with analytical thinking ability, it will have a positive impact on the students' performance. But, the result showed that integrated thinking (the science process skill and analytical thinking ability) was relatively low ie. 30.67% [5]. This is what drives the empirical study in this paper.

Student activities are related to chemistry learning, one of which is a buffer solution material. If students are actively involved in the learning process, they can better understand the learning material. Furthermore, it can increases achievement [6]. Students learn science best when the learning model enables them to get involved actively in class activities. So, they should participate actively in doing experiments, carrying out demonstrations, class discussion and other relevant learning experience.

Therefore, we need learning model that will help our students to develop their integrated thinking and increase activity. However, this cannot be done without giving the students the opportunity to express, shape and test their ideas. The use of various teaching approaches in a single lesson can create more opportunities for inculcation and acquisition of science process skills [7] also analytical thinking ability in the classroom.There are many model that can used such as inquiry based.The 7E learning cycle model is inquiry based learning that developed a constructivist theory by intellectual development. The constructivist theory is an educational theory that arouses students' thinking and makes students active and interactive during the learning process. Consequently, it concentrates on the students activity during learning and emphasizes meaningful learning based on understanding through the students' active role and effective participation in the activities they do, in order to build their concepts.

7E learning cycle model was divided into 7 phases as follows [8], (1) elicitation, the teacher asks students to express the knowledge they already have; (2) engagement, the teacher motivates students to be curious about learning that will take place;(3) exploration, students identify ways of exploring and checking, setting hypothesis, identifying the possible choices, practicing to collect some data to be a base for the next phase; (4) explanantion, after students have got enough information, then bring it to analyze, summarize and present in various formats; (5) expansion/elaboration, students bring the knowledge that students built up to cooperate with their own knowledge, or it is to bring the model or even conclusion to explain another case; (6) evaluation, teachers evaluate students by any technique to find out what students know after learning; and (7) extension, teachers prepare students knowledge after learning for applying using in theor daily life. Also, motivate students to use their knowledge so as to make a new one.

Consequently, we can say that the 7E learning cycle model makes meaningful learning so that it can improve integrated thinking skills and students' activities. This study aims to compare in integrated thinking ability and student activities of the eleventh grade toward chemistry learning using the 7-E learning cycle model and expository model. Thus, the study's problem is represented by the following main question: Are there differences in integrated thinking ability and stdent activities from the eleventh grade in Bantul District, Yogyakarta Province, Indonesia to chemistry learning using the 7-E learning cycle model and the expository model?

2. Methodology

The quasi experimental model with post-test only design was used as an research design. The sample consisted of sixty students were randomly selected from one public senior high school in Bantul regency, Yogyakarta Province, Indonesia then distributed in two groups: (1) an experimental group which was selected randomly and studied using the 7-E learning cycle model, and (2) a control group studied using the expository model. The sample size of the study shown in the Table 1.

Table 1. Sample size of the study

Groups	Number of students
Eksperiment	28
Control	32
Total	60

2.1. Research tools
The study implemented the following instruments:

2.1.1. Five items of integrated thinking test on Buffer. Its type was a essay question test. The test's difficulty index ranged from -2.90-4.26 and its reliability was 0.87.The reliability coefficient level is above 0.7, hence acceptable.This implied that there was a good internal constituency of items [9].

2.1.2. Questionnaire on students' activities toward chemistry learning, including 4 levels, 29 items.

2.2. Procedure
Teach students by Buffer lesson plans. The experimental group, learned by 7E learning cycle plans and the controlled group, learned with the expository plans. Each group was to spend 8 h to learn and also 2 h to posttest. Test of students after studying by integrated thinking test on Buffer and questionnaire on students' activities toward chemistry learning.

2.3. Data analysis
In order to compare the students' integrated thinking scores and students' activities toward chemistry learning, one-way MANOVA was used.

3. Results and disscussion
Since the researchers had studied and compared the students' integrated thinking ability and students activities toward chemistry learning entitled Buffer who learned with 7E learning cycle and expository, we got the conclusions as follows: the students learnt by using the 7E learning cycle model and the expository model shows differences in integrated thinking ability and students activities on Buffer toward chemistry learning at the 0.05 level of significance. In order to check this difference, one-way MANOVA was used. Table 2 shows the result of one-way MANOVA.

Table 2. Analysis of MANOVA of integrated thinking skill and students' activities[a].

	F	DF	Sig.	Partial eta squared
Model	13.974[b]	2.000	.000	.329

[b] Exact statistic.

It followed the hypothesis that the students learnt by using 7E learning cycle model and expository model had different integrated thinking ability and students activities at 0.05 level of significance. The different learning model would be an important variable for the students to have statistical differences of an integrated thinking ability and students' activities.

Student who learned with 7E learning cycle model got higher integrated thinking on Buffer than the model. This can be facilitated with 7-E learning cycle because the 7-E learning cycle can help students to understand the relationship of a concept [6]. Also added that inquiry-based learning can improve students' analytical thinking by involving students in science and laboratory classes [10]. Learning with the 7E learning cycle model causes students to have analytical thinking skills and attitudes towards chemistry learning rather than expository classes because these models provide students to build their own knowledge, especially in the exploration and elaboration phase [11]. Table 3 shows the means and standard deviation.

In addition to analytical thinking ability, in integrated thinking there are science process skills. Science process skills are very important for students in solving a problem. It is also facilitated by 7E learning cycle because the learning 7E learning cycle model is one of inquiry-based learning. Inquiry learning gives students thinking, communication, problem solving skills [12]. These three things are obtained from the stage of inquiry, namely asking, making hypotheses, investigating, and making conclusions. The results of several studies also show that science process skills can be developed using inquiry or investigative approaches to learning and learning science that give them the opportunity to practice their skills [13-16]. Inquiry-based instruction is a teaching strategy that aims to develop students' skills in dealing with problems by using methods used by scientists through research, inquiry, analysis and investigation in the classroom [17]. Inquiry-based learning has a positive influence on science process skills [18]. Consequently, the 7E learning cycle can improve students' integrated thinking skills.

Also, the students learned by the learning cycle model has higher learning activities on chemistry than the expository model. It was because the model was new, so the students would be enthusiastic to learn. Moreover, the LC 7E learning model is based on constructivism theory [6]. Constructivism theory assumes that students are active thinkers who build concepts and regard knowledge as a result of their thoughts and activity [19], it was concentrates on the learner and his/her activity. Thus, the 7E learning cycle can increase student activity, such as demonstrations, reading, experimenting, discussion, problem solving and problem solving, where this is done through seven stages. Table 3 shows the means and standard deviation.

Table 3. Comparison of means and standard deviation at two group.

	Model	Mean	SD	N
Integrated thinking	Eksperiment	42.4812	19.81905	28
	Control	22.1217	10.02319	32
	Total	31.6228	18.36995	60
Activity	Eksperiment	82.7665	10.69940	28
	Control	82.0502	13.94386	32
	Total	82.3844	12.43690	60

The students who learned using the 7E learning cycle and the expository model have statistical differences in integrated thinking toward chemistry learning at the 0.05 level of significance. Tests of between-subject effects in one-way MANOVA was used. The result can be seen on table 4.

Table 4. Tests of between-subject effects of integrated thinking.

Dependent variable	DF	SS	Mean square	F	Sig.
Integrated thinking	1	6190.000	6190.000	26.168	.000

But, the students who learned using the 7E learning cycle and the expository model were no statistical difference in students activities toward chemistry learning at the 0.05 level of significance. Tests of between-subject effects in one-way MANOVA was used. The result can be seen on table 5.

Table 5. Tests of between-subject effects of activities.

Dependent variable	DF	SS	Mean square	F	Sig.
Activity	1	7.662	7.662	.049	.826

Table 5 showed that the students who learned using the 7E learning cycle and the expository model were no difference in activities, but eksperimen group scored higher mean (82.7665) than control group. This is because students are not familiar with the 7E learning cycle model so students are still rigid and confused with the stages in learning. Consequently, the difference in average scores from the experimental class and controls is only a little. It can be seen on table 3.

The expository were model which deals with the teaching and learning process where the teacher plays a fundamental role and lead to a memorizing without observing, while 7E learning cycle were supported by the learner. 7E learning cycle was an active cognitive process, which the students explore their knowledge through various educational experience. Moreover, it was indeed a meaningful learning which is the opposite of the expository, which does not lead to meaningful learning and does not improve integrated thinking ability and activities of students. Finally, more researches should be conducted to test further effects of learning cycle as learning model with a larger number of students and the effect of this model on other instructional variables.

4. Conclusion

The students learnt by using the 7E learning cycle model and the expository model shows differences in integrated thinking ability and students activities toward chemistry learning on Buffer. The students who learned with 7E learning cycle had higher integrated thinking ability and activities than students who studied by using expository model. The reasons were that 7E learning cycle make meaningful learning. The students themselves were to arrange knowledge to learn and evaluate it. After that, they are continuously building and rebuilding understanding. Consequently, every phase at learning cycle encouraged students to develop their integrated thinking and activities. Thus, 7E learning cycle model can be used to improve integrated thinking ability and students' activities.

References

[1] Chebii R, Wachanga S and Kiboss J 2012 *Creative Education* **3** 1291-96
[2] Germann P J and Aram R J 1996 *Journal of Research and Science Teaching* **33** 773-98
[3] Karamustafaoglu S 2011 *J. Phys. Chem. Educ* **3** 26-38
[4] Anwar B and Mumthas N S 2014 *Int. J.Adv. Res.* **2** 455-8
[5] Irwanto, Rohaeti E, Wijayanti E and Suyanta 2017 *Students' science process skill and analytical thinking ability in chemistry learning* (4th ICRIEMS: AIP Publishing) p 030001-1
[6] Qarareh A O 2012 *Int. J. Edu. Sci.* **4** 123-132
[7] Rauf R A, Rasul M S, Mansor A N, Othman Z and Lyndon N 2013 *Asian Social Science* ***9*** 47-57
[8] Eisenkraft A 2003 *Sci. Teacher* **70** 57-59
[9] Fraenkel R J and Wallen E N 2000 *How to design and evaluate research in education* (4th ed) (San Fransisco: McGraw-Hill)
[10] Colburn A 2007 *The Science Teacher* **74** 14
[11] Siribunnam R and Tayraukham S 2009 *J. Social Sci.* **5** 279-282
[12] Cuevas P, Lee O, Hart J and Deaktor R 2005 *J. Res. Sci. Teach* **42** 337-357
[13] Leonard W H 1984 *An experimental test of an extended discretion laboratory approach for university general biology* (New Orleans: Paper Presented to the Annual Meeting of the National Association for Research in Science Teaching)
[14] Rissing S W and Cogan J G 2009 *CBE Life Sci Educ* **8** 55–61
[15] Marx J G, Honeycutt K A, Clayton S R and Moreno N P 2006 *The American Biology Teacher* **68** 140-147
[16] Ulu C 2011 *The effect of using inquiry based approach known as the science writing heuristic on concept learning, science process and metacognition skills in science teaching* (İstanbul: Marmara University) PhD Thesis
[17] Perry V R and Richardson C P 2001 *The New Mexico tech master of science teaching program: An exemplary model of inquiry-based learning* (*T*he 31st ASEE/IEEE Frontiers in Education Conference: Reno)
[18] Sen C and Vekli G S 2016 *Universal Journal of Education Research* **4** 603-612
[19] Wheatly G 1991 *Science Education* **75** 9-21

Acknowledgment

The author would like to express thanks to the DP2M DIKTI (Directorate of Higher Education), Ministry of Research, Technology and Higher Education Republic of Indonesia through "Tim Pascasarjana" Research Grant 2018 for the support.

IOP Conf. Series: Journal of Physics: Conf. Series **1156** (2019) 012017 doi:10.1088/1742-6596/1156/1/012017

Improving creativity of prospective chemistry teacher through chemoentrepreneurship oriented inquiry module on colloid topics

C A Dewi
Department of Chemistry Education, FPMIPA IKIP Mataram,
JalanPemuda No.59A Mataram 83126, Indonesia
Email: ayudewi_citra@ikipmataram.ac.id

Abstract. This study investigates the effect of Chemoentrepreneurship Oriented Inquiry Module (COIM) a colloid topic to enhance students creativity in teaching chemistry subject. The work was quasi-experimental, including pretest-posttest control group design. The used instrument is in the form of a description test, specially designed to determine the creativity of students. The results showed the use of COIM was effective as proven by 1) increase in student creativity index from 60 to 79, and 2) N-Gain level of attainment was in the range of 0.5 with a medium category. Overall, concluded that the use of Chemoentrepreneurship Oriented Inquiry Module effectively can enhance. Introduction The colloid is one of the learning topics in basic chemistry courses that are very important to be taught to students because it is a highly contextual topic. Sub-chapters in the colloid topic include dispersion systems, types of colloids, colloid properties and the application of colloids in everyday life.

1. Introduction

The colloid is one of the learning topics in basic chemistry courses that are very important to be taught to students because it is a highly contextual topic. Sub-chapters in the colloid topic include dispersion systems, types of colloids, colloid properties and the application of colloids in everyday life. As in the properties and application of colloids in various fields, it is needed to have teacher-centred learning [1]. Colloid concepts also include chemical characteristics that are summarized in three levels: macroscopic, microscopic and symbolic [2]. The macroscopic level is the easiest because it can be seen visually and more accessible, depending on the extent of sensory use. While the microscopic level requires analysis since everything that happens cannot be seen by the invisible or applied theory cannot be proven visually. Further symbolic is a description of a process[3]. To prepare the students in the 21st century must become the driving issues the improve of all pathways and levels of education[4]. A high demands to the qualified teacher an outcome of qualified education, so that students' thinking skills are provided by quality education.One of the skills students requiredthis time is creativity thinking [5].

The main problem faced by Most lecturers in Departement of Chemistry Education of FPMIPA IKIP Mataramis difficulty in teaching basic chemistry, especially on the colloidal system. One of the difficulties is to relate the concept to a student's daily life. In addition, teaching materials used by lecturers often does not contain the application of the concepts studied. It is necessary to prepare the college graduates to be qualified and skilled so that they can fulfil the market demand. Chemistry is one of the science that the application is very much in everyday life and has the potential to be developed into an entrepreneur in the field of chemistry especially on the colloidal system. Some of the examples are making shampoos from aloe vera, candy from seaweed, milk making from corn, from palm water and making jam from pineapple[6].

This study aims to improve the creativity of prospective chemistry teacher through Chemoentrepeneurship Oriented Inquiry Module (COIM) on colloid topic. Thus, in addition, students also have a chance to learn the process of processing a material into a useful product, economic value and foster entrepreneurship spirit.

2. Methods

2.1. General Background of Research

The pre-experimental design was used in this study. There were no control variables and the samples were not randomly selected. The form of pre-experimental design in this study was One Group Pretest-Posttest Design. The shape of the design is illustrated in table 1[7].

.

Table 1. Pre-experimental design

Subject	Pretest	Posttest
One Group	O_1	O_2

Information:
O_1 = Pretest value before learning using COIM.
O_2 = Posttest value after learning using COIM.

2.2. Sample of Research

The study was conducted at IKIP Mataram, Pemuda street No. 59A Mataram West Nusa Tenggara. The first semester students of Chemistry Education Study Program was being the subject of this study. The study instrument used was the creativity test using essay in the form of an essay with the aim to know the improving creativity of perspective chemistry teacher after learning using COIM on the colloid topics.

2.3. Instruments and Procedures

Instruments and Procedures Instruments used in this learning activity included the syllabus, lesson plans, Chemoentrepreneurship Oriented Inquiry Module, and test items. The instruments had been previously validated by two experts/science education experts and two education practitioners (chemistry education lecture). The validation was done by giving scores to each component on the validation sheet with the score range of 1–4. A score of 1 indicated "not valid," 2 indicated "less valid," 3 indicated "valid," and 4 indicated "very valid." The data analysis of the validation results was done by calculating the average scores from the experts and practitioners. The validation showed the following scores: 3.75 (valid) for the syllabus, 3.70 (valid) for the lesson plan, 3.80 (valid) for the Chemoentrepreneurship of Inquiry Module, and 3.89 (very valid) for the test items. The reliability test was also performed for the essay test instrument for creative thinking. The reliability test involved 70 students who were asked to finish 12 questions that had previously been validated. The score of the 96 students was then tested using Cronbach's alpha with SPSS statistics program 18.0 for Windows. The reliability test showed that the score was 0.84 (very high reliability, acceptable). The test instruments of creative thinking that were valid and reliable were used for collecting data on creative thinking. Data collection was done twice, pretest and posttest, on COIM (Chemoentrepreneurship Oriented Inquiry Module).

2.4. Data Analysis

The scores of creative thinking had converted into a range of 0–100. Then, normality and homogeneity tests were performed on the data. The study procedure was carried out, among others: a) The initial stage of the study, by conducting tests on creativity tests on students who have taken the course of Fundamental Chemistry. Furthermore, students are taught to using of chemoentrepreneurship Oriented Inquiry Module during the learning process takes place. After the learning activities are completed, posts

IOP Conf. Series: Journal of Physics: Conf. Series **1156** (2019) 012017 doi:10.1088/1742-6596/1156/1/012017

are conducted to measure students' creativity in the material being taught. b) The final stages of the study, by analyzing the data obtained and preparing the study report. This study used the N-Gain test with the aim to measure the increase of creativity of students before and after learning using Chemoentrepreneurship Oriented Inquiry Module on colloid topics.

3. Results

The learning model used is a Chemoentrepreneurship Oriented Inquiry consist of 4 stages of activity are problem orientation, formulating problems, submitting hypotheses, observing & collecting data through chemoentrepreneurship activities.These stages are presented inChemoentrepreneurship Oriented Inquiry Module (COIM) presented in figure 1.

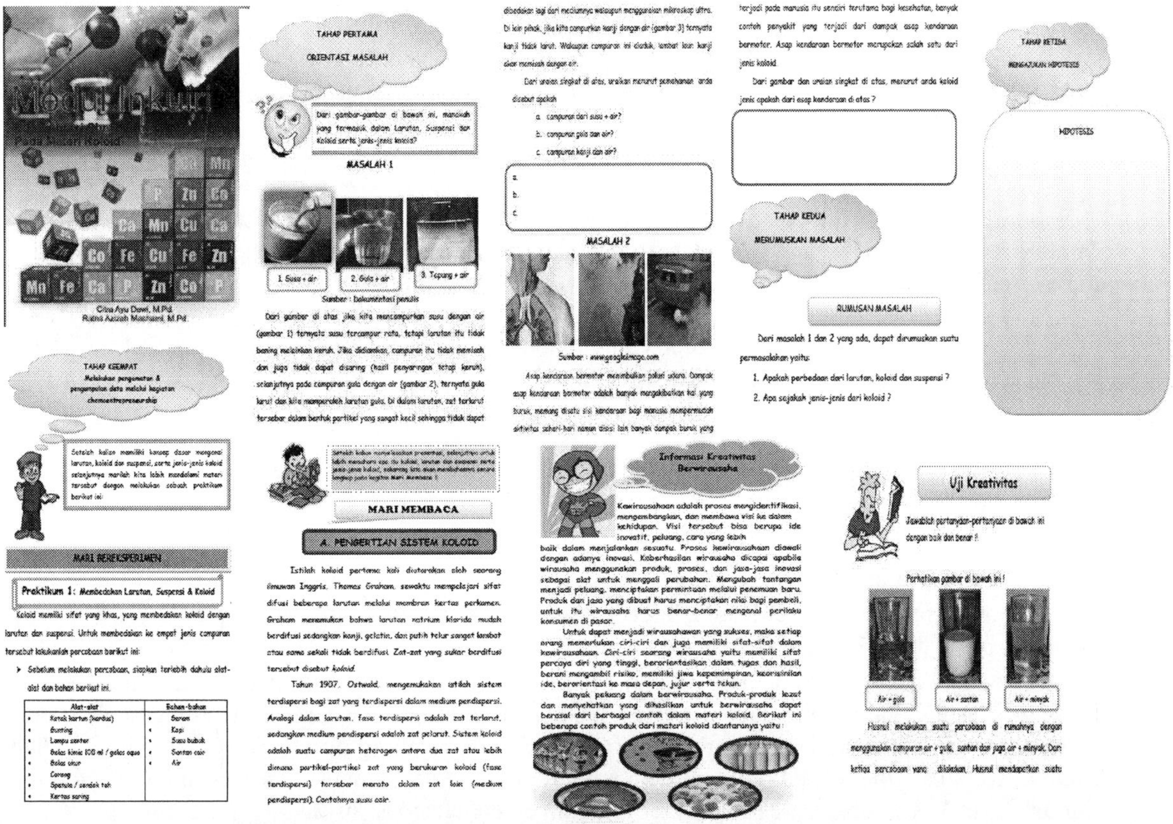

Figure 1. Chemoentrepreneurship Oriented Inquiry Module

Based on product effectiveness test that has been done so that the result of the final test of the students after the learning through the COIM. Implementation scenarios were pretest and final test (posttest). From this final test result (posttest) is used to know the improvement of student creativity on colloid topics as data that Chemoentrepreneurship Oriented Inquiry Module ineffective chemistry lesson or not. The data of the effectiveness of COIM can be seen in detail in table 2.

Table 2. Recapitulation Pretest-Posttest Results of Creativity

No	Group	n	Ideal score	Value		Average
				Minimum value	Maximum value	
1.	Pretest	25	100	58	62	60
2.	Posttest	25	100	75	82	79

Based on the calculation analysis of student creativity, the average value of student creativity of the effectiveness of COIM can be seen in figure2.

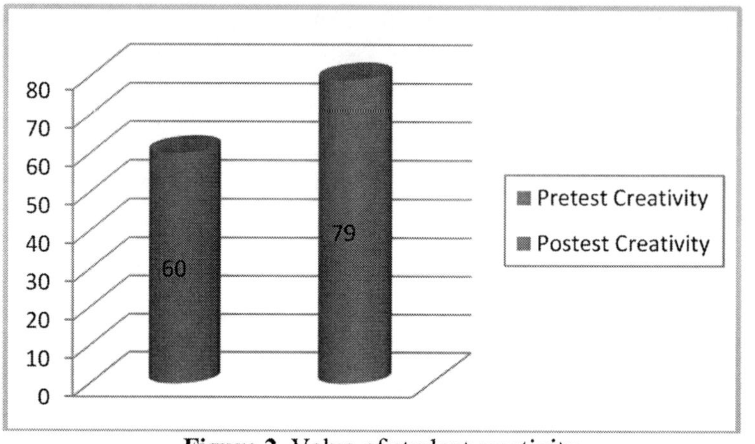

Figure 2. Value of student creativity

Figure2 shows that the creativity of students after using ChemoentrepreneurshipOriented Inquiry Module is obtained an average value of 79 thus if the data range is above 75% then it can be categorized to be effective.

Table 3. Recapitulation Results of N-Gain

No	Group	N	Ideal score	Value		N-Gain
				Minimum value	Maximum value	
1.	Pretest	25	100	58	62	0,5
2.	Posttest	25	100	75	82	

To know the existence of student creativity improving before and after usingChemoentrepreneurshipOriented Inquiry Module the procedure of using N-Gain testwas performed. It was found that the gain level of N-Gain is in the range of 0.5 with the medium categoryas shown in Table 3. The difference of students' average creativity onpretest and posttest is caused by using ChemoentrepreneurshipOriented Inquiry Module, where the learning steps used in ChemoentrepreneurshipOriented Inquiry learning are students doing observation to find the problem, students formulate the related problem with natural phenomena, students experiment and try to prove through hypotheses, observe and analyze and convey conclusions, in this learning also students are encouraged to be more active in learning, finding new ideas, always responding to opinions of friends and giving opinions in solving problems. So when working on the problem, students have enough information to do the test questions. This is in line with research conducted by Mursiti, *et al*[8]shows that learning using CEP approach and use of simulation game as CET media is very effective to improve student creativity. Learning using constructive and inquiry-oriented collaboration CEP can improve student learning outcomes[9]. Research states that CEP learning tools developed are valid, effective and can enhance students creativity and science process skills[10]. The result ofother research [11]stated that there is a significant between mastery of concept and creativity together with problem-solving ability.

4. Discussion

Based on this study it can be seen that learning by using ChemoentrepreneurshipOriented Inquiry Module can improve student creativity. In addition, ChemoentrepreneurshipOriented Inquiry Modules are eligible to be used as a learning resource and students provide a positive view of COIM.Boden[12]states that creative thinking is the ability to produce brilliant new ideasand values.Creative thinking about new things, the ability to create something, to implement new forms, to produce a lot of imaginative skills or to make something that is already new to something new[13]. Furthermore, Abraham[14] stated that creative thinking is a form of self-expression in a unique way.Students' creative thinking skills which are different from one another require a learning condition involving a learning experience so that the potential of creative thinking can develop [15]. Creative thinking can be incorporated into learning by the teacher, so the teacher must be able to carry out the development mandate the same skill to students. This is reinforced by the opinions of Wheeler *et al*[16], stated that the teacher's task is to provide the best teaching to get relevant thinking skills. Creative thinking skills are very important to considered [17]. Seyihoglu&Kartal[18] stated that to the challenges of a dynamic and uncertain modern life, it is necessary to develop the ability to think creatively. In fact, this ability to creative thinking is the foundation of science [19].

Various indicators of creative thinking have been expressed by several experts. According to Vlassi&Karaliota[20]there are five indicators of creative thinking, namely, (1) fluency, the ability to produce ideas, ways, suggestions, questions, and alternative answers smoothly within a certain time; (2) flexibility, the ability to produce ideas, answers, or questions, where ideas or answers are obtained from different points of view by changing the way of thinking and approach used; (3) authenticity, the ability to produce phrases, ways, or ideas to solve problems or create unusual and unique parts or elements that are not thought of by others; (4) elaboration, the ability to enrich, develop, improve, describe or specify the details of objects, ideas, products, or situations to make it more attractive; and (5) metaphorical thinking, the ability to use comparisons or analogies to make new connections.

The existence of modules made as interesting as possible to make students the spirit of reading colloids concepts let alone chemoentrepreneurship (CEP) oriented learning makes students more enthusiastic learning [21]. This is in line with Agustini's findings that chemistry learning model with chemoentrepreneurship (CEP) approach can improve learning motivation, entrepreneur interest, and student learning outcomes [22]. Meanwhile, Prayitno*et al*[23] indicated that the CEP-oriented SETS-oriented learning chemistry module is feasible to be used in chemistry learning with scores on the 95.00 aspects of graphics, presentation aspect 95,33, language aspect 95,00. And the results of module testing at MA Mu'alliminMu'allimat showed 20% motivation improvement, 25% entrepreneur interest, and 79% student learning result. Module trials in MAN Rembang improved students'learning motivation, entrepreneur interest, and students' experimental class learning outcomes by 27%, 17% and 66%, respectively, while control classes improved by 0.4%, 11%, and 24%, respectively.

The study that has been done by Rusilowati[24]showed that students' creativity can be grown by making IPA props interesting and liked by the students. Guided inquiry method can improve students' creativity [25-26]. In addition, students' creativity increases through project-based learning [27]. According to Vincentrisia[28]that student creativity can be improved through the LC 5E learning model. Guided inquiry methods have a significant effect on improving student achievement and participation in social learning better than conventional learning methods [29-31]. In addition, Özdemir&Işik[32]says that the use of inquiry strategies and science process skills has an effect on inquiry-based science activities in elementary school teachers. Meanwhile, Aktamiş[33]argued that inquiry-based learning had a significant effect on learning outcomes, science process skills and student attitudes compared with traditional learning methods. Tthere are differences in students' creative thinking skills for different models [34]. The highest creative thinking skills of the students are taught using a distinguished science investigation integrated with mind map model.

5. Conclusion

The effectiveness of using Chemoentrepreneurship Oriented Inquiry Module can be improving prospective chemistry teacher's creativity. This can be proven through the result of the statistical test of N-Gain that is obtained by a mean value equal to 0.5. Thus the ChemoentrepreneurshipOriented Inquiry Module was effectively categorized used in the learning process. The difference of students' average creativity on pretest and posttest is caused by using ChemoentrepreneurshipOriented Inquiry Module, where the learning steps used in Chemoentrepreneurship Oriented Inquiry learning are students doing observation to find the problem, students formulate the related problem with natural phenomena, students experiment and try to prove through hypotheses, observe and analyze and convey conclusions, in this learning also students are encouraged to be more active in learning, finding new ideas, always responding to opinions of friends and giving opinions in solving problems. So when working on the problem, students have enough information to do the test questions.

References

[1] Pradita Y, Mulyani B, and Redjeki T 2015 *JPK* **4(1)** 89-96
[2] Purtadi S and Sari R L P 2011 *Using Structured Clock Reaction Demonstration to Assess Students Understanding of Solution and Colloid Concepts* (Yogyakarta: Jurdik Kimia FMIPA UNY)
[3] Sunyono and Leni Y 2013 *JPP* **3(1)** 65– 79
[4] Mahanal S, Zubaidah S, Bahri A, and Dinnurriya M S 2016 *Asia-Pacific Forum on Science Learning and Teaching* **17(2)**
[5] Moon J 2008 *Critical thinking: an exploration of theory and practice* (New York: Taylor and Francis Group)
[6] Dewi C A and Mashami R A 2017 *Prosiding* Seminar Nasional Kimia FPMIPA UNESA Surabaya 7 Oktober 2017
[7] Sugiyono 2013 *Metode Penelitian Kualiatatif dan R&D* (Bandung: Alfabeta)
[8] Mursiti S, Wahyukaeni S, and Sudarmin 2008 *JIPK* **2(2)**
[9] Supartono, Saptorini, and Asmorowati D S 2009 *JIPK* **3(2)**
[10] PrahastutiW, Supartono and WidodoA T 2013 *IJCET* **2(1)**
[11] Silaban B 2014 *JPBP.* Volume **20(1)** 65 - 75
[12] Boden M 2001*Creativity in education* London Continuum
[13] Greenstein L 2012 *Accessing 21-century skills: To guide to evaluating mastery and authentic learning* (USA: Corwin)
[14] Abraham A 2016 *Brain Imaging and Behavior* **10(2)** 609-618
[15] Yusnaeni, Corebima A D, Susilo H, and Zubaidah S 2017 *Int. J. Inst.* **10(2)** 245-262
[16] Wheeler S, Bromfield C, and Waite S J 2002 *JCAL* **18(1)** 367-378
[17] Baker M and Rudd R 2001 *JSAER* **51(1)** 173-188
[18] Seyihoglu A and Kartal A 2010 *Educational Science*: *Theory & Practice* **10(3)** 1637-1656
[19] Hadzigeorgiou Y, Fokialis P, and Kabouropoulou M 2012 *Creative Education* **3(5)** 603-611
[20] Vlassi M, and Karaliota A 2013 The Comparison Between Guided Inquiry and Traditional Teaching Method. A Case Study for The Teaching of The Structure of Matter to 8th Grade Greek Students. Priced–Social and Behavioral Science **93** 494-497
[21] Lestari E and As'ari AR 2013 *Pengembangan Modul Pembelajaran Soal Cerita Matematika Kontekstual Berbahasa Inggris Untuk Siswa Kelas X* (Malang: Universitas Negeri Malang)
[22] Agustini F 2007 *Prosiding Seminar Nasional Implementasi Pembelajaran Tematik dalam Mengoptimalisasi Kurikulum 2013* Semarang.
[23] Prayitno M, Agus, Dewi, Kusuma N, Wijayati, Nanik 2016 *JIPK* **10(1)** 1617-1628
[24] Rusilowati A2009 *JPFI* **(1)**102–107
[25] Kurniawan AD 2013 *JPII* **2(1)** 8–11
[26] Deta U A and Widha S 2013 *JPFI* **9(1)** 28–34
[27] Kind P M and Kind V 2007 *Studies in Science Education* **43(1)** 1-37
[28] Vincentrisia A 2015 *JPII* **4(1)**56–66
[29] Ifeoma OE and OgeEK 2015 *BJESBS* **3(3)** 206-222
[30] Nworgu, LN and Otum V V 2013 *JEP* 4(27) 35-40
[31] Matthew B M and KennethI O 2013 *International Studies* **2(1)** 134-140
[32] ÖzdemirO and Işik H 2015 *J. TUSED* **(1)**
[33] Aktamiş H, Hiğde E, and Özden B 2016 *J. TUSED* **13(4)**
[34] Zubaidah S, FuadNM, Mahanal S, and Suarsini E 2017 *J. TUSED* **14(4)** 77-91

Acknowledgements
Special thanks to DRPM DitjenRisbangKemenristekDikti of Indonesia government that funding this research and development project.The Rector of IKIP Mataram for the support and encouragements.In addition, the Chairman of LPPM of IKIP Mataram for the writing and research guidance.

The effect of discovery learning on students' integrated thinking abilities and creative attitudes

D F Syolendra[1,*] **and E W Laksono**[2]

[1] Postgraduate of Chemistry Education, Universitas Negeri Yogyakarta, Indonesia
[2]Department of Chemistry Education, Faculty of Mathematics and Natural Sciences, Universitas Negeri Yogyakarta, Indonesia
*E-mail: dwifinna_syolendra.2017@student.uny.ac.id endang_widjajanti@uny.ac.id

Abstract. The aims of this study was to investigate the effect of discovery learning model on students' integrated thinking abilities in the form of analytical thinking skills and science process skills, and students' creative attitudes in the buffer solution lesson. The population was students of eleventh grade students of science class in Yogyakarta, academic year of 2017/2018. The sample of this study consisted of two classes in Yogyakarta public high school taken using a random sampling technique. Each experimental and control class was 32 students. The treatments for experimental class was the use of discovery learning, and expository model for the treatment on control class. Data collection used in this study was an integrated thinking ability test and students' creative attitudes questionnaire. The data of students' integrated abilities and creative attitudes were analyzed by observing the results of significance value and partial eta squared on MANOVA test. The results of multivariate analysis showed that there was an effect of discovery learning on students' integrated thinking abilities and creative attitudes of 12%. This means the integrated thinking abilities and creative attitude of students applying discovery model is better than integrated thinking abilities and creative attitudes of students with expository model.

1. Introduction

The chemistry learning process cannot be separated from cognitive development and students' skills development. This is due to the relationship between cognitive and process skills. Process skills are essential in understanding and applying chemical concepts [1]. Process skills are very important in learning process to acquire a knowledge [2]. Actamis and Ergin [3] are revealed that science process skills are skills that can be used by students in everyday life in order to improve students' understanding and essence of science. Science process skills also emphasize students' ability to find their own knowledge based on learning experiences, laws, principles and generalizations, so it can provide students more opportunities to enhance their thinking skills [4]. Also clarified that science process skills are considered to have a correlation with higher thinking skills.

Higher thinking skills in Bloom cognitive domain is at the analytical, evaluation and creating level [5]. Other perspectives defined high-level thinking skills as thinking abilities that not only require the ability to memorize or know, but also require other higher abilities such as the ability to think critically and creatively and to solve a problem [6]. Through creative learning, students are expected to be able to process abstract and complex chemical information into new and original knowledge [7]. Analytical thinking ability is the ability of students in visualizing, articulating and solving complex problems to make logical decisions based on information obtained [8]. Analytical thinking is part of high-level thinking that includes the ability to hone and apply information that can provide knowledge [9].

Analytical thinking in Bloom's taxonomy revision is part of the C4 cognitive domain which consists of distinguishing, organizing, deducing, and attributing [10]. Irwanto et al. [11] described that process skills students relate with thinking activity and require reasoning. Science process skill is a tool to generate and to use scientific information on scientific research and to resolve problems, and the students are able to get these skills through science learning activities [2]. Science process skills are not only used in the process of science learning, but also needed in solving every problem in the 21st century. Mansila and Jackson [12] argued that there are four global competences of dimention skills that the students need to have in order to acquire the process skills, namely investigate the world, recognize prespective, communicating idea, and taking action. Further, Mansila and Jackson [12] explained that the global competences can enhance students' thinking skills including critical thinking, high-level thinking and analytical thinking

Creativity is the affective aspect of learning outcomes (attitudes) as a result of creative thinking in order to understand a thing or the ability of processing information. Kenneth [13] revealed effective instructional strategies that creative thinking is thought of putting information together to come up with a whole understanding, concept, or idea. The four stages generally identified with the development of creative thought are preparation, incubation, illumination, and verification. Numerous thinking are used during each of these stage of the creative process. In fact, the greater the flexibility in thinking, the greater the possibility for developing student's creative thinking ability. Dessailly [14] identified key elements of creativity, namely (1) generating new ideas, (2) applying the known skills and ideas in different contexts, (3) taking other people's ideas or starting points and moving them on or personalizing them, (4) communicating ideas in interesting or varied ways, (5) putting different or disparate ideas together to make something new, (6) working towards a goal or set of goals, (7) evaluating their own or others' work, and (8) adapting and improving on their work in the light of their own or others' evaluations. This study used six aspects of creativity, namely curiosity, feeling the challenge, imaginative, original thinking, detailed thinking and willingness to take risks.

One of the solutions to improve students' science process skill, analytical thinking skill and creativity is the selection of appropriate learning models, such as using discovery learning models. Discovery learning is thought to increase the ability of students to transfer information, they construct to other areas, as it allows the students to independently explore broader issues [15]. Discovery is the mental process of understanding the concepts and principles, and is a study of how to use the mind to find. With discovery learning, the students construct knowledge based on new information and data collected by them in exploratory learning environment [16]. Learning discovery, making children can learn too think analysis and try to solve their own problem encountered. Discovery learning consists of six syntax, namely stimulation, problem statement, data collection, data processing, verification and generalization [17]. The main targets of discovery learning models are maximum students engagement in teaching and learning activities and develop self-belief about what is found during the learning process.

The purpose of this study was to determine the effect of discovery learning model on students' integrated ability and students' creative attitude. As for benefit of this study, it is expected that it can improve integrated thingking ability and creative attitudes of students on eleventh grade chemistry subject, as one input for educator to improve integrated thinking ability and creative attitudes in yogyakarta high school, can be applied in learning and learning increase knowledge and ability to realize

2. Methods

This research is a quasi-experimental research with post-test only control group design as presented in Table 1. The population of this research is eleventh grade of science students of high schools in Yogyakarta, academic year of 2017/2018. The research sample consists of two classes selected by random sampling. Learning in experimental class used discovery learning models while learning in control class used expository learning models. Student worksheets are used to assist the learning process. Student worksheets based on discovery learning model are used in the experimental class, and students worksheet are based on expository learning model in control class.

Table 1. Research design

Class	Treatment	Post-test
Experiment	X	Z, A
Control	Y	Z, A

Note:
X = Discovery learning model
Y = Expository learning model
Z = Result of integrated thinking abilities
A = Students' creative attitudes questionnaire

The study used an integrated thinking ability test developed by Sukmasari [18] with the Aiken index value is 0.87 and reliable if the instrument was tested to the participants with the ability (Θ) between - 3.70 to 2.95. of the twelve integrated thinking ability test items developed by Sukmasari, five items wereselected to be used in this study. While for the data of creative attitudes students were collected using a questionnaire consisting of 27 item and covered aspects of curiosity, feeling challenges, imaginative, detailed thinking and willingness to take risk. Data of students' creative attitude questionnaire were ordinal data which have no absolute value on an object, so it must be converted to interval data. The data analysis technique used was Multivariate Analysis of Variance (MANOVA) at a significance level of 0.05. Before analyzing the data, there are several prerequisites that must be carried out in conducting MANOVA test. They include two variables, two independent groups, no interclass correlations, having minimum sample size, no multicollinearity, a linear relationship, multivariate normality, multivariate outlier test, and homoscedasticity test [19]. Of the nine tests above, the researchers particularly considered normality and homogeneity test in this study. The researchers measured multivariate normality through Mahalanobis vs chi-squared test. The homogenity test of variance was carried out through Box's test of equality of covariance matrices. Besides, Hotelling's Trace test was used for multivariate analysis and partial eta squared test in order to see the effect of discovery learning models on students' integrated thinking abilities and creatives attitude.

3. Result and discussion

Data of the research were the results of integrated ability test and students' creativity questionnaire. Multivariate normality test using Mahalanobis vs chi-squared test. The results show that Mahalanobis value which is under the chi-square value from the experimental class is 15 out of 32 students with a percentage of 46.88%. Similarly, students from control class who had a smaller Mahalanobis value than chi square were 19 out of 32 students (53.13%). Based on the percentage of Mahalanobis and chi-square values in both samples, the results indicated that the distribution of data in each class meets the requirements of multivariate normality. The significance value obtained from Box's M test was 0.807, meaning that the significance value of Box's M test is greater than 0.05, thus it can be concluded that the samples were from homogeneous population. The summary of the result Box's test of equality of covariance matrices is presented in table 2. After testing multivariate normality and variances of matrix homogeneity test, MANOVA test (difference testing) was measured as the further test.

Table 2. Result Box's test of equality of covariance matrices

Box's M	F	df1	df2	Sig.
1,010	0,325	3	691920,000	0,807

To find out whether there are differences in students' integrated thinking abilities and students' creative attitude in the experimental class and control class, a multivariate test was conducted using Hotelling's Trace test. Hotelling's Trace test was selected based on the results of the prerequisite MANOVA test, in which the data was normally distributed and had a homogeneous variant matrix. Significance value obtained from multivariate test for the two dependent variables was 0.02 which was smaller than 0.05. This indicated that there are differences in integrated thinking abilities and creative attitude between students who learn with discovery learning model and students who study with expository learning model. Therefore, the effect of discovery learning model on students' integrated ability and students' creative attitude was 12%. Result of multivariate test and partial eta square value is presented in Table 3.

Table 3. Result of multivariate test and partial eta square value

Effect	Value	Hypothesis df	Sig.	Partial eta squared
Hotelling's Trace	0,136	2,000	0,020	0,120

The results of this study indicated that discovery learning models not only influence the integrated abilities, but also influence both students' integrated ability and students' creative attitude at once. This is because discovery learning models emphasize the importance of structures or ideas towards a discipline, namely through active students' involvement in the learning process. Through the syntax of discovery learning models, students were guided by the teacher to find the concepts and principles presented by the teacher that was not fully comprehensive. Catronova [20] stated that "in discovery learning, students are active. Learning is not defined as simply absorbing what is said or read, but actively seeking new knowledge. Students are at least engaged in hands-on purpose for finding answer and learning more". Jamil [21] added that in discovery learning students are encouraged to actively learn, through their own active involvement to concepts and principles, and teachers encouraged students to have experience and conduct an experiment that enables students to discover principles and concepts for themselves. With the discovery of principles and concepts by students, students may not be easy to forget, because the understanding of principles and concepts has been recorded in the long-term memory of the students. In addition, discovery learning model can improve students' reasoning, the ability to think freely, and train students' cognitive skills to find and solve problems without the help of others [22]. Kutsiyah [23] argued that discovery learning models have a significant effect on learning. Using module based on discovery learning in the learning process is effective to increase generic science skill [24]. Hence, the significant difference in this research is the discovery learning model with the help of students' creative attitudes.

4. Conclusion

Based on the results of the research and discussion, it can be concluded that discovery learning model affects 12% to students' integrated abilities and creativities. This means the integrated thinking abilities and creative attitude of students applying discovery model is better than integrated thinking abilities and creative attitude of students with expository model.

References

[1] Marasigan A C and Espinosa A A 2014 *Int. J. Learn. Teach. Educ. Res.* **1** 35-72
[2] Rauf R A, Rasul M S, Mansor A N, Othman Z, and Lyndon N 2013 *Asian Soc. Sci.* **9** 47-57
[3] Aktamis H and Ergin O 2008 *Asia-Pac. Forum Sci. Learn. Teach.* **9** 11-12
[4] Chebii R, Wachanga S, and Kiboss J 2012 *Creative Education* **3** 1291-96
[5] Iskandar D and Senam 2015 *Jurnal Inovasi Pendidikan IPA* **1** 65-72
[6] Lissa 2012 *Jurnal Lembaran Ilmu Pendidikan* **41** 27-32
[7] Unnanantn T and Boonphadung S 2015 *International Journal of Educational and Pedagogical Sciences* **9** 515-9
[8] Ramos J, Dolipas B and Villamor B 2013 *International Journal of Innovative Interdisiplinary Research* **1** 48-60
[9] Anderson L W and Krathwohl D R 2001 *A taxonomy for learning, teaching and assessing* (New York: Longman)
[10] Rezba R, Sprague C, McDonnough J T, and Matkins J J 2007 *Learning and assessing science*

process skills 5th edition (Lowa: Hunt Publishing Company)
[11] Irwanto, E Rohaeti, E Widjajanti and Suyanta 2017 *AIP Conf. Proc. Of ICREAMS* **1868** 030001-4
[12] Mansilla V B and Jackson 2011 *Education for global compentence: preparing our youth to engage the world* (New York: Asia Society)
[13] Kenneth L C and Jane P L 2009 *Management information system: managing the digital firm* (Bergen: Prentice-Hall)
[14] Desailly J 2015 *Creativity in the promary classroom (2nd ed)* (London: Sage Publication)
[15] Klahr D and Nigam M 2004 *Psychol. Sci.* **15** 661-7
[16] Nadira S Wouter R V J and Hout-Woltres VB H A M 2005 *Br. J. Educ. Psychol.* **75** 603-21
[17] Syah M 2004 *Psikologi pendidikan dengan pendekatan baru* (Bandung: PT Remaja Rosdakarya)
[18] Sukmasari H 2016 *Pengembangan instrumen penilaian terintegrasi untuk mengukur kemampuan berpikir analitis dan keteramilan proses sains kimia peserta didik kelas XI SMA/MA pada materi larutan penyangga* (Yogyakarta: Universitas Negeri Yogyakarta) Thesis
[19] Tabachnick B G and Fidel L S 2007 *Using Multivariate Statistical analysis (5th ed)* (US: Pearson Education)
[20] Castronova J A 2001 *Discovery learning for the 21st century: an action research study* (Valdosta GA: Valdosta State University)
[21] Jamil S 2013 *Pengaruh model pembelajaran berbasis masalah (PBM) pada mata pelajaran fisika terhadap keterampilan berpikir kreatif dan konsep diri siswa kelas X SMAN 1 Kandangan Kalimantan Selatan* (Yogyakarta: Universitas Negeri Yogyakarta) Thesis
[22] Hosnan 2014 *Implementasi saintifik dan kontekstual dalam pembelajaran abad 21* (Bogor: Ghalia Indonesia)
[23] Kutsiyah 2017 *Pengaruh model discovery learning terhadap motivasi belajar, pemahaman konsep, dan creative thinking siswa kelas XI SMA N 1 Sentolo pada materi sistem pernapasann* Thesis (Yogyakarta: Universitas Negeri Yogyakarta)
[24] Khabibah E N, Masykuri M and Maridi 2017 *Journal of Education and Learning* **11** 146-153

Acknowledgments
This work received financial support from the Kemenristek Dikti (Indonesian Directorate General of Higher Education) through "Tim Pascasarjana" Research Grand 2018.

The computer-assisted testlet assessment instrument to measure students' learning difficulties in chemical bonding

E Lutviana*, S B Rahardjo, E Susanti, S Yamtinah, S Mulyani, and S Saputro

[1]Master Program of Chemistry Education, Universitas Sebelas Maret, Jl. Ir. Sutami No.36A, Jebres, Surakarta, Jawa Tengah,Indonesia
*E-mail:luthfiana.eva92@gmail.com

Abstract. Many students assume that the chemistry subject is difficult. We create a Testlet Aassessment Instrument (TAI) as a diagnostic tool that can be used to detect students' level of difficulty in learning chemistry. Hence, this study aims to analyze the computer-assisted TAI to measure students' learning difficulties in the chemical bond subject matter for class X students of the senior high school in Madiun of East Java Indonesia. This study used descriptive-quantitative analysis methods and15 students from 3 different schools as participants. The computer-assisted TAI consists of 10-question stems, that's stem 1 until stem 10. The result showed that for stem 8 only 4% students can answer correctly and thus the stem is considered the most difficult. Scoring of the TAI used graded scoring with 4 categories. The implication for applying the computer-assisted TAI for measuring students' difficulties in chemical bonding subject was discussed.

1. Introduction

An examination of studies shows that most of the basic concepts that are less studied in learning basic physical and chemical concepts[1-2]. Identify what difficulties students face in learning this concepts more research needs to be done. More research needs to be done to identify what difficulties students face in learning because teaching and learning difficulties are important. Both science educators and cognitive researchers agree that efforts to understand and improve science education must focused on the essential of knowledge [3-4].

The assessment and teaching of 21st century skill projects were sited very much in this space, of establishing frameworks within which to link assessment and teaching. The project closed normally in 2012 with development and delivery of a conceptual framework for twenty-first-century skills [5]: a focus on two skills, areas-collaborative problem solving and ICT literacy in digital networks, an approach to formative assessment, and teacher professional development modules [6]. Assessment and teaching of 21st century skills provided an approach to assess students with online and collaborated tasks. Student navigation of digital networks and problem-solving behavior was captured electronically in activity log files for synchronous scoring and reporting against developmental progressions. The progressions provided guidelines for the teacher about how students might demonstrate the skills at increasing levels of sophistication and competence such that these could be integrated into their teaching. How the insights into these complex twenty-first-century skills might be translated into classroom practice is the challenge for national education systems globally [7].

Modern testing is beginning to move away from thinking that focuses on items as the molecular construct out of which tests are built. Test builders are beginning to use as the fundamental units of which tests are manufactured. The psychometric foundation for such an idea was provided and

IOP Conf. Series: Journal of Physics: Conf. Series **1156** (2019) 012019 doi:10.1088/1742-6596/1156/1/012019

expanded [8]. They called this new construct the computer-assisted TAI [9]. Thus a test developer can take pains to avoid the context effects elaborated upon previous [10].This means that test developers can focus on what test structure measures best the construct of interest and trust that there will be an appropriate scoring model available for it. This has not been the case in the past [9].

Chemical bonding is one of the keys and basic concepts in chemistry. Clearly, many of the concepts thought in chemistry in both secondary schools as well as in the colleges are highly based on understating the fundamental ideas related to this concept. Nevertheless, the concept is perceived both by the teachers as well as by learners as difficult in teaching-learning. Very often, when considering the teaching of chemical bonding, the arguments are focused on pedagogical issues rather than scientifically related disagreements [1]. There is no doubt that chemical bonding is one of the key concepts in chemistry and therefore is one of the most fundamental ones. It is also one of the areas in the physical sciences where understanding is developed through diverse models – which are in turn built upon the range of physical principles and when learners are expected to interpret a disparate range of symbolic representations for chemical bonds [1]. Among the students' difficulties, the abstract nature of the chemical concepts was a common theme. This is also recognized by the lectures. The other difficulty which is related to the nature of the subject, or general chemistry, is the mathematical content of the course. Some students refer to chemistry as too mathematical. The study presented here allows for some solutions in certain diagnose the extent to which students' difficulties in learning chemical bonds might be solved.

2. Method

The method of the study used descriptive-quantitative analysis. We designed an assessment instrument in the form of the computer-assisted testlet where this instrument is used to see how far the level of difficulty of students in learning chemical bonds. This computer-assisted TAI consists of 10 stems and 30 question in the form of multiple choice that is arranged hierarchically according to the problem. Each stem of chemical bond subject matter covers sub-topics namely, electron configuration, type of ionic bond, the tendency of bonding an element, type of covalent bond, the correct statement of covalent bonds, and determining the number of lone electron pairs on the Lewis structure. Examples of diagnostic instrumentsof difficulty learning in the form of testlet is shown in figure1.

STEM 1
In a data according to the stem it is known that an element ${}^{32}_{16}X$ has an electron configuration below:
$1s^2 2s^2 2p^6 3s^2 3p^4$

1. The X element is in a group ....
   a. IIA (2)
   b. IVA (14)
   c. VIA (16)
   d. IVB (4)
   e. VIB (6)
2. Possibilities that occurs if the element X is bound to element Y is ....
   a. Release 2 electrons so that the charge is +2
   b. Catch 2 electrons so that the charge is +2
   c. Release 2 electrons so that the charge is -2
   d. Catch 2 electrons so that the charge is -2
   e. Release 3 electrons so that the charge is +3
3. The statement below according to the element X above is ...
   a. Can form ionic compounds with sodium
   b. Element X is a metal
   c. Can only react with metals
   d. Have 6 protons on each atom
   e. Can release electron

Figure 1. Diagnostic instruments of difficulty tetslet

This research was conducted in the Madiun, East Java, Indonesia. The respondents were 15 students from three different school. Each student was asked to work on this computer-assisted TAI to completion so that the pattern of answers to questions that had been done by students, could be recorded by the system so that it could be analyzed based on the TAI. scoring used graded with 4 categories is shown in table 1.

Table 1. Graded scoring on testlet instruments

No.	Assessment Aspect	score
1	Wrong answer in the whole steps	0
2	The correct answer in the first step, but wrong or nor answer in step 2 or 3	1
3	The correct answer in the first and second steps but wrong in step 3	2
4	Correct answer at all steps	3

3. Results and Discussion

The use of TAI was tested to 15 students. It is expected that the instrument can be used to measure the student's difficulty in chemical bonding percentages with these indicators as shown in table 2.

Table 2. The indicators of chemical bonding in testlet assessment instrument

Stem	Indicators stem of chemical bonding	%
1	Electron configuration	28%
2	Type of ionic bond	35%
3	The tendency of bonding an element	55%
4	Type of covalent bond	7%
5	The correct statement of covalent bonds	13%
6	Determination the number of lone electron pairs on the Lewis structure	7%
7	Type of metal bond	13%
8	Molecular Form	4%
9	The value of the electronegativity of an element	7%
10	Electron domains	22%

After the students conducted TAI on the chemical bonding subject matter, the result showed its stem of instrument difficulty. About 28% studentscan answer correctly on stem 1 with indicator electron configuration, 35 % on stem 2 with indicator type of ionic bond, 55% on stem 3 with indicator of tendency of bonding an element, 7% on stem 4 of type of covalent bond, 13% on stem 5 of the correct statement of covalent bonds, 7% on stem 6 with indicator determination the number of lone electron pairs on the Lewis structure, 13% on stem 7 of type metallic bond, 4% on stem 8 of molecular form, 7% on stem 9 the value of the electronegativity of an element, and 22% on stem 10 electron domain. The highest percentage is in stem 3 with 55% of students who can answer correctly and lowest at stem 8 only 4% of student can answer correctly. After the students completed the TAI problem, students are given a questionnaire to find out the reasons why many of the students cannot answered correctly on the problem of stem 8. From here it can be seen how far this instrument can measure a student's ability on chemical bond material and how the feasibility of this instrument is generated from questionnaires.

Therefore, interviews are needed for every student who cannot completed the stem 8 appropriately. Many students argue that students do not understand the material presented at stem number 8, there is still an overlap of understanding between bonding electron pairs and lone pairs of electrons in a bond so that students are still confused in determining it, one of the factors students feel unsure of they have done because of lack of learning motivation in this chemical bond material. Students assume the material of this chemical bond is difficult to imagine in the minds of students, such as the determination of molecular shape and the force produced in a bond that requires analysis in the form of illustrations in order to find the molecular shape of a compound developed because the teacher can easily analyze where the difficulties of students in learning a material. The result of students develop can be analyzed so that they can be found where their difficulties in learning are related to the material of this chemical bond.

The student fills out a prepared questionnaire. The result of questionnaires and interviews stated that 80% of students had never done the test in the form of testlet, they often get questions in the form of multiple choices and essays that are not hierarchical. 67% of students thought that the testlet was difficult to understand because they didn't understand it as a whole related to chemical bonds. 80% of students thought that learning about chemical modelling was difficult because the teacher always explained by lecture [11] or with a limited understanding of the previous models[12-13], 20% of students assume that problems with STEM such as in computerized testlet are needed because they assume that by solving problems with a testlet form like this. They must understand the basic concepts first before stepping in to completed the second question so they know where level of understanding. 20% of students finding it difficulties to understand the language used in the testlet questions. 67% of student stated this computer-assisted TAI is very practical and more efficiently. 40% of student stated that for a computerized look still needed improve to be attractive and interesting. 100% according to the computer-assisted TAI students can be used to knowing the ability of students to master chemical bonding material.

In another study, it was found that the majority of students considered science lessons very difficult. Only 1% of students thought that chemistry was easy. This indeed could be a major reason as to why pupils choose not to pursue the study of Chemistry even though they may find it enjoyable and interesting [14]. Many high school and university students experience difficulties with fundamental ideas in chemistry [15]. Many researchers, teachers and science educators argue that the abstract nature of many chemical concepts includes: teaching styles applied in classrooms, lack of teaching aids and lack of chemical language difficulties [15]. Many middle school students and universities have difficulty with basic ideas in chemistry [11]. Most students emerge from introductory courses with limited subject understanding despite the importance of chemical foundations [16] which have been considered difficult subjects for students.

The assessment instruments used by researchers are computerized testlet. Where in this testlet consist of 10 stems, each stem consists of 3 questions. For instance, the cyber school makes teaching and learning environment more visual and concept formation is enhanced. There is a need to integrate computer technologies into learning and teaching [16]. In the chemistry education literature, there have been numerous studies reporting positive effects of the use of computers on students' achievement [17]. Computer-assisted curricula also provide opportunities for inquiry-based approaches to the learning of chemistry and it seems they discouraged rote memorization and algorithmic problem solving while encouraging conceptual understanding and critical thinking.

Besides giving easiness in correcting, testlet can also result from the cognitive indicator that can be used to measure indicators [18] of chemical bonding. Testlet is designed with multiple choices. Chemical bonding is an abstract topic, something far removed from the daily experiences of secondary school students. As students cannot see an atom, its structure and how it interacts with other atoms, it is difficult for them to understand the concepts involved in the topic of chemical bonding and there is great potential for the formation of alternative conceptions [19]. There are still many students who have difficulty in working on this computer-assisted TAI, this is caused by several factors such as many students who have not really understood the material of chemical bonding so that they cannot solve the problem correctly, the network is in school so that it can inhibit students in working on, and some students to understand questions that would be presented as evaluation material for researchers as an improvement to this computer-assisted TAI being tested back to the next stage [19]. More research needs to be done to identify what sort of difficulties students face in the learning of physical and chemical concepts [15]. Learning difficulties are important for both teaching and learning, both science educators and cognitive researchers agree that efforts to understand and improve science education should be focused on fundamentally important knowledge domains [4].

4. Conclusion

It can be concluded that the computer-assisted TAI used to diagnose students' difficulties in learning chemistry in the chemical bond subject matter is found in stem 8 which not all students can answer correctly and only 4% of students can answer correctly. This TAI can determine the consistency of students in answering questions and can detect student difficulties in learning chemical bonds. There are several factors that cause students unable to answer questions presented in the form of testlets. One of them is due to a lack of understanding of the basic concepts of students, where the ability to describe bonds is still low because it is difficult to find in real life. therefore, the teacher needs an evaluation tool that can be used to evaluate students' level of understanding. one of them is the teacher can use the computer-assisted TAI as an evaluation tool and also as an alternative concept that will be delivered to students so that they become more aware of the material delivered.

5. References

[1] Nahum T L, Mamlok-Naaman L, Hofstein L, and Taber K S 2010 *Stud. Sci. Educ* **46** 179–207
[2] Sozbilir M and Bennett J M 2007 *J. Chem. Educ.* **84** 1204–1208
[3] Browman K, Ekborg M and Johnels D 2011 Chemistry in crisis? Perspectives on teaching and learning chemistry in Swedish upper secondary schools *NORDINA* **1** 43–60
[4] Eilks I and Hofstein A 2013 *Teaching Chemistry – A Studybook* (The Netherlands: Sense Publisher)
[5] Griffin P, McGaw B, and Care E 2012 *Assessment and Teaching of 21st Century Skills* (Victoria Australia: Springer Science)
[6] Piquer C F 2015 *Journal of Technology and Science Education* **5** 107–121
[7] Care E, Griffin P, and Wilson M 2018 *Assessment and Teaching of 21st Century Skills* (Springer International Publishing)
[8] Wainer H, Bradlow E T and Wang X 2007 *Testlet Response Theory and Its Applications* (Cambridge: Cambridge University Press)
[9] Wainer H and G L Kiely 1987 *J. Educ. Meas.* **24** 185–201
[10] Nimmermark A, Ohrstrom L, Martensson J, and Davidowitz B 2014 *Chem. Educ. Res. Pract.* **17** 985–1005
[11] Daniel K C, Khang N, and Sai L 2016 *J. Sci. Math. Educ. S.E. Asia* **24** 40–50.
[12] Chwee K, Tan D, Goh N K, Chia L S and Treagust D F 2009 *Multiple Representations in Chemical Education* (The Netherlands: Springer)
[13] Sheehan M 2010 Identification of Difficult Topics in the Teaching and Learning of Chemistry in Irish Schools and the Development of an Intervention Programme to Target Some of These Difficulties Learning *Thesis* ulir.ul.ie/bitstream/10344/432/3/MSpthesis.pdf
[14] Woldeamanuel M M , Atagana H and Engida T 1998 *African J. Chem. Educ.* **4** 1–43
[15] Gudyanga E 2014 *International Journal of Secondary Education* **2** 11
[16] Windschitl M and Andre T 1998 *J. Res. Sci. Teach.* **35** 145–160
[17] Ratna I, Yamtinah S, Ashadi M, Masykuri M and Shidiq A 2017 *Studies in Science Education* **46(2)** 179-207
[18] Hanson R 2015 IJSRST **4** 115–122
[19] Keng L and Ho T 2008 *A Comparison of Item and Testlet Selection Procedures in Computerized Adaptive Testing* 1–42 (Austin: Pearson)

Acknowledgments

Authors grateful to the *Sebelas Maret* University's community research and development institute (LPPM UNS) for contributing to the funding of this study.

A need analysis in developing virtual laboratory according to the chemistry teachers

F Solikhin[1*], J Ikhsan[2] and K H Sugiyarto[2]
[1]Postgraduate of Chemistry Education, Universitas Negeri Yogyakarta, Indonesia
[2]Department of Chemistry Education, Faculty of Mathematics and Natural Science, Universitas Negeri Yogyakarta, Indonesia

*Email: febrians2802@gmail.com

Abstract. Technology-based learning was always done to face up the growth of technology in the modern era. It attracted students to understand the subject matter. The development of the technology-based practicum media could facilitate students in practicum without the real tools and materials. The media could save the tools and materials that were expensive. However, need analysis were required to find out what it needed in the chemistry class according to the chemistry teachers. This research is descriptive research with quantitative and qualitative data. This need analysis used a questionnaire as an instrument that contain 6 open-ended questions and 2 yes-no questions. A virtual laboratory is one of the practicum media that can resolve the limitation of tools, materials and learning time. Respondents from this research were 37 chemistry teachers who have been teaching chemistry for at least 5 years. The data were analyzed using analysis of percentages. The results of this study obtained 91.89% of respondents stated to need an integrated virtual chemistry laboratory in hybrid learning. While practicum-based matters that were perceived to require a virtual chemistry laboratory were chemical equilibrium, electrochemistry and reaction rates. This research can be used as a criterion for developing the technology-based media.

1. Introduction

Research and development (R&D) is a step to developing the learning media. The form of this media is physical media and digital media. Physical media has a real shape, while digital media is a digital-based media that can only be opened through computers, laptops, tablets or smartphones. Currently, digital media is developed by many researchers. One example of learning media is the virtual laboratory [1-3]. A virtual laboratory is a digital-based learning media that can facilitate students in practicum [4]. This media is used in overcoming the limitations in the school laboratory. These limitations include the limited time of learning, practicum tools, and chemicals. The virtual laboratory produces a higher learning outcome, learning independence and the ability to understand concepts in their implementation [5]. Learning with virtual laboratory also improves effectiveness in learning [6-9].

Hybrid learning or blended learning consists of face-to-face learning and online meeting [10]. This learning is done in overcoming the distance of students with school. Hybrid learning has a positive effect on students [11]. In a previous study, hybrid learning could facilitate accessibility in the following learning and become an attraction in learning [12]. Therefore, hybrid learning should be developed for the utilization of existing technology. Implementation of technology in this learning can make students more interested and happy in following the lesson.

Need analysis is done to determine the problems that exist in the environment [13]. This analysis can be used in knowing the different status in the field with a status that should exist [14]. Need analysis is the first stage of the ADDIE model [15]. Stages in the ADDIE model are Analysis, Design, Development, Implementation, and Evaluation. These five steps are used in the process of developing the media. In the research and development, the analysis phase is performed to find out the problems that exist in the learning and find solutions to overcome them [16]. A learning requires media to achieve educational goals, including practicum-based media. This media allows overcoming the limited tools, materials and learning time at school. Before the development of the media, it is necessary to do an analysis phase to find out the situation in the field. The purpose of this analysis is to determine the needs of the development of integrated practicum media in hybrid learning, specifically a virtual laboratory.

2. Methods

2.1 Research Design and Procedure

This research adopted a descriptive study with a qualitative approach through the need analysis. Need analysis is done through 2 ways, namely the analysis of previous studies and analysis of the questionnaire. Analyze previous studies by analyzing the journal concerned with virtual laboratory and hybrid learning. While the analysis of needs questionnaire begins by giving the questionnaire to the chemistry teachers forum of Bantul Regency, Yogyakarta, Indonesia.

2.2 Subjects

The research respondents amounted to 37 chemistry teachers with varying teaching and schooling. All of these teachers had experience of teaching chemistry for at least 5 years so have known the ins and outs of chemistry learning in high school. While the school of origin of teachers is divided into private schools and public schools.

2.3 Instrumentation and Data Analysis

The instruments used are journaled analysis tables and need analysis questionnaire. Journal analysis table is used to find out the results of the analysis of articles related to virtual laboratory and hybrid learning. These articles are analyzed in terms of methods, objectives and research results. The result of this articles analysis is qualitative data. The data is compared each other to find the new thing about the virtual laboratory and hybrid learning. It is synthesized to determine innovation in this development of the media. Questionnaire of need analysis is a mixed questionnaire between 2 yes-no questions and 6 open-ended questions. The instrument is explained in table 1.

The results of this questionnaire are mixed data between quantitative and qualitative data. The qualitative data are produced by question number 4 and the quantitative data are produced by 7 number others. It is analyzed quantitatively with the percentage. The calculation percentage of questionnaire used formula :

$$\% = \frac{number\ of\ teachers\ who\ agreed}{total\ respondents} 100\%$$

However on the material is chosen, the percentage was calculated using the formula :

$$\% = \frac{number\ of\ subject\ matters\ that\ chose\ by\ teachers}{total\ subject\ matters} x100\%$$

The total subject matters that mentioned by the teachers were 156. It's because not all teachers choose 5 materials. There are only 3 or 4 materials.

Table 1. Indicators of the Questionnaire

Number of Question	Type of the Question	Indicators
1	Yes-No	Teachers state whether or not practicum-based matter is considered difficult for students
2	Open-ended	If teachers say yes (question number 1), teachers must be mentioned at least 5 practicum-based matters that considering difficult for students.
3	Yes-No	Teachers state whether or not practicum-based matters done completely
4	Open-ended	If teachers say no (question number 3), teachers must be explained the strategies to replace the practicum activities.
5	Open-ended	Teachers explained the matters always needs the practicum to master the matters.
6	Open-ended	Teachers explained what the virtual laboratory can help students to an understanding of the matters
7	Open-ended	Teachers explained what the forum out of learning time is needed.
8	Open-ended	Teachers explained the integrated virtual laboratory in hybrid learning is needed or not for the reason.

3. Results and Discussions

This study was a preliminary study for the development of a virtual chemistry laboratory integrated hybrid learning. It is an analysis step of the ADDIE model. The analysis phase of ADDIE development model can be done in various ways. Among them are by interview, questionnaire or pre-survey, class observation, teacher or student observation, article analysis, and so on. The process is to strengthen existing research. However, in this study, the analysis phase is only done with 2 processes only, that is articles analysis and giving of the questionnaire. It because this ways is easy and need a little time to know the situation of many schools.

The preliminary study process or often called need analysis began by analyzing the journals on virtual laboratory and hybrid learning. This article analysis was useful to investigate the existence of technology in education and learning. The components analyzed were researched methods, the purpose of the research, and research results.

The results of the analysis of previous studies, several studies have been carried out the development and use of virtual laboratories in the study of science, especially chemistry. The use of virtual laboratories has more effective results than conventional laboratories [2,6,8]. The study was a research on the development and implementation of media using pretest-posttest design. This virtual laboratory media can increase students' abilities and motivation in chemistry practicums with simulations [17-18].

Learning using hybrid learning mode can significantly facilitate students in accessing learning [12]. This hybrid learning is also effective in practical preparation [19]. Hybrid learning mode is one example of student-centered learning. Learning like this is highly desired by students.

The analysis of the journal stated that there has been a lot of virtual laboratory development and implementation in the classroom. But in many research, the implementation in the classroom compared just 2 classes with real laboratory and virtual laboratory. This comparison clearly showed that the use of a real laboratory will have a good effect. There makes students gain real experience. Students with a virtual laboratory only get visualization. For that, it is necessary to compare the two treatments coupled with the use of virtual laboratory as support.

The virtual laboratory that will be developed is a laboratory based on hybrid learning. Hybrid learning is used in a learning process that not only depends in the classroom but also uses online learning. In its implementation, this laboratory is used in 2 types. The first type is the use of a laboratory as a substitute for conventional laboratories, and the second type is the use of the laboratory as a supporter of the conventional laboratory. Both types of virtual laboratory use will be compared with conventional laboratory usage as usual. The three treatments used have the same learning time and the same teacher. Different learning times and teacher will make a difference in the outcomes later on. The use of a virtual laboratory as a supporter is used in online learning.

The result of an open-ended questionnaire given to chemistry teachers stated that 94.6% of respondents stated practicum-based matters were difficult to understand by students. These matters should be understood through practicum, whereas according to them the limitations of tools, chemicals, and time make practicum activities minimized. Each respondent mentions a maximum of 5 materials, the percentage of the materials were stated in figure 1.

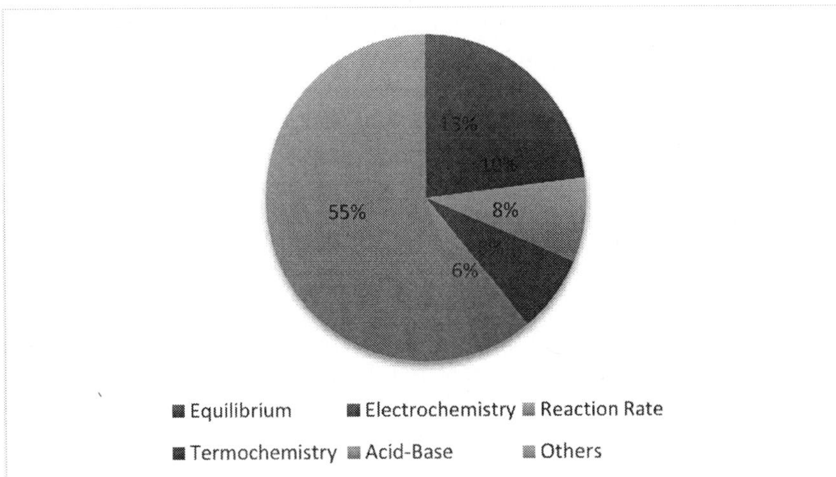

Figure 1. Percentage of Practicum-based Matters that were Considered Difficult

The five materials become the matters that required a practicum in learning. Limitation of the time is the main reason for eliminating practicum activities. From the open-ended question about strategies to replace the practicum, real practicum takes a lot of time. They often only provide a finished video or practicum data for analysis. This activity did not give a practical experience to students. Students just receive data without experiment. As much as 97% of respondents stated the matters not always needs the practicum to master it. Another result, 89.19% of respondents stated that virtual laboratory can help in the process of learning-based practice. A virtual laboratory may allow students to gain experience in the practicum of such materials [3].

For forums outside the school time, 64.86% of respondents stated that the forum was needed to increase knowledge. Activities in this lesson make students more relaxed but serious in following the lesson. This activity is still given the presence to know the students who follow this lesson. Learning involving technology make students more interested in following the learning. This is because today's students have relied on technology, especially smartphones. This dependency makes students comfortable in learning with technology rather than learning as usual. Learning using multimedia is effective in improving the quality of learning for the better [20]. Media that will be developed later is expected to improve the quality of chemistry learning.

In the development of virtual laboratory integrated hybrid learning, as many as 91.89% of respondents said they need this media. Learning by using a virtual laboratory as much as possible can resolve the limitations of time, tools and practicum materials. This learning can streamline the time so that the aims of learning can be achieved completely. Hybrid learning is an efficient mode in chemistry learning [21]. This learning is expected to face up the technological developments.

The virtual laboratory becomes a learning media that can facilitate students in practicum. Practicum can virtually stimulate students with visualization of the tools, materials, and practicum. Practicum with this laboratory does not require a long time, because we do not need to prepare tools and materials manually. The tools and materials used already exist in this laboratory. While conventional laboratory requires considerable preparation time. In addition, the laboratory process will also take time. The results achieved in conventional laboratories are uncertain according to existing theory. We need to match the results of that practicum theory.

In previous research, the stage of analysis is the stage of finding the problems that exist in an environment [22]. The research is to determine the indicators in the design of e-learning in Vocational School. The difference with this research is that the research is only determining the indicators to be developed in a learning design. Another study suggests that this phase of need analysis is to find out the needs of students and teachers in the learning process, especially in the needs of instructional media [16]. In the study, the analysis phase by pre-survey at school. The result of this stage is to know the need for developing learning media in the form of mind-mapping. The difference with this research is the process undertaken. In this study, the analysis phase was done by giving an open-ended questionnaire to the chemistry teachers, whereas in the previous research, the analysis phase was done by observation to the school. Giving an open-ended questionnaire to the chemistry teacher will be more meaningful than observation to school. The chemistry teachers involved came from more than 20 schools so the problems that exist in these schools can be known.

4. Conclusion

The results of this study are needed to develop a virtual chemistry laboratory. The developed laboratory is an integrated hybrid learning. The selected material is electrochemical. It is a material of grade 12th. This analysis can help the researchers to find out the needs of developing the media, especially integrated virtual laboratory in hybrid learning. Then they can develop the media in other matters to resolve the problems in the chemistry class. So that, students can understand the matters with simulation in the virtual laboratory. Furthermore, implementation of hybrid learning can be developed to resolve the limitation of the learning time.

References

[1] Arista F S and Kuswanto H 2018 *Int. J. Instruction.* **11** 1–16
[2] Hidayat A and Utomo V G 2015 *Int. J. Com. App.* **120** 14–18
[3] Tatli Z and Ayas A 2013 *Educ. Technol. Soc.* **16** 159–170
[4] Donnelly D, Reilly J O, and Mcgarr O 2013 *J. Res. Sci. Educ.* **43** 1571–1592
[5] Balid W, Abdulwahed M, and Alrouh I 2014 *Int. J. Electrical Engineering. Educ.* **51** 341–353
[6] Bakar N, Zaman H B, Kamalrudin M, Jusoff K, Khamis N, and Jaya H T 2013 *Aus. J. Basic and App. Sci.* **7** 78–84
[7] Elsunni and Abdelwahed H 2014 *American Int. J. Soc. Sci.* **3** 166–171
[8] Hawkins I and Phelps A J 2013 *Chem. Edu. Res. Prac.* **14** 516–523
[9] Popovic N and Naumovic M B 2016 *Int. J. Electrical Engineering Educ.* 1–14
[10] Chigeza P and Halbert K 2014 *Aust. J. Teach. Educ.* **39** 132–146.
[11] Wu D, Yu X, Shi Y, Yang H H, and Lu C 2014 The effect of hybrid learning in vocational education based on cloud space: taking the vocational education cyber-platform as an example. In S. K. S. Cheung, J. Fong, J. Zhang, & R. Kwan (Eds.), Hybrid Learning Theory and Practice (pp. 23–35). (London: Springer)
[12] Heirdsfield A, Walker S, Tambyah M, and Beutel D 2011 *Aust. J. Teach. Educ.* **36** 1–16
[13] Almomen R K, Kaufman D, Alotaibi H, Al-Rowais N A, Albeik M, and Albattal S M 2016 *Int. J. Clinical Medicine* **7** 538–546
[14] Dick W, Carey L, and Carey J O 2005 *The systematic design of instruction* (8th ed.) (Florida: Pearson Education Inc)
[15] Branch R M 2009 *Instructional Design: The ADDIE Approach* (New York, NY: Springer)

[16] Sari S A and Sakdiah H 2016 *J. Edu. and Learning* **10** 53–62
[17] Tsovaltzi D, Rummel N, McLaren B M, Pinkwart N, Scheuer O, Harrer A, and Braun I 2010 *Int. J. Technol. Enhanced Learning* **2** 91–110
[18] Sun D, Looi C, and Xie W 2014 *Educ. Technol. Soc.* **17** 390–403
[19] Gregory S and Trapani G D 2012 *Int. J. Innovation in Sci. Math. Educ.* **20** 57–70
[20] Aloraini S 2012 *J. King Saud University - Languages and Translation* **24** 75–82
[21] Antonoglou L D, Charistos N D, and Sigalas M P 2011 *Chem. Educ. Res. Pract.* **12** 454–468
[22] Triyono M B 2015 *Procedia – Soc. and Behavioral Sci.* **204** 54–61

Acknowledgment
This research was supported by the Ministry of Research and Higher Education, Indonesia in financial.

Chemistry in context: The development of hydrocarbon chemistry and petroleum module based on vehicle case

Febrianto[1*], A Wiyarsi[2], C F Partana[2] and B Sulistyo[3]
[1]Study Program of Chemistry Education, Faculty of Mathematics and Natural Sciences, Universitas Negeri Yogyakarta, Indonesia
[2]Department of Chemistry Education, Faculty of Mathematics and Natural Science, Universitas Negeri Yogyakarta, Indonesia
[3]Automotive Engineering Education Department, Faculty of Engineering, Universitas Negeri Yogyakarta, Indonesia

[*]E-mail: febrianto9696@gmail.com

Abstract. This study aims to analyze the characteristic, feasibility, teachers' and students' responses of Hydrocarbon Chemistry and Petroleum (HCP) module based on the context of vehicle case on automotive engineering. Exploratory mixed method has been employed in this research. The research procedures used four steps including qualitative, development, quantitative, and interpretation step. The instruments used to collect the data consisting the three questionnaires, one open questionnaire to obtain the validation and two close questionnaires to collect the teachers and students' responses toward the HCP module. HCP module based on the context of vehicle case on automotive engineering program was yielded as a result of this reseach. The characteristic of research product was integrating between chemistry and automotive subject based on basic competence of curriculum and integrated by vehicle context in first chapter, concept map, activities, examples, and exercise. A group of experts consisting two lectures were confirmed that the HCP module has a correct chemistry content. A very good category was given as the response of the teacher. Moreover, the students response were very interest toward the HCP module. The students and the tachers could be utilize the HCP module as learning sources reference to gain a succesfull chemistry learning.

1. Introduction

Vocational school is emphasizing at practices than theory, but it was conducted by vocation subject only, was not be the other subject learning. The students weren't interest of some subject matter, for example in chemistry. Chemistry is ones of subject matter that the students were not interest, because they feel that it difficulty lies in the number of formulas and terms that must be memorized [1]. Actually, chemistry is basic subjects in the vocational curriculum to support vocational subjects, prepare basic student ability in expand expertise and good chemistry literacy implementation [1-2]. Vocational student likes practice to find the concept than memorized only. However, practice method in chemistry lesson not yet applied to vocational schools because the limited time allocation in the school.

Vocational learning was fundamentally different from any other type of learning except in one aspect, namely context. The context is integrating basic competence vocation when process, objective, and outcome would be wished in learning [3]. Context-based learning is one approach to chemistry learning that can be used to increase student's interest in learning chemistry. Context-based learning will increase knowledge to connect chemistry learning to the community, because contain applications required by students [4]. According to [3] vocational school context is a broader concept to form the analysis and students' critical thinking on science. However, context-based learning has not been widely practiced in vocational school because a teacher was being difficult to connecting application in a particular skill program [1]. Thus, teachers needed reference to apply context-based learning in chemistry learning to optimize knowledge about applied chemistry and student skills.

Chemistry based contexts module would be needed because chemistry module is relevance's of everyday life [5], competence, learning objectives [6], and literacy sciences [7], it easier for students to master chemistry. In the study of [9], the sequence of context-based chemistry module was presenting the desired context of learning in the beginner of each chapter, then presenting problems to be solved and solving science issues by combining some scientific knowledge. Context-based chemistry module also contains contact phases, questioning or cultivating curiosity, elaboration, exploring and connecting [10]. Content in context-based chemistry module includes experimental activities, inquiry activities, scientific perspectives and chemical history [11]. Accoding to [8] that explain a component of chemistry module covering science, technology, and social elements such as context, content, scientific processes or skills and attitudes. Thus, context-based chemistry module has a good quality if the chemistry module contains material that presents the problem according to the context that can be solved with inquiry activities and scientific perspectives, fosters curiosity, and develops student skills.

Automotive engineering is one of the program in vocational school, which the chemistry learning is general that chemistry applications for automotive engineering are lacking. It confirmed the previous study conducted by [1] that the learning of chemistry in automotive engineering program has not connected with the material on automotive technique so that the chemistry matter is difficult to understand and student is not interested in chemistry learning. Chemistry based context module is need to developed in automotive engineering program. It will increase the students' interest because the content of chemistry was correlate to the automotive engineering conten. Thus, students will be enthusiastic in following the chemistry learning in the classroom.

Vehicle context is principle subject in the automotive engineering learning. Vehicle is a unity of complex systems, so it needs for an understanding from various points of knowledge view, such as chemistry. Applications of chemistry that can be applied in vehicles quite a lot, such as fuel oil and lubricating oils [1], fuel oils have contained chemical compounds, the many types of fuel oils had used for vehicles, and it needs students knowing. Therefore, students need an appropriate learning resource to understand what is in the vehicle not only on material in automotive engineering but rather from other scientific point of view such as chemistry and able to increase student interest in chemistry learning.

In relation to the development of chemistry module, previous research has been conducted on feasibility testing of work sheets oriented approach to contextual material of chemical rate with the result that context-based chemistry module is feasible to be used because it can improve students' critical thinking skill [12]. The development of chemistry with a context-based is enough to give positive results for learning. Thus, chemistry module with a vehicle context for current automotive engineering students is indispensable. Therefore, the research development is expected to help meet the needs of chemistry module in accordance with the context of vocational automotive engineering and the fulfillment of the need for chemical materials vocational school is also expected to improve the quality of chemistry learning in vocational school.

The following research question guide this study. How are the characteristic, feasibility, teachers' and students' responses about the HCP module based on vehicle case for vocational school student on automotive engineering program?

2. Method

Exploratory mixed method design was employed in this research. The aims of this research method were the exploration of the problem and developing the product. The scheme of this research design according to [13] showed in figure 1.

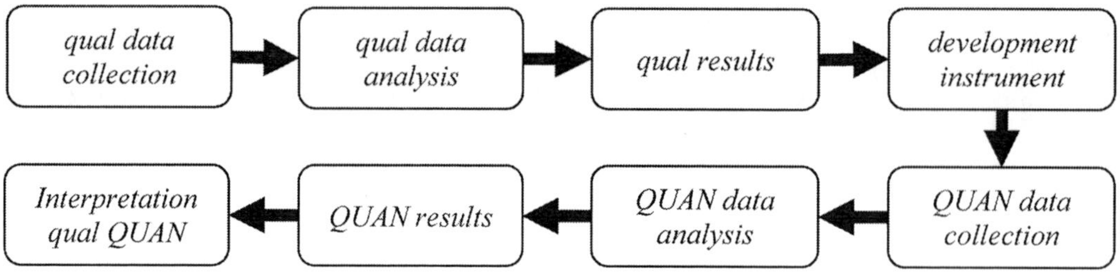

Figure 1. Design Research Chart

Based on figure 1, it can be explained that product development systematic begins with qualitative data collection. Firstly, the exploratory problem was collected by interview with the teachers of chemistry in vocational school. The topic of the interview was about chemistry learning in vocational school on automotive engineering program. In addition, the basic competence in vocational school curriculum and basic competence of chemistry in vocational school on automotive engineering program were analyzed through the sheet analysis. The sheet analysis is a table connecting the basic competence on chemistry and automotive engineering, which can be integrated on chemistry learning. Hence, the chemistry content could be integrated according to the automotive engineering context. The result of the qualitative analysis showed that arranging teaching material on chemistry topic based on the context of automotive engineering program was necessary.

The principles of instrument development were adopted from [8] with the steps consisting the preparation, arranging and validation of teaching material. The step of preparation, firstly is determine the source of the books that used as a reference on arranging the teaching material to arrange teaching material, there are chemistry and automotive engineering books. After that, determine content framework with considering of consistent and content aspects as well as determining context on automotive engineering will use the teaching material of chemistry. The chemistry module have arranged by making layout, arranging content of chemistry, and development content. Finally, a validation step was conducted by asking the judgment from chemistry education experts toward HCP module. They were review about some component to judgment feasible or not the product. Aims to collect the responses from the user in the term of the assessment from teacher and student responses about product. The teacher responses was knowing the product to using in learning chemistry and student responses was knowing to interest them about the product. The results of these steps were the average scores on each component of assessment. Finally, the interpretation of qual-quan step was score conversion based on ideal conversion from [15].

2.1. Research subject

The research subject consisted two expert judgments by validation, five teachers and ten students by user responses on automotive engineering program. The expert judgments were the lecture of chemistry education department and teaching the hydrocarbon and petroleum subject. All teachers were having teaching experiences more than 10 years and minimum graduate of bachelor degree. The teachers ware The students in automotive engineering program enrolled on hydrocarbon chemistry and petroleum course. The students came from the school which belongs to the same district of Yogyakarta.

2.2. Instrument

The interview guidelines had been used to teachers' interview. It was including five aspects, there are: (a) curriculum chemistry in vocational school, (b) learning sources, (c) vocational context integration, and (d) Integration of chemistry subject with vocational context. The validations step was used an open questionnaire with specific declaration for each chapter. For example: *Is chemistry content had been adjusted in automotive engineering?*. Its aim to knowing in a proper manner and judgments of content correctness. The content validation questionnaire was consisting of 31 items in four components; there are vocational content integration, content proper, presentation, and Indonesian grammar rules.

The closed questionnaire have used in this quantitative steps. The teacher response questionnaire had arranged with the adaptation of assessment module from Education National Standard Institution (BSNP) and the development module based context journal from [5, 8, 9, 10, 11]. The adaptations were resulted 29 items indicator from five aspects; there are content, presentation, leanguage, integration of vocational context and readability. Then, the student's response questionnaire had arranged adaptation from reading literacy in Programme for International Student Assessment (PISA) [14]. The adaptations were resulted 18 items statement response student about product. The teacher and student responses were resulted qualitative data. It had been converted become quantitative data with ideal conversion, thus obtained average scores each component.

3. Results and Discussion

3.1 Qualitative Data

Qualitative data was collecting information about problems in student learning in automotive engineering program to knowing needs analysis from interview of the chemistry teacher. The results of the interview showed that some problems in learning chemistry for automotive engineering program, due to the result focused on the topic only.

Firsly, the curriculum chemistry in vocational school. The teachers' opinion about curriculum chemistry in vocational school had been appropriated with student necessary, but each chapter have not relating with subject in vocational program and it has limited time in practice plan allocation besides many matter which must to learned. Student centers' learning is demand of curriculum to teaching chemistry in the class. The teacher was difficultly, because students are less motivation and interest about chemistry. Second, learning sources. It was using self-made modules with reference derived from several high school books and internet, because special chemistry book for the automotive engineering program was not available. The impact given by students was study chemistry subject is general, not specific.Third, the Integration of chemistry subject with vocational context. It had been used to open lesson but not all of chemistry subject can be integrated. The teachers have difficulty in relating chemistry content with the context of a student's program, because chemistry teachers in vocational schools are less good at automotive engineering subjects and not just teaching one program. Finally, chemistry learning in vocational school. Students are less motivated in chemistry learning, because chemistry was not related to vocational subject. However, the effort of teacher had been related each chemistry subject in opening lesson and example chemistry in vocational subject if it's be present. In addition, they have difficult to study chemistry, because consider many calculating in chemistry. Hydrocarbon chemistry and petroleum subject had taught with paper task, a glance, and not practice in laboratory. Due to the teacher reasons which category of chemistry subject which memories only, nothing calculating.

According to interview result, it is founding that the vocational students has less motivated in the chemistry lesson, because the teachers has difficulty to relating chemistry learning with the student's program so chemistry learning in vocational school is general. The case is underlying researcher in development HCP module based on vehicle case.

3.2 Development Instrument

The product of HCP module based on vehicle case was resulted in development instrument. Firstly, it was analyzed core and basic competences of chemistry subject in vocational school with basic competence subjects of automotive engineering program. The analysis aims are find out basic competence in automotive engineering vocational subjects that can be integrated with basic competence chemical subject. The result is integrated matrixes of chemical content with automotive engineering vocational context, for example was shown in table 1.

IOP Conf. Series: Journal of Physics: Conf. Series **1156** (2019) 012021 doi:10.1088/1742-6596/1156/1/012021

Table 1. Integration Matrix of Chemical Content with Context of Automotive Engineering

Basic Competence of Chemistry	Content able Chemistry Content Related	Content Vocational Lessons Chemistry	Integrated Chemistry Content to Vocational Engineering Automotive
3.9 Analyze structure, properties and combustion impact of hydrocarbon compound	- Type of hydrocarbon compound - Naming of hydrocarbon compound - Reaction and life application of hydrocarbon compound - Combustion impact of hydrocarbon compound for surrounding and healthy	• Energy conversion process at machine • Work process of heat exchanger/ radiator • Treatment of diesel fuel system • Emission contents in throw gas.	- Introduction of carbon and hydrocarbon compound - Grouping and naming of hydrocarbon compound - Physic properties of hydrocarbon compound - Isomerization of hydrocarbon compound - Reaction of hydrocarbon compound - Combustion impact of hydrocarbon compound - Heat energy concept

The HCP module was arranging base integrated matrix. The draft of materials had been produced consist of 149 pages with several components. The component is determined by reference analysis, discussion, and integrated matrixes. The product had been showed in figure 2.

Figure 2. The Product of Hydrocarbon Chemistry and Petroleum Module Based on Vehicle Case

The HCP module based on vehicle case have eight components, there are introduction books, matrix integration, concept map integrated automotive engineering, content modules, reference, glossary, index, and elements periodic table. The HCP module characteristic was integrating between chemistry and automotive subject based on basic competence of curriculum and integrated by vehicle case in first chapter, concept map, activities, examples, and exercise. First chapter is general introduction, which contains descriptions of hydrocarbon chemistry and petroleum present in vehicles. It begins with a vehicle-related image or scheme to stimulate initial knowledge and attract students to learn more about the subject. For example, ...*The combustion process was occured because fuels of motorcycle has liquid phase changed to gases phase. Actually, what is the fuel containing? If you not know them, Let's learn later in 3rd chapter about petroleum fraction*. The concept map is sketch of hydrocarbon chemistry and petroleum linkages with automotive engineering. It is the order basic of content presentation in module. Activities were containing discussion or practicum chemistry had connected to automotive engineering subject. The matter of hydrocarbon chemistry and petroleum had completed with example chemistry in vehicle and the objectives of the exercise construct in HOTS.

The product has been compiled and consulted to supervisor lecturer, then validated by a group of experts consisting two lectures for review of chemistry learning experts. A group of experts was comment is there are still many words that are typo, less sentence writing and punctuation in accordance with the correct General Regulation of Spelling Indonesian (PUEBI) in the HCP module. The statement was showing in some introductory chapters is still less clear, images and captions are still unclear. The logo or brand on the cover and the lubricant products in chapter trird is better omitted. The writing of a phase on the chemical reaction equation is skewed. Some material concepts are not quite right. Each sub-section is bold because it's still not clear even though the colors are different. The sentence structures, punctuation, numbering, and some chemical concepts are still less precise. The comments from the group of experts was used to revison of HCP module, but it has been proper.

3.3 Quantitative data

Products have been revised of the validation stage, then assessment teacher and student responses to HCP module. The teacher's responses were using closed-questionnaire, which contains five aspects, and free critical. The teacher responses were resulted form of qualitative value will then be converted into quantitative value at the interpretation stage by using Likert scale. Then analyzed data and the result is average score of each aspect and converted into qualitative criteria according to the category of aspect assessment, it had showed in table 2.

Table 2. The teachers response result

Aspect	**Teacher**					**Total**	**Average**	**Maximum Score**	**Response Category**	**Ideals (%)**
	A	**B**	**C**	**D**	**E**					
Content	15	16	15	12	16	74	14.8	16	Very Good	92.5
Presentation	42	44	34	37	43	200	40	44	Very Good	90.9
Language	22	24	18	18	23	105	21	24	Very Good	87.5
Integration of vocational context	13	13	13	12	16	67	13.4	16	Good	83.8
Readability	15	14	12	13	14	68	13.6	16	Very Good	85.6
Total	**107**	**111**	**92**	**92**	**112**	**514**	**102.8**	**116**	**Very Good**	**88,6**

The teacher response had been obtained of average score on overall component is 102.8. Based on the ideal scoring criteria, it entered in the range $X \geq 93.75$ with the category of very good. The result was indicating that HCP module can have used to support of learning chemistry and student interest will increase.

3.3.1 Content aspect. According to the result, accuracy content indicator has lowest value response compared to the other indicator in content aspect. All people was representing to each chemistry subject to certain level. The research from [6] suggested that students have difficult to represent every material of science, so the teacher is expected to be able to represent chemistry subject into certain levels. Therefore, the accuracy aspect has lowest value, it possible for the teacher to represent truth of content according to their perception; even though according to the expert who is more mastering the accuracy content judges that, it is feasible.

Some teachers were adding suggestion about HCP module. Firstly, each start is better given the page so readers are easier to searching them. Then, the concept of heat energy had been already existing on thermochemistry so cannot be included again, the students need wished course only. Mixed separation subject is not usually included in the hydrocarbon compounds and petroleum, but in the solution. Display of the repaired content restored the layout.

The teachers critical are not used as a consideration of the revision, because that the concept of thermal energy already exists on the thermochemistry so that it is not necessary to be included again. The students need to wished course only, this is not revised because in the new curriculum, thermochemistry is not in basic competence, so put on this material is due to the combustion of the hydrocarbon compound to produce heat. It is directly connected with the heat conversion on radiator by addition of thermochemistry. Then, mixed separation is not normally included in hydrocarbon and petroleum chemicals, but on the solution matter. Chemical separation is incorporated in hydrocarbon and petroleum chemicals as basis of the petroleum separation process and other solvent separation processes as enrichment or reference for students.

The other indicators on content aspect have the same value so that in terms of completeness, breadth and depth content, this HCP module has been very good category. Module had been complemented by student involvement activities, practicum activities and questions about the development of scientific literacy that relate directly to the context of student life. Therefore, this module is expected to increase students' motivation and interest in learning chemistry [10-11].

3.3.2 Presentation aspect. The presentation aspect has the most indicators than other aspect. It had been concluded that very good, because HCP module was presenting of overall indicator aspect in each chapter. However, student involvement activities and development of HOTS question indicators have the lowest value response. HOTS question is stimulating students' ability to apply their knowledge and solve problems related to everyday life, so that the development of HOTS requires the latest context in real life problems [16]. Thus, the development of HOTS questions in the context of automotive engineering were needed in order to master automotive engineering subject and produce good HOTS questions.

3.3.3 Language aspect. HCP module was using grammar instruction from General Regulation of Spelling Indonesian (PUEBI). The teacher response was finding some wrong sentence structure; it was causing to lowest value in accuracy sentence structure indicators. The overall of language aspect have very good category response.

3.3.4 Integration of vocational context aspect. The integration of vocational context aspect has lowest value compared to other aspects. This integration had caused that learning of chemistry in vocational schools is less interest to students, because the chemistry subject taught by teacher is not related to the vocational learning [1]. In other words, the teacher does not understand to integrated chemistry subject of automotive engineering, so that in assessing of HCP module cannot provide a maximum assessment of HCP module.

3.3.5 Readability aspect. According to the result, layout and font indicators have lowest value in this aspect. It was caused that some table and picture in HCP module was wrong arranging on some page. The teacher also was critical that some page layout have very full and less space between sentences, so the reader was not comfortable to read this module. The critical will used to improving the HCP module quality and overall the readability have very good category.

The students' response had been conducted assessment after teacher response. It was using closed-questionnaire, which contains 18 item of interest statements about HCP module. The interest statement in closed-questionnaire contains from three aspects, there are readability, content, presentation. The process analyze is same between teachers' response. The result had showed in table 3.

According to the table 3, readability indicators has lowest value in this aspect. It was caused when review the HCP module, the student conditions were not interest to chemistry. They have opinion that chemistry is difficult. The overall student response has included of very interest category. Thus, it can be concluded that students' responses to HCP module based on vehicle case are excellent.

Table 3. The student response result

Aspect	Students										Average score	Maximum Score	Category	Ideals (%)
	1	2	3	4	5	6	7	8	9	10				
Readability	22	23	24	18	19	24	17	14	18	17	19,6	24	Interest	81.7
Content	36	35	36	27	28	36	29	34	25	30	31,6	36	Very Interest	87.8
Presentation	12	12	12	9	8	12	10	9	9	12	10,5	12	Very Interest	87.5
Total	**70**	**70**	**72**	**54**	**55**	**72**	**56**	**57**	**52**	**59**	**61,7**	**72**	**Very Interest**	**85.7**

4. Conclusion

This research had been successfully developed of HCP module Based on the context of vehicle case on automotive engineering program. The characteristic of research product is integrating between chemistry and automotive subject based on basic competence of curriculum and integrated by vehicle context in first chapter, concept map, activities, examples, and exercise. HCP module had a correct of content aspect according to the expert judgments. The teacher response included in very good category and student response included in very interest category toward HCP module. The HCP module based on vehicle case could be used as learning sources for students and to support a successful chemistry learning on automotive engineering program

References

[1] A Wiyarsi, H Pratomo and E Priyambodo 2017 Proc. of *3rd International Seminar of Science Education* **3** 359-366 Yogyakarta
[2] Eilks I and Hofstein A Relevant Chemistry Education – From Theory to Practice (Rotterdam: Sense Publishers)
[3] Faraday S, Overton C, and Cooper S 2011 *Effective Teaching and Learning in Vocational Education* (London: LSN)
[4] Blondera R, Zemlera E, and Rosenfelda S 2016 *Chem. Educ. Res. Pract.* **17** 1-13 doi:10.1039/c6rp00177g
[5] Ikävalko V M and Aksela M 2015 Contextual, revelant and practical chemistry teaching at upper secondary school level textbooks in Finland (Lumut)
[6] Kapici H and Funda S 2015 *Chem. Educ. Res. Pract.* 518-536
[7] Bybee R and McCrae B 2011 *International Journal of Science Education* **3(1)** 7-26.
[8] Abualrob M M and Daniel E G 2013 *International Journal of Science Education* **35(15)** 2538–2558. doi:10.1080/09500693.2011.610381
[9] Vaino K, Holbrook, J, and Rannikmae M 2012 *Chem. Educ. Res. Pract.* **1(13)** 410-419 doi:10.1039/c2rp20045g
[10] Vos M A, Taconisa R, Jochemsa W M, and Pilotb A 2011 *International Journal of Science Education* **33(10)** 1407–1432
[11] Chen B and Wei B 2015 *Science Education International* **26 (2)** 195-216
[12] Arifin F F and Nasrudin H 2014 *Unesa Journal of Chemical Education* **3(3)** 93-99.
[13] Creswell J and Clark V 2012 *Educational research: Planning, conducting, and evaluating quantitative and qualitative research* (United States of America: Pearson Education Inc.)
[14] Organitation for Economic Coperation and Development (OECD) 2006 *Assessing Scientific, Reading, Matematical Literacy: A Freamework for PISA 2006* Retrieved at htpps://oecd.org/pisa
[15] Sudijono A 1987 *Pengantar Statistik Pendidikan* (Jakarta: Raja Grafindo)
[16] Mohamed R and Lebar O 2017 *International Journal of Academic Research in Business and Social Sciences* **7(2)** 466-476

Impact of student-initiated green chemistry experiments on their knowledge, awareness and practices of environmental sustainability

Taha H[1*], Suppiah V[2], Khoo Y Y[3], Yahaya A[1], Lee T T[1] and Muhamad Damanhuri M I[1]

[1]Faculty of Science and Mathematics, Sultan Idris Education University 35900 Tanjong Malim, Perak, Malaysia.
[2]SMK Jalan Pasir Puteh 31650 Ipoh, Perak, Malaysia
[3]Faculty of Management and Economics, Sultan Idris Education University 35900 Tanjong Malim, Perak, Malaysia

[*]E-mail: hafsah@fsmt.upsi.edu.my

Abstract. This study investigated the impact of student-initiated green chemistry experiments on their level of knowledge, awareness and practices of environmental sustainability; and on their achievement in the topic of rate of reaction. The quasi-experimental study involved 46 secondary school students with 23 students each in control and treatment groups. A set of questionnaire and a set of test were employed in the study; and data was analysed using descriptive and inferential statistical techniques. Results showed significant difference in awareness, between the groups; but a weak insignificant correlation between knowledge and awareness on green chemistry. It was also indicated awareness on green chemistry as a strong significant predictor for practice. However, results revealed no significant difference in the achievement between students undergoing traditional chemistry experiments and the green chemistry experiments. Nevertheless, students participated in designing green chemistry experiments have higher level of awareness and practice of environmental sustainability. This study recommends that teachers should empower students with suitable mechanisms to address sustainable environment issues in classroom or laboratory instruction to promote sustainable practices among them.

1. Introduction

In Malaysia, the government had first integrated environmental consideration into the sixth Malaysian plan (1991-1995) [1]. Since then, environmental education has been infused into primary and secondary schools' curriculum. Malaysian secondary school chemistry syllabus stated,

"The curriculum also aims to develop a dynamic and progressive society with a science and technology culture that values nature and works towards the preservation and conservation of the environment. Realise that scientific discoveries are the result of human endeavour to the best of his or her intellectual and mental capabilities to understand natural phenomena for the betterment of mankind. Be aware of the need to love and care for the environment and play an active role in its preservation and conservation.") [2].

The initiation of education for sustainable development (ESD) was officially launched with a guidebook for teachers titled; Teachers' Guide for Environmental Education developed by the

Curriculum Development Centre. Teachers are supposed to integrate the topics of sustainability in all subjects, with varying presence, depending on their relevance to the learning) [3]. However, the implementation of Environmental Education (EE) in primary and secondary schools is still unclear and seemed to have no significant impact on Malaysian green practices) [4]. It was reported Malaysia only scored 10.5% recycling rate, far too behind compared to some other countries with some almost 70% rate and we produce 0.8 kg of waste daily individually; 30000 to 33000 metric tons of waste daily. Malaysian disposed recyclable waste worth RM476 million two years ago; where primary source of waste is paper (17%) and second is plastic (9%) [5].

Recently Malaysian bauxite miners in the state of Pahang had been suspended from their mining activities for three months due to serious pollution issues. The rivers and sea water near the mines turned red, with reddish dust all over several villages and towns. Analysis carried out by the Environment Pollution Department on rivers for instance; Sungai Riau, Mabuk and Pinang found levels of aluminium, iron and mercury were at a dangerous level for mankind and will have very bad impact on aquatic organisms and water resources) [6]. If teachers could embrace ESD in the teaching and learning of chemistry, it could serve as a platform for learners in the rural areas to examine some of the environmental issues rampant in the community.

2. Literature Review

Some studies [7-8] indicated teachers' knowledge of sustainability issues were not well-developed. Many teachers seemed knowledgeable when asked about sustainability and ESD, however, their ideas were shallow and they were unable to relate those knowledge into practises in classroom. It was also revealed that teachers lack the skills and knowledge to implement ESD [9-11]. Borg et al. [12] pointed out that by using traditional instructional methods teachers assumed learners will display appropriate sustainable actions. It seems mere campaigning or teaching sustainability in schools does not seemed to be effective in instilling sustainable practices among students.

A study [13] comparing student learning of chemistry content and experimental design skills following completion of one of the LRT (Laboratory Report Templates) experiments or a traditional experiment on identical chemistry content was conducted. Study results indicated students who completed the LRT experiment learned significantly more content and experimental skills than did students who completed the traditional experiment. This study demonstrates changing one lab from traditional to research-based has positive effects on students' learning.

Environmental education is a way for teachers to expand science into the real world [14]. Outdoor education is place-based, involves experiential learning and fieldwork; thus could be employed to encourage learners to participate actively when learning about the environment [15]. Another study [16] showed recycling project positively affected students' attitudes towards chemistry learning and environmental literacy. However, they also acknowledged changing their own behaviour is still difficult even when they fully understood the importance of environmental sustainable practices. Nevertheless inquiry-based teaching of life-cycle thinking could still be seen as an effective approach to nurture and enhance sustainable practices in children.

Revell [18] suggested local authority in the country should enforce harder penalties to address unsustainable practices and to meet their national emission reduction targets. This shows green practice must also be enforced rigorously by legal means and not only depends on people's voluntary actions. If this were to be applied in Malaysia, recycling and systematic waste disposal, should be among the firsts to be enforced fiercely. This is also seconded by Dzulkifli [17] (ex-vice chancellor of USM, Malaysia) in his talk, posited people should be empowered with appropriate mechanisms of how they could participate to care for the environment.

Green Chemistry (GC) is rather new in Malaysia. Some chemistry teachers could not even tell what GC is exactly and how to practice it! Green chemistry is sometimes mistakenly understood as environmental chemistry. Maybe because it started as an academic research that later became noble practice championed by academia, government and NGOs. It is an approach which reduce pollution and waste from chemical reactions [19]. This approach could convey environment sustainability; as well as, enhance students' understanding of chemistry [20]. Karpudewan and her team [20] tried out green chemistry experiments by downsizing the apparatus and amount of chemicals used in experiments. Their study found significant difference in students' awareness on the environment however found no significant differences in students' green practices. A recent study by Tuğçe, Nalan & Şenol [21] uses problem-based learning (PBL) to help students understand green chemistry better. They found students took more prominent role since the experiments were related to daily life. Hence, PBL seemed to enhance students' level of understanding of green principles; and helped them to be more proactive in pro-environmental practices.

Mellor, Coish, Brooks, Gallagher, Mills, Kavanagh and their team [22] involved students to develop a green chemistry educational game where students need to apply green principles as they design a hypothetical, chemical product. Students' seemed to enjoy it very much and positive impacts were observed on their awareness of sustainable practices. Another study done by Ribeiro & Machado [23] analysed the effects of reaction conditions on several green chemistry mass metrics for reactions with 1:1 and 1:3 ratios. Although there were some practical limitations to the greenness due to the complex nature of the chemistry, it was still very crucial to pay more attention to quantitative details of chemical syntheses to increase the *greenness* of the chemical reactions. On the other hand, Tsakeni [24] posited contemporary environmental issues offer teachers opportunities to disseminate ESD in practical work. However this could be easier said than done. For instance, Kosovo's researchers, Spahiu, Korca & Lindemann-Matthies [15] investigated the integration of EE by 244 high school teachers. They found teachers focused on pollution and hazards of pollutants, however teachers did not seem to effectively relate them to sustainability issues. They also reported EE was also aimed at capacity-building. However, this approach was hard to implement, due to insufficient teacher resources, unsuitable class size, and very limited time. Nevertheless, interviewees insisted ideal EE should involve outdoor education, field work, and practical first-hand experiences.

3. Statement of Problem

Schools would be able to manage chemical waste more effectively if students and teachers were more knowledgeable and committed towards green chemistry practices. Solid and chemical waste from schools' laboratory, should comply with environmental preservation principles. Therefore integrating the principles of green chemistry in a school's chemistry laboratory plays significant role to raise awareness and understanding of all members of the school community on improving environmental sustainability practices. Hence, to play an important role as environment sustainability educationist, chemistry teachers need to equip themselves with ample knowledge and skills to disseminate the effective green practices to their students. Since the nature of chemistry lessons is very unattractive to students, teachers need to manipulate investigative and research-like approaches to trigger and enhance students' awareness of the environment. Armed with green chemistry as the platform, chemistry teachers could involve students in designing green chemistry experiments, encouraging them to embark on their own research and seek information on green chemistry practices. This would in turn enhance students' practical knowledge and increase their pro-environment awareness and practices. Thus, based on what has been stated above, and students as the major player of the project; the objectives of this study were formulated as follows:

1. To identify differences in students' knowledge, awareness and practices of environmental sustainability between students who did the conventional chemistry experiments from the textbook with those who developed their own green chemistry experiments.
2. To seek relationship between students' awareness and practices of environmental sustainability.
3. To identify best predictor that contribute to students' practices of environmental sustainability.
4. To identify differences in students' performance in the topic of rate of reaction between students who did the conventional chemistry experiments from the textbook with those who developed their own green chemistry experiments.

4. Method

4.1. Sample

A total of 46 secondary school students were involved in the study. 23 students were undergoing traditional chemistry experiment (control group) and 23 students developed and carried out green chemistry experiments (experimental group).

4.2. Instruments

A set of questionnaires of 48 items was used to elicit students' knowledge, awareness and practices of environmental sustainability, while a paper and pencil test in the topic of rate of reaction was employed to gauge students' grasp of the topic. The questionnaire was constructed by the research team based on extensive literature analysis and some items were adapted from Tee [25], Mohd Zubair [26], Nurbaity [27] customized to Malaysian secondary school students. The test on rate of reaction (ROR) was constructed based on Malaysian secondary school chemistry syllabus. A pilot study conducted on 30 respondents found the validity and reliability of the instruments were high (Cohen's kappa = 0.94 and Cronbach's alpha = 0.91 for the questionnaire; and content validity agreement from another three experts for the ROR's test was 0.99 and inter-rater reliability r = 0.98).

4.3. Research Procedures

Pre-test on the ROR's topic were first given to both groups. The control and experimental groups were formed from intact classes which were equivalent in terms of performance and background. The control group then proceeded with direct teaching by a teacher (selected from the research team) using the conventional chemistry experiments from the text book provided by the Malaysian Ministry of Education. This group managed to finish off the topic in six school periods. At the same time, the treatment group started to design green chemistry experiments of the topic under the supervision of another member of the research team. Development of these experiments was based on students' research on the 12 principles of green chemistry; such as using benign materials used in daily life and by downsizing the amount of chemicals used; to study factors affecting the ROR. A series of brain-storming and debate sessions took part throughout the project. Students first tried out their ideas during the chemistry lab periods before finalizing the green experiments. The treatment group finished 4 teaching periods later than the control group. Both groups were guided to carry out research, to manage information, develop critical and creative thinking and build their coherent analysis of the experiments. Table 1 shows brief descriptions of some of the green chemistry experiments developed by the treatment group.

Table 1. Conventional and green chemistry experiments.

Factor affecting ROR	Conventional experiment	Green chemistry experiment
Total surface area	Zinc metal and hydrochloric acid	Vinegar and eggshells
Concentration	Sodium thiosulphate, sulphuric acid	Lemon juice , Alka Seltzer pills
Temperature	Sodium thiosulphate, sulphuric acid	Vinegar , eggshells

Once treatments were completed, a post-test on ROR was conducted to both groups intended to determine the effectiveness of green chemistry experiments on student achievement in the topic and after that the post-treatment questionnaires was administered to study the impact of the GC experiments on students' knowledge, awareness and practices of environmental sustainability.

5. Results and Discussion

First, differences in students' compatibility was examined by comparing the pre-test scores of both groups on the topic of ROR.

Table 2. Independent t-test of the pre-test on the topic.

Group	Mean	Mean diff.	Sd. diff.	t	df	Sig. (2-tailed)
Experiment	15.00	0.44	2.534	0.868	44	0.394
Control	14.56					

*Significance level 0.05

Table 2 shows the mean of the experimental group was 15.00, while the control group was 14.56 ($t = 0.868$ and $p > 0.05$). This shows there was no significant difference of both groups' performance in the pre-test. Thus the two groups were equivalent in terms of prior knowledge. Next, table 3 shows there was no significant difference in the achievement between students undergoing traditional chemistry experiments and the green chemistry (GC) experiments.

Table 3. Independent t-test of the post-test on the topic.

Group	Mean	Mean diff.	Sd. diff.	t	df	Sig. (2-tailed)
Experiment	18.88	0.52	2.12	5.933	41	0.289
Control	18.36					

*Significance level 0.05, N = 43

The post-test results demonstrates there was no significant difference in the post-test mean scores between the experimental and control groups, ($t = 5.933$, and $p > 0.05$). The results indicates the achievement of the experimental and control groups were not significantly different, although the mean scores for the experimental group (mean = 18.88) was slightly higher than the control (mean = 18.36).

Below, table 4 shows the comparison of the post-questionnaire results for both groups on the knowledge of green chemistry (GC). The mean value shows a significant difference between the groups. The treatment group shows a much higher mean score, indicating a significant increase of their knowledge of GC after completing the green chemistry experiments project. Students seemed to gain more knowledge when the topic was taught using green chemistry experiments. They were compelled to look for more information on green experiments, had many brain-storming sessions and heated discussions on how to ensure the experiments could really be done realistically. They even tried out the experiments several times before finalizing their report. These exercises of active learning might be the contributing factors of the present results of this study.

Table 4. Mean score of post-test for knowledge GC.

Group	N	Mean	SD	t	df	Sig
Experiment	23	15.13	1.46	10.62	41	.000
Control	20	10.30	1.52			

*Significance level 0.05, N = 43

Table 5 shows the pre and post-test comparison for awareness of GC for students who were involved in the green chemistry experiments. Result shows there was significant difference in awareness, with the mean value showed substantial increase after the exposure towards green chemistry experiments. These seemed to support research done by Cullifer [28] that reported students' understanding of green chemistry was significantly enhanced when presented with two real-world scenarios and applied benefits-costs-risks analysis that emphasizes a decreasing dependence on limited non-renewable resources and an increasing focus on preventing pollution by-products of the chemical industry.

Table 5. Mean scores for experimental group on awareness on GC.

Group	N	Mean	SD	t	df	Sig
Pre-test	23	61.56	5.84	1.58	44	.021
Post-test	23	84.34	5.92			

*Significance level 0.05, N = 46

Next, correlation between students' knowledge and awareness on green chemistry was also examined. Table 6 shows the correlation between knowledge and awareness on green chemistry among the participants. The results shows a weak insignificant correlation ($r = 0.08$ and $p > 0.05$) between students' knowledge and their awareness of green chemistry. This result seems to support findings of a research done by Aminrad, Syed Zakaria, Hadi & Sakari [29] and Ahmad, Md Noor & Ismail [30] that also revealed weak relationship between knowledge and awareness of environmental issues.

Table 6. Correlation between students' knowledge and awareness on GC.

Variables	1	2	Sig
Attitude		0.080	0.462
Knowledge			

*Significance level 0.05, N = 46

Although the literature suggests knowledge could be a viable predictor for environmental attitude of GC [31], this study does not seem to yield similar results. Next, Table 7 seems to indicate awareness on green chemistry as a strong significant predictor for practice. The percentage of the predictor is 60.5 percent towards the criterion variable and however, as expected, knowledge does not significantly predict practice of GC ($p > .05$).

Table 7. Multiple regression analysis for the predictor for practice of GC.

Model	B	Standard error	t	Sig.
Constant	27.894	4.802	5.83	.000
Knowledge GC	.149	.151	.984	.450
Awareness GC	1.78	.156	11.44	.000

$R^2 = .614$, $AR^2 = .605$

Though we might logically think and believe that when one is more knowledgeable about the environment, they will, in turn, be more inclined to be more 'green' or act toward the environment more responsibly, this is not apparent in this study. The pro-environmental behaviors have to come from people's true understanding and commitment. The global change in the 21st. century changes with fast-paced technological advance, plus emerging life styles and values, are now exposing nature with serious environmental damage. Pollution from industrial emissions, waste management, and sustainability of water supply, and energy efficiency; need to be addressed now, in order for younger generation to be able to inherit a clean and beautiful Malaysia. Prevention is always better than cure. Thus, preventing the environmental catastrophe in the future is a better option before it is too late to regret.

6. Conclusion

This study revealed student-initiated green chemistry experiments have positive impact on students' knowledge, awareness and practice of green chemistry. It is evident from the study that encouraging students' participation in designing green experiments enhances their sustainable practice. Integrating green chemistry in school chemistry experiments triggers students' sense of responsibility towards the environment. It would be excellent if every school subjects called for students' involvement in such a way that they would feel responsible for environment in whatever they do. Integrating sustainable practices in lessons actively would be an integral parts to cultivate the awareness among students. Furthermore, it is also recommended proper training for teachers so that they will be able to teach their students effectively. Teachers need to be equipped with more 'hands on' skills rather than mere theories and facts to deliver effective green chemistry lessons. If teachers were not environmentally literate, it will be almost impossible for them to produce environmentally literate students. ESD in schools should call for more collaborative participation and engagement from all school staff, parents and local community to instill pro-environmental practice into children and community at large.

References

[1] Economic Planning Unit 1991 Ministry of Economic Affairs, Malaysia http://www.epu.gov.my/en/rmk/sixth-malaysia-plan-1990-1995
[2] Ministry of Education Malaysia 2006 *Integrated Curriculum for Secondary Schools Chemistry Syllabus* (Curriculum Development Centre, Ministry Of Education)
[3] Kementerian Pendidikan Malaysia 1998 *Buku Panduan Guru Pendidikan Alam Sekitar Merentas Kurikulum KBSR dan KBSM* (Kuala Lumpur: Pusat Perkembangan Kurikulum)
[4] Pudin J, Tagi K, and Periasamy A 2005 *Environmental Education in Malaysia and Japan: A Comparative Assessment Int. Conf. on Ed. for Sustainable Future* (Ahmedabad, India) pp 18-20 Jan 2005 www.ceeindia.org/esf/download/paper20.pdf
[5] Abas M A and Wee S 2014 *Municipal Solid Waste Management in Malaysia: An Insight Towards Sustainability* http://dx.doi.org/10.2139/ssrn.2714755
[6] Ibrahim K 2016 Perlombongan bauksit usah jejas alam sekitar *Berita Harian* http://www.bharian.com.my/node/111114
[7] Riess W and Mischo C 2010 *Int. J. of Sci. Ed.* **32** 705-25
[8] Burmeister M and Eilks I 2013 *Sci. Ed. Int.* **24** 167-94
[9] Garner N, Siol A and Eilks I 2015 *Sustainability* **7** 1798-1818
[10] Makoni M 2013 *Int. J. of Asian Soc. Sci.* **3** 971-91
[11] Burmeister M, Rauch F and Eilks I 2012 *Chem. Ed. Res. Pract.* **13** 59-68
[12] Borg C, Gericke N, Höglund H, and Bergman E 2012 *Research in Sci. & Tech. Ed.* **30** 185-207 doi: 10.1080/02635143.2012.699891
[13] Carciatore K L 2010 *Development and Assessment of Green, Research-Based Instructional Materials for the General Chemistry Laboratory* (ProQuest Dissertations UMI Dissertation Publishing) UMI3420092
[14] Warner B P and Elsier M 2015 *The J. of Environmental Ed.* **46** 1-22
[15] Spahiu M H, Korca B and Lindemann-Matthies P 2014 *Int. J. of Sci. Ed.* **36** 2750-71
[16] Juntunen M and Aksela M 2014 *Chem. Ed. Res. Pract.* **15** 639-49
[17] Dzulkifli A R 2015 *Sustainability and Leadership ProSPER.* Net Programme Sabah, Malaysia https://www.slideshare.net/HannaatUNU/sustainability-and-leadership-dzulkifli-abdul-razak
[18] Revell K 2013 *J. of Integrative Environmental Sci.* **10** 199-218
[19] Ahluwalia V K 2008 *Green Chemistry: Environmentally Benign Reaction* (New Delhi India: Ane Books India)
[20] Karpudewan M, Ismail Z, and Roth W M 2012 *Environmental Ed. Research* **18** 375-89
[21] Tuğçe G, Nalan A, and Şenol A 2017 *Research in Sci. & Tech. Ed.* **35** 500-20
[22] Mellor K E, Coish B P, Brooks W, Gallagher E P, Mills M, Kavanagh T J, Simcox N, Lasker G A, Botta D, Voutchkova-Kostal A, Kostal J, Mullins M L S M, Nesmith J C, Kristofco L, Saari G, Steele W B, Melnikov F, Zimmerman J B and Anastas P T 2018 *Green Chem. Letters and Reviews* **11** 103-10
[23] Ribeiro M G T C and Machado A A S C 2013 *Green Chem. Letters and Reviews* **6** 1-18
[24] Tsakeni M 2018 *African J. of Research in Maths., Sci. and Tech. Ed.* **22** 125-36 doi: 10.1080/18117295.2018.1446498
[25] Tee S T 2017 *Green* Chemistry Awareness among Form 4 Student in Kuala Selangor *Unpublished thesis* (Malaysia: Universiti Pendidikan Sultan Idris)
[26] Arifin M Z A 2015 Tahap kesedaran teknologi hijau dalam kalangan guru-guru teknologi kejuruteraan zon utara *Unpublished thesis* Universiti Tun Hussien Onn Malaysia http://www.eprints.uthm.edu.my/7069/
[27] Nurbaity N 2011 *Jurnal Riset Pendidikan Kimia* **1** 13-21
[28] Cullipher, S G 2015 *Research for the Advancement of Green Chemistry Practice: Studies in Atmospheric and Educational Chemistry* Graduate Doctoral Dissertations Paper 203 http://scholarworks.umb.edu/doctoral_dissertations
[29] Aminrad Z, Zakaria S Z S, Hadi A S, and Sakari M 2013 *World Applied Sci. J.* **22** 1326–33

[30] Ahmad J, Md Noor S, and Ismail N 2015 *Asian Soc. Sci.* **11** 284-93
[31] Sadik F and Sadik S 2014 *Procedia–Soc. and Behavioral Sci.* **116** 2379–85

Acknowledgments
The authors wish to thank UPSI for the University Research Grant (GPU2016) that supported the study and thus made it possible for us to experience this great academic endeavor.

IOP Conf. Series: Journal of Physics: Conf. Series **1156** (2019) 012023 doi:10.1088/1742-6596/1156/1/012023

Automotive engineering chemistry module: Exploring acid base and electrochemistry topic in vehicle context

H Febiana[1], C F Partana[2], A Wiyarsi[2] and B Sulistyo[3]

[1]Study Program of Chemistry Education, Faculty of Mathematics and Natural Sciences, Universitas Negeri Yogyakarta, Indonesia
[2]Chemistry Education Department, Faculty of Mathematics and Natural Science, Universitas Negeri Yogyakarta, Indonesia
[2]Automotive Engineering Education Department, Faculty of Engineering, Universitas Negeri Yogyakarta, Indonesia

*E-mail: handarifebiana8@gmail.com

Abstract. The research purposed to analyze the characteristics of chemistry module on acid-base and electrochemistry topic based on vehicle context, to analyze the feasibility of module according to chemistry learning experts and to analyze the response of chemistry teachers and students of Automotive Engineering Program (AEP) about this module. This research using mixed methods and exploratory model design. The data collected from expert validation sheets also teachers and students response questionnaire. Technique of data analysis begins by converting qualitative data into quantitative data using Likert scales. Furthermore, the total and average score of each component can be determined from the quantitative data. The result of the average scores were converted into qualitative feasibility criteria according to the rating category. The product of the development was a chemistry module on acid-base and electrochemistry topic based on vehicle context for students in AEP. The experts considered feasibility of the module. This module had assessed by 5 teachers and 10 students. The module assessed by teachers showed a very good response while the module assessed by students showed a good response. Therefore, this module is feasible and can be used as a learning resource for students of AEP to enhance quality of chemistry learning.

1. Introduction

Vocational education more emphasized for students who want to work after graduation. Vocational High School curriculum is more practical than theory. Vocational High School has various programs, student can choose according to interests and talents. Graduates of vocational school ready to work based on their expertise. They learn not only in school but also in workplace.

One of the programs in Vocational High School is the Automotive Engineering Program (AEP). Student of AEP not only accepts subjects that related to automotive engineering skills courses but also gets other general subjects. Chemistry is one of the subjects in Vocational High School that must be followed by teacher especially in AEP.

The aim of the chemistry subject in vocational schools are to prepare the basic skills of vocational students in developing their skills, to support vocational subjects, as a basic knowledge for higher education, and to provide the basic knowledge so that students have good science literacy [1]. In fact, according to vocational school students', chemistry subjects are not important subjects so students tend to underestimate. It is not surprising that many vocational school students in Indonesia do not like chemistry subjects. Many chemistry teacher and science teacher researchers note that chemistry

IOP Conf. Series: Journal of Physics: Conf. Series **1156** (2019) 012023 doi:10.1088/1742-6596/1156/1/012023

widely felt by students as a difficult subject at several levels of education [2]. They also assume that the chemistry subject is not beneficial for their next career. Many students are not too interested in the subject matter of school chemistry. They do not understand why they need to know the chemical concepts [3].

Based on the results of interviewed with teachers, chemistry learning was given to vocational students are still very common, it is no different from chemistry learning in senior high school. This is due to not available of chemistry modules for vocational schools especially for AEP. Chemistry modules for vocational high school should be differentiate with chemistry modules for senior high school. The appropriate chemistry module for AEP can be created by developing the chemistry learning that integrated with vocational context. The context-based modules can increase students' motivation intrinsically and can improve students' chemical literacy [4]. The chemistry learning has the potential to increase students' motivation, and can reinforce both the meaning of learning chemistry and its perceptions [5]. The students tend to be more interested in studying big ideas of contextual discipline than basic concepts [3]. Consequently, with this concept, learning outcomes are expected to be more meaningful for students.

Meaningful learning can be gained from connecting subject content to specific contexts. Learning chemistry subjects in Vocational School should be associated with the vocational context. Context-based learning aims to enable students connecting between scientific discourse and real-world contexts to enhance curricular relevance without reducing conceptual comprehension [6]. However, it is still constrained by the limited context-based modules. Some Vocational School teachers say that there is no context-based chemistry module for AEP.

One context that is very close to the students of Automotive Engineering is the vehicle. Every day students of AEP always discuss about vehicle. The vehicles body is influenced by the acid-base factor in the environment. Many acid and base products are used in vehicle maintenance. Solution in battery is an acid which is one of the important components in battery. Battery is an application of electrochemistry cells. Acid base and electrochemistry are chemistry topic that taught to vocational students majoring in Automotive Engineering. In this topic easy to find automotive engineering context. Therefore, this research was concerned to develop chemistry modules that can relate chemistry content to automotive contexts, that is acid-base and electrochemistry based vehicles.

The context-based modules was developed can help students to understanding chemistry topic especially for vocational schools. Context-based modules can be used as student learning resources. Context-based chemistry learning is more meaningful and not boring. The problem of this research are: How the characteristics of chemistry module on acid-base and electrochemistry topic based on vehicle context? How the feasibility of module according to chemistry learning experts? And how the response of chemistry teachers and students of Automotive Engineering Program (AEP) about this module?

2. Method

2.1 Research Design

This research used exploratory mixed methods design. This design was appropriate with it aims because can explore problems and develop products.

2.2 Procedure

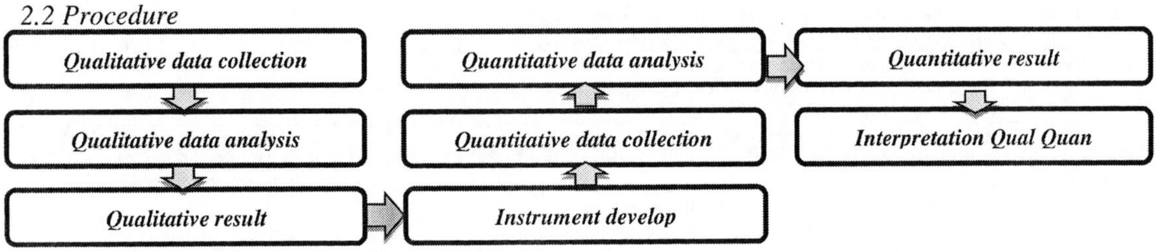

Figure 1. Exploratory design mixed method

Based on exploratory design mixed method that showed on the figure 1 [7], the research procedure can be divided into 4 phase including qualitative, development, quantitative, and interpretation phase. In the qualitative phase, data was collected by analyzing the problems of chemistry learning in AEP, analyzing basic competence of chemistry subject in AEP, and analyzing basic competence of vocational subject that exist in AEP of Vocational School. In the development phase, the first steps were to determine the learning resources to be used as references, make a framework of content, and determine the vehicle context that appropriate with acid-base and electrochemistry topic. Then, arrange the layout and develop the content. In the validation phase was done by feasibility test of experts. The validation was conducted by asking the review of the module in the term of vocational content integration, content feasibility, content presentation, language and grammar toward two experts from chemistry education department. In the quantitative phase, collection of data conducted by responses test of teachers and students. Teachers responses test was conducted by 5 chemistry teachers of Vocational School AEP. Students responses test was conducted by 10 students of Vocational School AEP.

2.3 Research Instrument

The instruments that used to collect the data are validation instrument and qualitative data collection instrument. Validation instrument is chemistry learning expert validation sheets. Quantitative data collection instruments are teachers' responses questionnaire, and students' responses questionnaire. Chemistry learning expert validation sheets were adapted from Education National Standard Agency [8], Programme for International Student Assessment (PISA), and relevant journals. These validation sheets consist of 5 aspect, there are vocational content integration, content feasibility, content presentation, language and grammar aspects. Teachers' responses questionnaire were adapted from Education National Standard Agency [8] and some journals [4] [9-12]. These teachers' responses questionnaire present in the table 1. Teachers' response questionnaire consist of 29 indicators and 5 components. On the other hand, students responses questionnaire was adapted from PISA [13] . Students response questionnaire consist of 18 indicators and 3 components showed in the table 2.

Table 1. Instrument of teachers responses questionnaire

Component	Instruments were adapted	Indicator number	Proposed by
Content Feasibility	Completeness, breadth, Depth, and accuracy of content [8]	1, 2, 3, and 4.	Education National Standard Agency 2012
Content Presentation	Systematic consistency in every chapter, coherency of presentation, suitability the picture with content, addition of question example, addition of concept maps, addition of exercise in every chapter, addition of evaluation question in every chapter, and identity consistency of table/picture [8]	5, 13, 11, 9, 7, 10, 14, and 12.	
	Development of HOT question in evaluation [4]	15	Holbrook, Rannikma and Vaino (2012)
	Student involvement activities [4] [9]	8	Holbrook, Rannikma and Vaino (2012) & Chen and Wei (2015)
Readability	Entirety attraction of module, cover, layout, and font [8]	29, 26, 27, and 28.	Education National Standard Agency
Content Presentation	Completeness of content presentation (content was integrated with context, table of contents, concept maps, core competence and basic competence, glossary, index, and reference) [9]	6	Chen and Wei (2015)
Language	Language use, accuracy of sentences structure, linkages between paragraphs, suitability of punctuation, spelling accuracy, and consistency of terms [8]	16, 17, 18, 20, 19, and 21.	Education National Standard Agency
Vocational Content Integration	Accuracy of integration results with content in module [10]	22	Abualrob and Daniel (2011)
	Suitability of experiment with vehicle/automotive context [11] [9]	23	Ikavalko and Aksela (2015) & Chen and Wei (2015)
	Content application was integrated with context [10]	24	Abualrob and Daniel (2011)
	Accuracy of narrative at the beginning of chapter with context [4] [12]	25	Holbrook, Rannikma and Vaino (2012) & Vos, Taconis, Jochems and, Pilot (2011)

Table 2. Instrument of students responses questionnaire

Component	Indicator Number	Statement
Readability	1	I feel interested when learning acid-base and electrochemistry topic using this module.
	2	I am more excited learning acid-base and electrochemistry topic using this module.
	3	I know the benefits of acid-base and electrochemistry in automotive engineering through this module.
	4	I think easy to understand the concepts of acid-base and electrochemistry using this module.
	5	I know the benefits of acid-base and electrochemistry in everyday life through this module.
	6	I think this module can increase the reading interest about acid-base and electrochemistry topic.
Content	7	I think that acid-base and electrochemistry topic is important to learned because it can be helpful to support my job in future.
	8	I think this module can help the students understanding acid-base and electrochemistry topic in vehicle context.
	9	I know the alternative treatment to respond chemicals (battery, H_2 gas, etc.) in everyday life.
	10	I think the problem in vehicles related to acid-base and electrochemistry can solved by learning this module.
	11	I can understand acid-base and electrochemistry topic easily using this module.
	12	I think this module appropriate with automotive engineering student needs.
	13	I get some new knowledge about acid-base and electrochemistry from this module.
	14	I think this module can give other knowledge out of major for careers in the future.
	15	I feel motivated to do chemistry practical activity because this module contain practical worksheets integrated with vehicle context..
Content Presentation	16	I think more easy to understand acid-base and electrochemistry topic because there are illustrations and application in this module.
	17	I feel this module can motivate me to discussion with friends because the presentation of module content support to discussed.
	18	I can test my understanding of acid-base and electrochemistry topic through evaluation test in the end of chapter.

2.4 Data Analysis

The assessment result of teachers and students' responses tests were qualitative data and some suggestions. Then, qualitative data was converted into quantitative values with Likert scale while the results in the form of suggestions were used as consideration for revision. Revisions were aimed to improve the modules and increase the quality of modules. Then, quantitative values were analyzed and the average score for each instrument could be determined.

The final average scores were obtained from this quantitative data analysis was used to determine the feasibility category of chemistry module. The results of quantitative data analysis were interpreted with ideal conversion guidelines .The categories that used are very good, good, average, less, very less [14].

3. Result and Discussion

Based on the results of interviews, could be known some problems that occur in the process of chemistry learning in Vocational School AEP. Using basic competence analysis sheets could be gotten acid-base and electrochemistry topic that integrated with automotive engineering vocational basic competence. The analysis result of it was a matrix of integration between chemistry basic competence especially acid-base and electrochemistry topic and vocational basic competence. Based on the matrix, it has been arranged the chemistry module of acid-base and electrochemistry topic for AEP.

Validation result that conducted by chemistry learning expert test were some suggestions. The result of validation by chemistry learning expert, there are some mistakes about chemical nomenclature and some typo, but the experts said that the module have been feasible. The suggestion of validation results was used for consideration of revision.

Quantitative data were gotten from analysis result of teachers and students assessment. Teachers responses of module completely are shown in table 3. In addition, the students' response of module are shown in table 4.

The chemistry module products including cover, introduction, table of contents, introductory sections including a concept map of the integration of chemistry content with the vehicle context, the matrix of integration of chemistry content with the automotive engineering context, content, glossary, indexes, and references. Furthermore, analysis of basic competence of chemistry subject on acid base

IOP Conf. Series: Journal of Physics: Conf. Series **1156** (2019) 012023 doi:10.1088/1742-6596/1156/1/012023

and electrochemistry topic and basic competence of Automotive Engineering Vocational subjects in were conducted according to Decree of Director General of Primary and Secondary Education Number 330 / D. D5 / KEP / KR / 2017 [15].

Table 3. Data of teachers responses of chemistry module on acid-based and electrochemistry topic based on vehicle context completely

Component	*Teachers*					Total Score	Average Score	Maximum Score	Category	Ideals (%)
	1	2	3	4	5					
Content Feasibility	16	14	15	14	16	75	15	16	Very Good	93.75
Content Presentation	43	38	43	37	42	203	40.6	44	Very Good	92.27
Language	24	19	21	18	21	103	20.6	24	Very Good	85.83
Vocational Content Integration	16	14	14	14	16	74	14.8	16	Very Good	92.50
Readability	16	14	16	14	16	76	15.2	16	Very Good	95.00
Total Score	115	99	109	97	111	531	106.2	116	Very Good	91.55

Table 4. Data of student responses of chemistry module on acid-based and electrochemistry topic based on vehicle context completely

Component	Students										Total Score	Average Score	Maximum Score	Category	Ideals (%)
	1	2	3	4	5	6	7	8	9	10					
Readability	18	17	24	19	19	20	20	19	23	19	198	19.8	24	Good	82.50
Content	27	30	36	29	31	29	29	30	36	29	306	30.6	36	Good	85.00
Content Presentation	9	10	12	10	9	9	9	8	12	10	98	9.8	12	Good	81.67
Total Score	54	57	72	58	59	58	58	57	71	58	602	60.2	72	Good	83.61

Based on the analysis of teacher responses on the table 3, an overall average score is 106.2 from a maximum score of 116 is obtained, so it has a very good category with an ideal percentage of 91.55%. Ideal percentage of teachers' responses each component showed on the figure 2.

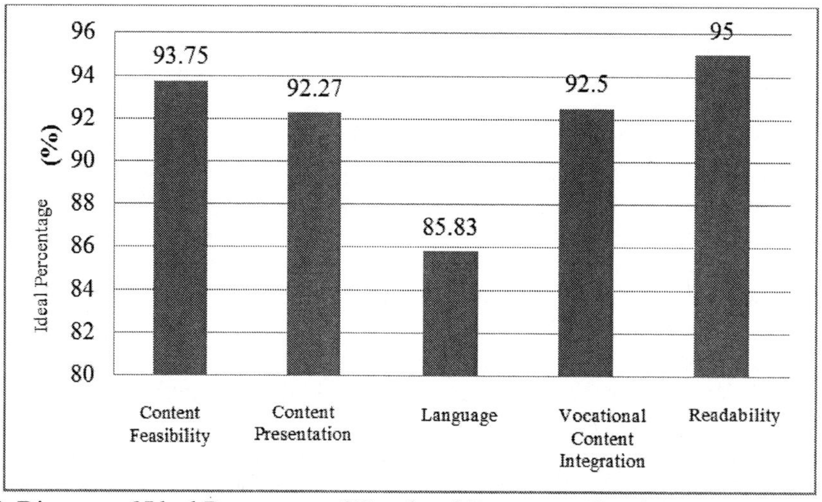

Figure 2. Diagram of Ideal Percentage of Teachers' Responses Teacher responses assessment instrument consists of 5 components

3.1 Components of Content Feasibility

Based on the results of the analysis that showed in table 3,it can be concluded that the components of content feasibility in the module have been very good. The highest average score is in indicator 1 about content completeness while the lowest average score is in indicator 3 and 4 about depth and accuracy of content. In indicators 3 and 4 although the average score is the lowest but it is still in very good category. This shows that of the four indicators in the content feasibility component in terms of completeness, breadth, depth, and accuracy of modules content has been very good.

3.2 Components of Content Presentation

Based on the results of the analysis that showed in table 3, it can be concluded that the content presentation component in the modules has been very good. The highest average score is in indicator 12 about consistency of table/ image identity while the lowest average score is in indicator 8 about student involvement activities. This shows that indicator 12 viewed from the consistency of the identity of the table / picture of the modules have been very good while the indicator 8 about the student involvement activities fall into good category. Other indicators in the content presentation component are also very good because the average score is more than 3.4.

Indicator 8 about student involvement activities has the lowest score because activities in module are less interesting. Activities in modules should encourage students to do something. Other indicator, the development of HOT questions in modules is very good. One of HOTS (High Order Thinking Skill) realization is able to cultivate critical, creative, and innovative thinking skills [16].

3.3 Component of Language

Based on the results of the analysis showed in table 3, it can be concluded that the language component in the modules has been very good. The highest average score is in indicator 18 while the lowest average score is in indicator 19 and 20. This shows that when viewed from indicator 18 that the linkage between paragraphs of modules has been very good while the indicators 19 and 20 regarding the spelling accuracy and the suitability of punctuation in good categories.

Indicators 19 and 20 have the lowest average score because there are spelling of words in modules that have not complied with general guidelines for Indonesian spelling and the use of some precise punctuation are not exact. Such errors may cause the reader to be unable to understand the content of the module. Sentences become ineffective.

The indicator 21 viewed from the consistency of term also got a low score. Some terms in the module have not been consistent. There are the use of foreign words and Indonesian words on the same terms. It must be used only foreign terms or Indonesian terms only, not both.

3.4 Components of Vocational Content Integration

Based on the results of the analysis showed in table 3,it can be concluded that the component of the integration of vocational content in modules has been very good. The highest average score is in the indicator 24 about content application while the lowest average score is in the indicator 22, 23, and 25 about accuracy of the integration result, suitability of experiment, and the accuracy of the beginning narrative of the chapter. The indicators are 22, 23, and 25 although the lowest average score is still in very good category. This indicates that of the four indicators in the vocational content integration component have been very good. The context-based modules designed to capture students' intrinsic motivation and increase chemical literacy to become fast-response students [4]. To develop competence in an area of inquiry, student must understand fact and ideas in the context of conceptual framework and organize knowledge in ways that facilitate retrieval and application [17]. The vocational content integration component of modules is so good that the modules are expected to increase the motivation of vocational students in AEP to study chemistry.

3.5 Components of Readability

Based on the results of the analysis showed in table 3, it can be concluded that the components of readability in modules have been very good. The highest average score is in indicator 26 about cover of module while the lowest average score is in indicator 28 about font type. In indicator 28 even though the lowest average score but still in the very good category. This shows that when viewed from the cover and the font type of modules are very good. Other indicators in the readability components that are indicators 27 and 29 seen from the layout and entirety attraction of the modules have also been very good.

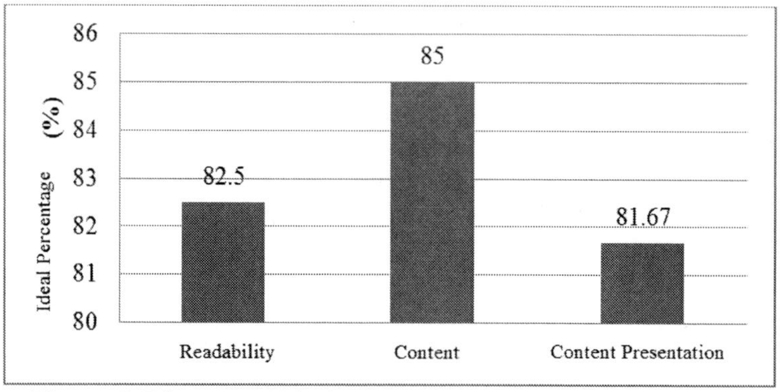

Figure 3. Diagram of Ideal Percentage of Students' Responses

Based on the results of data analysis showed on table 4, obtained the overall average score is 60.2 from a maximum score of 72 so that it has good category with an ideal percentage is 83.61%. Thus, it can be concluded that students' responses to module are good. Ideal percentage of students' responses each component showed on the figure 3.

4. Conclusion

Based on the research, the chemistry module has a characteristic, that is the module was the result of integration between basic competence of chemistry subject on acid-base and electrochemistry topic and basic competence of Vocational Automotive Engineering subject. Based on chemistry learning expert, this module is feasible to be tested to students of AEP but with some suggestions. Based on teachers' responses, this module is included in very good category. Based on students' responses, this module is included in good category. Hence, the module should be used as a learning resource for students of AEP in order to support the learning of chemistry.

5. Reference

[1] Wiyarsi A 2017 *Kurikulum dan Konten Pembelajaran Kimia di SMK untuk Guru dan Calon Guru Kimia* (Yogyakarta : UNY Press)
[2] Johnstone A H 2000 *Chem. Educ. Res. and Prac. in Europe* **1**(1) 9-15
[3] de Jong O and Talanquer V 2015. *Relevant Chemistry Education* ed I Eilks and A Hofstein (Rotterdam: Sense Publishers) p 11
[4] Holbrook J, Rannikmae M and Vaino K 2012 *Chem. Educ.* **4** 410-19.
[5] Sevian H and Bulte A M 2015 *Relevant Chemistry Education* ed I Eilks and A Hofstein (Rotterdam: Sense Publishers) h 7
[6] King D T and Ritchie S M 2013 *Inter. J. Sci. Educ.*, **35** 1159-82.
[7] Creswell J and Clark V 2007 *Designing and Conducting Mixed Methods Research* (Thousand Oaks, CA: Sage Publications)
[8] Kemendikbud 2014 *Penilaian Buku Teks Pelajaran oleh Badan Standar Nasional Pendidikan (BSNP)* www.kemdikbud.go.id
[9] Chen B and Wei B 2015 *Sci. Educ. Inter.* **26**(2)195-216.

[10] Abualrob M M and Daniel E G 2013 *Inter. J. Sci. Educ.*, **35**(15) 2538-58
[11] Ikavalko V and Aksela M 2015 *LUMAT: International Journal on Math, Science and Technology* **3**(3) 304-15
[12] Vos M A, Taconis R, Jochems W M and Pilot A 2011 *Inter. J. Sci. Educ.*, **33**(10) 1407-32
[13] Organization for Economic Cooperation and Development (OECD) 2006 *Assessing Scientific, Reading, Matematical Literacy : A Framework for PISA 2006* Accessed at https://oecd.org/pisa
[14] Sudijono A 1987 *Pengantar Statistik Pendidikan* (Jakarta: Raja Grafindo)
[15] Kemendikbud. 2017 *Decree of Director General of Primary and Secondary Education Number 130 / D5 / KEP / KR / 2017* www.kemdikbud.go.id
[16] Faraday S, Overton C and Cooper S 2011 *Effective Teaching and Learning in Vocational Education* (London: LSN)
[17] Wilson S M and Peterson P L 2006 *Theories of Learning and Teaching What Do They Mean for Educators?* (Washington : NEA)

Analyzing scientific approach and problem solving in *salt hydrolysis* topic

H N Pramesthi*, A Ashadi and S Saputro
Master Program of Sciences Education, Universitas Sebelas Maret,
Jalan Ir. Sutami No. 36A, Jebres, Surakarta, Jawa Tengah 57126, Indonesia
*E-mail: hilda.nita19@yahoo.com

Abstract. This study aimed to analyze the suitability of materials teaching with aspects found in the scientific approach, and the steps of problem solving model. The sample of this research was three textbook materials of class XI chemistry. It was used by some high schools in Klaten Regency. The analysis was conducted on three teaching materials that were coded X, Y, and Z. The result of the research showed that the compatibility toward the scientific approach of textbook X and Y was sufficient category, textbook Z was very poor category. Conformity to problem solving model step sample X was poor category, sample Y was sufficient category, and sample Z was very poor category. These chemistry book materials which used in the high school should be added according to aspects of the scientific approach and problem solving steps that have not been listed.

1. Introduction

Education in Indonesia currently uses curriculum 2013. The curriculum suggests that students are subjects who have the ability in learning, actively seek, process, construct, and use knowledge. In order to truly understand and apply knowledge, students need to be invited to solve problems, find everything for themselves, and strive to realize their ideas. Students are expected to be more active, creative, and innovative in solving any problems encountered.

Chemistry includes product (cognitive), process (psychomotor), formation of scientific attitude (affective), and application. To understand chemistry requires the ability to describe three representations of macroscopic, microscopic, and symbolic [1]. The presentation of chemical concepts with three levels of representation simultaneously is an important aspect to be considered by teachers in the chemistry learning process. But in reality, many students do not understand and cannot use the three representations (macroscopic, microscopic, and symbolic) in explaining a phenomenon [2]. Therefore, chemistry is often referred to as one of the difficult subjects.

The teachers tend to focus more on the procedural learning in calculation aspect than the conceptual aspect in explaining the chemistry [3]. As a result, students have difficulty understanding the concepts in chemistry correctly. Chemistry contains concepts that are sequential and tiered [4]. If students do not understand the basic concept, then students will have difficulty in understanding the concept further [5]. Student-centered learning can be implemented by applying a scientific approach. The scientific approach has five aspects that can be applied in the learning process. Aspects of a scientific approach include, observing, questioning, collecting information, processing information, and communicating results. Indicators of the five aspects of the scientific approach used in this study are in accordance with the descriptions of the scientific approach aspect of the approach by Harlen (2006) and Permendikbud 81 A 2013 about Implementation of Curriculum.

Problem Solving model is one of the learning model that suits the scientific approach. Problem Solving learning is more emphasized on the students activity, while the role of teachers as mentors who direct students to solve problems. Problem-solving syntax includes: (1) Defining the problem; (2) Diagnose the problem; (3) Formulating alternative strategies; (4) Determining and implementing strategies; (5) Evaluate the success of the strategy [6].

Learning resources or teaching materials that are often used in teaching learning are textbooks such as textbook lessons, modules, and student worksheets (LKS). The book is an instructional aid in the teaching and learning process and should be in accordance with the curriculum of the objectives, content, and teaching methodology of each subject [7]. Therefore, teaching materials should be in accordance with the applicable curriculum for the purpose of learning to be achieved.

This study aims to analyze the suitability of chemical resources used in several high schools in Klaten District with a scientific approach and Problem Solving learning model. The process of analysis of teaching materials was done on *salt hydrolysis* as teaching material for class XI. The analysis was conducted on three teaching materials that were coded X, Y, and Z.

2. Methods

The selection of teaching materials analyzed in this study was adapted to the teaching materials which used in five high schools in Klaten District. Based on the results of interviews and observations, there are three teaching materials used by teachers which is mostly owned by students in the five high schools. The three teaching materials are coded X, Y, and Z.

Method of this research was content analysis, which used a coding qualitative and quantitative analysis. The data collecting technique used documentation method. Data analysis is the assessment of compatibility of teaching materials according to scientific approach and according to the steps in problem solving learning model. The research instrument is the form of table consisting of elements of scientific approach or problem solving learning models in the first column, the second column shows the characteristics of elements, third and fourth column contains textbook contents which is compatible or not. Qualitative data obtained from the checklist on the form. Quantitative data is derived from the score of the calculation of the percentage of the teaching materials calculated using the following equation:

The compatibility of teaching materials with aspects of a scientific approach

$$\%PS = \frac{KM}{K} \text{ x } 100\%$$

The compatibility of teaching materials with Problem Solving model step

$$\%MPS = \frac{SS}{S} \text{ x } 100\%$$

Information:

% PS : Percentage of scientific approach
KM : Total of the compatibility aspects
K : The total number of aspects
% MPS : Percentage of Problem Solving model
SS : Total of the compatibility syntax
S : The total number of syntax

As the provision in giving meaning and decision-making above calculation results can be interpreted with a range as in table 1.

Table 1. Categories of Compatibility Assessment of Materials

Percentage	Quality Categories of compatibility
$81 < \% \leq 100$	Very Good
$71 < \% \leq 80$	Good
$41 < \% \leq 70$	Sufficient
$21 < \% \leq 40$	Poor
≤ 20	Very Poor

3. Results and Discussion

The result of compatibility of teaching materials according to scientific approach can be found in the table. There are four learning indicators of *salt hydrolysis* in the first column and five aspects of scientific approach in the next column. This research consisted of five aspects of scientific approach which analyzed its compatibility with teaching materials. The five aspects of a scientific approach include observing, questioning, experimenting, associating, and communicating. The results of the analysis on materials teaching X, Y, and Z are respectively shown in table 2, table 3, and table 4.

Table 2. Result of Compatibility of textbook X with Scientific Approach

Learning Indicators	Scientific Approach					%
	A	B	C	D	E	
• Grouping salt types based on their acid-base formers	√	√	√	√	√	100
• Determine the type of hydrolysis reaction that occurs in a salt	-	-	-	-	-	-
• Determine the character of the hydrolyzed salt solution	√	√	√	√	√	100
• Calculate pH of hydrolyzed salt solution	-	-	-	-	-	-
					Average	50

Information:
A: Observing
B: Questioning
C: Experimenting
D: Associating
E: Communicating

Textbook X there are only 2 learning indicators that compatibility with five aspects of scientific approach. Aspect observed that students are given the opportunity to make observations through activities: viewing, listening, and reading. Teachers facilitate students to make observations. The observing aspect is an object image as well as a clue for students to observe the object. The aspect of the questioning can be done with the teacher guiding the students to ask questions related to the observation. Aspect of experimenting or collecting information is to experiment and collect as much information as possible related problems. Experimenting aspect is contained in textbook X in the form of instructions or tasks for students to conduct an inquiry to answer questions. Associating aspect or processing information is to find patterns of information interconnected with each other and take the conclusions of the patterns found. The activities of communicating aspect is writing or telling the results found and assessed by the teacher as a result of learning. Communicating aspect is also contained in the form of commands for the task of preparing reports and presenting them. Based on the calculation, the percentage of compatibility of textbook X with scientific aspect is 50%, sufficient category.

Table 3. Result of Compatibility of textbook Y with Scientific Approach

Learning Indicators	Scientific Approach					%
	A	B	C	D	E	
• Grouping salt types based on their acid-base formers	√	√	√	√	√	100
• Determine the type of hydrolysis reaction that occurs in a salt	-	√	-	√	√	60
• Determine the character of the hydrolyzed salt solution	√	√	√	√	√	100
• Calculate pH of hydrolyzed salt solution	-	-	-	-	-	-
					Average	65

Three learning indicators in book Y are in accordance with the aspects of the scientific approach. The introduction to this textbook is the question "what if acid or alkaline or salt substances are dissolved in water?" then continued with observation activities. Indicator "Grouping salt types based on their acid-base formers" is presented with a table of some salt then students are asked to guess the forming acid and base. Indicator "Determine the character of the hydrolyzed salt solution" is the investigation of the character of some salt solutions using litmus paper. Experimenting, associating, and communicating aspects in textbook Y appear in instructions or assignments for students to conduct investigations and discussions then answer questions and make reports. Unfortunately on indicator "Determine the type of hydrolysis reaction that occurs in a salt" there are no aspects of observation and experimentation. On this indicator only preliminary questions are given but there is a full explanation and students are asked to complete and conclude. Indicator "Calculate pH of hydrolyzed salt solution" of textbook Y only presents formulas and sample questions without involving students directly. The percentage of compatibility of teaching materials Y is 65% including the sufficient category.

Table 4. Result of Compatibility of textbook Z with Scientific Approach

Learning Indicators	Scientific Approach					%
	A	B	C	D	E	
• Grouping salt types based on their acid-base formers	-	-	-	-	-	-
• Determine the type of hydrolysis reaction that occurs in a salt	-	-	-	-	-	-
• Determine the character of the hydrolyzed salt solution	√	-	√	√	√	80
• Calculate pH of hydrolyzed salt solution	-	-	-	-	-	-
					Average	20

In textbook Z shows that the application of the scientific approach is less compared to textbook X and textbook Y. This textbook presents material without preliminary questions that can stimulate students to active. Associating aspect is present in the three materials teaching, but the materials teaching X and Y have a more complete command than the teaching material Z. The materials teaching X and Y have commands to create patterns against the observations, explain them and make conclusions and link the results with relevant concepts, while the materials teaching Z there are only commands to make a conclusion. The percentage of compatibility of teaching materials Z is 20% including very poor category.

Materials teaching that are still not in accordance with aspects of a scientific approach should be added to the appropriate aspects. Materials teaching that fit the scientific aspect are expected to help teachers in the learning process, because through the application of this materials teaching teachers have implemented the curriculum 2013. Use of materials teaching with constructivism (scientific approach) proved better when compared with the traditional approach [8]. Learning by scientific approach on chemicals is effective in improving argument analysis skills, learners will be more active in asking, expressing opinions, and increasing students' curiosity about learning materials [9]. Chemical modules based on an effective scientific approach improve student achievement from aspects of students' knowledge, attitude, and skills [10].

Table 5. Results of Compatibility of Textbook X with Problem Solving

Learning Indicators	**Problem Solving**					**%**
	A	**B**	**C**	**D**	**E**	
• Grouping salt types based on their acid-base formers	√	√	-	-	-	40
• Determine the type of hydrolysis reaction that occurs in a salt	-	-	-	-	-	-
• Determine the character of the hydrolyzed salt solution	√	√	√	√	√	100
• Calculate pH of hydrolyzed salt solution	-	-	-	-	-	-
					Average	35

Table 6. Results of Compatibility of Textbook Y with Problem Solving

Learning Indicators	**Problem Solving**					**%**
	A	**B**	**C**	**D**	**E**	
• Grouping salt types based on their acid-base formers	√	√	√	√	√	100
• Determine the type of hydrolysis reaction that occurs in a salt	-	-	-	-	-	-
• Determine the character of the hydrolyzed salt solution	√	√	√	√	√	100
• Calculate pH of hydrolyzed salt solution	-	-	-	-	-	-
					Average	50

Table 7. Results of Compatibility of Textbook Z with Problem Solving

Learning Indicators	Problem Solving					%
	A	B	C	D	E	
• Grouping salt types based on their acid-base formers	-	-	-	-	-	-
• Determine the type of hydrolysis reaction that occurs in a salt	-	-	-	-	-	-
• Determine the character of the hydrolyzed salt solution	-	-	-	√	√	40
• Calculate pH of hydrolyzed salt solution	-	-	-	-	-	-
					Average	10

Information:
A: Define the problem
B: Diagnosing problems
C: Formulate alternative strategies
D: Determine and implement strategy
E: Evaluate the success of the strategy

The syntax in the learning model is the learning steps that must be fulfilled and executed in order. Based on the results of the suitability of teaching materials with problem solving steps there are still steps that have not been fulfilled.

The first step in the problem solving model is to define the problem. Problem solving depends heavily on understanding the problem itself. Defining the problem can be done by presenting the event containing the problem according to the subsequent subject learners are asked to formulate the problem. Teaching materials X and Y each provide a dish of problems that can be understood learners and there are also commands and questions that stimulate learners formulate the problem. Teaching material Z there is no problem presented clearly so that students difficult to understand the problems encountered.

The second step, diagnose the problem that learners are asked to form a small group to discuss the causes of the problem. Teaching materials X and Y there is this step in the form of commands to discuss and seek related information that causes problems. The teaching material Z does not contain this step.

The third step, formulating an alternative strategy that the group finds some alternative problem solving. Teaching materials X and Y provide an opportunity to find an alternative problem solving one of them by asking learners to plan their own experiments.

Step four, define and implement strategy. The three materials teaching there is this step that there are experimental steps and tables of observations and related questions relevant to the material.

The last is to evaluate the success of determining the effectiveness and truth of the experimental results and discussions with the related material. The three materials teaching there is this step is an invitation to create reports and class presentations so that teachers can guide learners to evaluate.

Based on the result of percentage calculation, it is found that the teaching material X has the suitability with Problem Solving step 35% including the poor category, the teaching material Y has the percentage of suitability 50% including the sufficient category, while the teaching material Z has 10% suitability including the very poor category. Materials teaching that are not in accordance with the steps of the Problem Solving model should be completed with steps and done in sequence. Applying the Problem Solving model to the learning will build students problem solving skills so that they are accustomed to apply the scientific step when finding the problem and can solve it well [11]. Problem solving skills can be applied to various learning materials so it is best to model Problem Solving applied to school learning [12].

Application of learning through scientific approach and Problem Solving model is expected to improve the learning process of learners so that will improve learning outcomes and learning objectives in the 2013 curriculum can be achieved. Competence to be achieved is a balanced competency between attitude, skills, and knowledge. Teaching materials in accordance with the curriculum is expected to form a generation that characterize and reflect the personality of the nation. Students are expected to be more active, creative, and innovative in solving any problems encountered.

4. Conclusion

Based on the analysis that the conformity of scientific approach to the teaching materials X was the sufficient category, the percentage of conformity of teaching materials Y was sufficient category, and the conformity of teaching materials Z including the category is very poor. Conformity of Problem Solving model to teaching materials X including poor category, teaching material Y has sufficient category, while teaching material Z has very poor category.

Suggestion that can be submitted from this research is the need of addition of aspect and syntax at instructional material which not yet according to scientific approach and model of problem solving so that teacher can apply learning according to curriculum 2013 by following step on the instructional material. The need for the development of materials teaching in accordance with the curriculum 2013.

References

[1] Gilbert J K and Treagust D 2009 *Multiple Representations in Chemical Education* (The Netherlands: Springer)
[2] Talanquer V 2011 *International Journal of Science Education* **33(2)** 179-195
[3] Orgill M and Sutherland A 2008 *Chem. Edu. Res. Prac.* **9** 131-143
[4] Kean E and Middlecamp C 1985 *Panduan Belajar Kimia Dasar* (Jakarta: Gramedia)
[5] Nakhleh M B 1992 *J. Chem. Educ.* **69(3)** 191-196
[6] Gulo W 2008 *Strategi Belajar Mengajar* (Jakarta: PT.Grafindo)
[7] Seguin R 1999 *The Elaboration of School Textbooks* (Unesco)
[8] Khalid A and Azeem M 2012 *International Journal of Humanities and Social Science* **2(5)** 170–177
[9] Saputri, Rosilawati and Sunyono 2015 *Jurnal Pendidikan dan Pembelajaran Kimia* **4(2)** 416-428
[10] Astuti D R, Saputro S, dan Mulyani S 2015 *Jurnal Inkuiri* **5(2)** 71–78
[11] Armağan F Ö, Sağir Ş U, and Çelik A Y 2009 *Procedia–Soc. and Behavioral Sci.* **1(1)** 2678–268
[12] Nfon N 2013 *Journal of Mathematics Education* **6(1)** 38–55

Implementating guided inquiry: The influence towards students' activities and communication skill

L A Lungan[1*] and E W Laksono[2]
[1]Graduate Programme, Yogyakarta State University
[2]Department of Chemistry Education, FMIPA Yogyakarta State University

*Email : lidwina_bi@yahoo.com

Abstract. The aim of this research is to identify the effect of guided inquiry on students' activities and communication skill in chemistry. The population was all eleventh grade high school students in SMAN 2 Wates. The samples of this study were obtained by using random sampling technique. The samples consist 48 students, with 21 students in class XI MIPA 2 as an experimental group and. 27 students in class XI MIPA 4 as control group. While the students of experimental group were taught using guided inquiry and the students of control group were taught using expository learning. Data of this research were collected by using questionnaire and observation for students' activities and communication skill. The data were analyzed by using Multivariate Analysis of Variance (MANOVA) test whereas the significant level used was 0,05. The results show that there were significant differences between students who were taught using guided inquiry and students who were taught using expository. This study demonstrates that guided inquiry is significantly more effective to improving student's activity and communication skill than the expository learning which are usually applied by teachers in class.

1. Introduction

The learning process in the curriculum applied in Indonesia has changed from teacher centered to student centered. The teaching and learning process in the 2013 curriculum provides opportunities for students to build and develop their own knowledge. But in reality, there are still schools that implement teacher-centered learning. Teachers still believe that students will become smarter if more and more know about the facts so that learning is only seen as an activity to remember / memorize facts. In such learning, students tend to emphasize the process of transferring knowledge rather than building knowledge [1]. This causes students to become less involved in building their knowledge.

The teacher has a role to create learning that is fun and interesting so that students can be actively engaged in learning activities. For example using appropriate learning methods or models. Inquiry is learning where students find and use various sources of information and ideas to improve understanding of a problem. Students do not merely submit and answer questions but are supported by investigation, exploration, search, quest, scientific research and study [2]. Muslich[3] reveals the characteristics of inquiry learning emphasize on the activities of students maximally to search and find, so that students act as the subject of learning. There are several types of inquiry, such as confirmation, structured inquiry, guided inquiry and open inquiry[4]. Guided inquiry used in this study is learning based on constructivist theory where knowledge is built into the minds of students[5]. Guided inquiry learning is carried out with the teacher as a facilitator to guide and direct students in classroom activities.

Communication skills is one skill competencies that must be owned by human resources in the 21st Century (21st Century Learning Partnership Framework). Communication is an activity of exchanging

IOP Conf. Series: Journal of Physics: Conf. Series **1156** (2019) 012025 doi:10.1088/1742-6596/1156/1/012025

information both verbally and nonverbally [6]. Communication skills can be developed through collaboration and discussion in learning. In the steps of inquiry learning, communication can be trained through a discussion process in finding problem formulations, conducting investigations and conclusions. In this step, students are trained to be able to interact well with friends or teachers.

Several studies that have implemented guided inquiry in learning have been conducted by several researchers. Mohammed, et al. [7] showed that there was an effect of guided inquiry learning and modified free inquiry on learning achievement. Conway [8] pointed out that the final exam score in the group that used guided inquiry was higher than the group using half guided inquiry and non-guided inquiry. Bilgin [9] conclude that students have develop their performance of acid and base concepts and positive attitude toward guided instruction better than students are studies with alone. Guided inquiry learning was indicated as a strong positive predictor for students achievements, and its effect was also positively associated with student interest [10].

The main purpose of this study was to investigate the effects of guided inquiry learning toward students' activity and communication skill. The study was done by following research question: (1) Is there a significant mean difference between experimental group and control group on students' activity and communication skill? (2) Is there an effect of guided inquiry on studens' activity and communication skill ?

2. Method

Post-test only control group quasi experimental design was used in this study. This study aims to determine the effect of inquiry learning on the activities and communication skills of students.

This research was conducted at a high school in Kulon Progo district in the second semester of 2017. Population in this study was all eleventh grade students in SMAN 2 Wates. With randomly technique sampling, forty eight students from two classes were used as samples, namely the XI MIA 2 class as the experimental group using the guided inquiry learning and the XI MIA 4 class as a control group using expository learning. The research design is shown in table 1.

Table 1. Research design

Group	Treatment	Postes
Experiment	X	A_1B_1
Control	-	A_2B_2

Description :
X : Guided inquiry learning
A_1 :Communication skill questionnaire after applied guided inquiry learning
A_2 :Communication skill questionnaire after conventional learning
O_1 : activity questionnaire after applied guided inquiry learning
O_2 : activity questionnaire after conventional learning

The study was conducted for 4 weeks in each class. The experimental class is carried out using guided inquiry learning, where students find knowledge/information related to the subject independently with the guidance of the teacher. While in the control class that uses expository learning, students listen to the teacher's explanation of the subject in front of the class. Data collection was carried out using questionnaires of communication skills and activities provided as post test. In addition to using questionnaires, observation sheet is used to see students' activities and communication skills were also carried out during the learning process. The activities of students observed in this study are visual activity, listening activities, writing activities and motoric activities. Based on the observation, at the initial meeting (first meeting), the activities of students in the control and experimental classes were not much different. But at the next meeting, the activities of the two classes looked different. This difference can be seen from the activeness of students when conducting group discussions, practicing, and looking for reference sources. In practicum activities, experimental students found their own procedures for

IOP Conf. Series: Journal of Physics: Conf. Series **1156** (2019) 012025 doi:10.1088/1742-6596/1156/1/012025

conducting lab work from various sources/references. Whereas in the control class, practicum procedures have been prepared.

The data obtained were analyzed using MANOVA to see whether or not there were differences in activity and communication skills between the control class and the experimental class. Subsequently, partial eta-squared calculations were performed to see how much influence inquiry learning had on activities and communication skills.

3. Discussion and Result

After the datas collected, it will be proceed with manova statistical analysis. However, before the manova test is carried out, multivariate normality test and homokedasticity must be tested first as a prerequisite test. Normality test in this study using mahalanobis vs chisquared statistics using SPSS 20 for windows. The percentage of mahalanobis values from both classes was 59% for the experimental class and 52% for the control class. This values shows that data from both classes is normally distributed. Then, the homogeneity test using the box's test of equality of covariance matrix, obtained a significance value of 0.32 ($p < 0.05$). this value indicates that the data is homogeneous.

Table 2. Descriptive Statistics

Group	Variable	Mean
Experimental	Activity	69,6653
	Communication skills	36,206
Control	Activity	65,2192
	Communication skills	32,3502

The results of the analysis using manova techniques with the Hotelling's trace test can be seen in table 3.

Table 3. Summary of MANOVA

	Effect	F	Sig.
Model	Pillai's Trace	3.485^b	.039
	Wilks' Lambda	3.485^b	.039
	Hotelling's Trace	3.485^b	.039
	Roy's Largest Root	3.485^b	.039

Based on the results in the table, Hotelling's Trace shows that the significance value is 0.039 ($p < 0.05$). These results indicate that at a 95% confidence level, there is a significant difference between experimental class that use guided inquiry and control classes that use conventional learning in terms of activity and communication skills. Furthermore, to see the influence of the learning model on each dependent variable (activity and communication skills) can be seen in the results of the Test of Between Subject *Effects* on the table 4.

Table 4. Test of Between Subjects

Dependent Variable	Mean Square	F	Sig.
Activity	233,503	4,550	,038
Communication	175,626	5,601	,022

The table shows that the significance value for student activity is smaller than the significance level (0.05). This shows that there is a significant difference between the control class and the experimental class in terms of the activities of students. The same thing can be seen in communication skills that show a significance value of 0.022 ($p < 0.05$), which shows that there are significant differences between experimental class that use guided inquiry learning and control class that use conventional learning. To see how much influence the learning model has on each variable, Partial Eta Squared is used. Table 5 shows the partial eta squared values for activity variables and communication skills, respectively 0.090 and 0.109, meaning that inquiry learning affects student activity by 9% and communication skills by 10%.

Table 5. Partial Eta Squared

Source	Dependent Variable	Sig.	Partial eta-squared
Intercept	Activity	.000	0.989
	Communication	.000	0.975
Model	Activity	.038	0.090
	Communication	.022	0.109

Activity data and communication skills were obtained from questionnaires given to both classes after being given treatment. The results of the data analysis showed that there were significant differences in activity and communication skills between classes using guided inquiry and classes that were taught using expository. These results are supported by observations on communication activities and skills that are carried out during the learning process. The difference in the average student activity between experiment and control class can be seen in Table 2. Students who use expository learning are less involved because in learning activities, the teacher is more dominant in directing lessons and solving a problem. Guided inquiry learning is developed based on constructivism learning theory [11-13]. Student-centered learning in constructivism theory allows students to be more active in the process of discovering their knowledge. Teachers are advised to use student-centered learning such as problem based learning, peer lead team learning, or process oriented guided inquiry learning [14]. Learning process that involves students can minimize passive students so that students can express their ability to develop optimally.

Communication skills in this study are verbal communication in expressing opinions, questions and presentations. Students in the experimental class look more active and enthusiastic in presenting results, expressing opinions and giving responses to other groups. The average communication skills of the experiment and the control class can be seen in Table 5. These results are consistent with the research conducted by Iswatun, et al.[15] which shows the inquiry learning model improves communication skills as one of the aspects of science process skills. In the inquiry learning, students are trained to communicate through discussion activities so that communication skills can be improved.

The implementation of guided inquiriy learning in this study is not optimal because students and teachers are not used to applying it in daily learning. Guided inquiry learning is a good approach for teachers who have not experienced implementing open inquiry [16]. Koksal & Berberoglu [17] revealed that guided inquiry learning is the transition from teacher centered to student centered. So to get optimal learning, students and teachers must use guided inquiry learning periodically.

4. Conclusion

Based on the results and discussion, it can be concluded that guided inquiry learning affects students' activities and communication skills. In more detail, the results of the study indicate that there are significant differences in communication activities and skills between the experimental class and the control class.

Researchers suggest that guided inquiry learning can be done periodically so that it can be applied more optimally in order to enhance students'activities and communication skill. In addition, further research needs to be carried out using guided inquiry models to improve other aspects of the learning process.

References

[1] Winkel W S 2004 *Teaching Psychology* Jakarta PT Gramedia
[2] Kuhlthau C C, Maniotes L K, Caspari A K 2007 *Guided Inquiry: Learning in the 21st Century*

Westport Libraries Unlimited
[3] Muslich M 2008 KTSP *Pembelajaran Berbasis Kompetensi dan Kontekstual* Jakarta Bumi Aksara
[4] Staver J R and Bay M 1987 *J.l of Res. in Sci.Teach* .**24** 629-643
[5] Bodner G 1986 *J. Chem. Edu.***63** 873-878
[6] Cook R and Weaving H 2012 *Key Competence in School Education in Europe* Brussels European Schoolnet
[7] Mohammed A. Baskoro A P and Widha S 2015 *J. Inquiry* **4**3 127-135
[8] Conway C J 2014 *J. of Chem. Edu* **91** 480-483
[9] Bilgin I 2009 *Sci. Research and Essay* **49** 1038-1046
[10] Kang J and Keinoen T 2017 *Research in Sci.Edu.* **48** 865-885
[11] Bruner J S 1961 The act discovery *Harvard Edu. Re.***31** 21-32
[12] Dewey J 1997 *How We Think* New York NY Dover Publications
[13] Vygotsky L S 1962 Thought and Language Ambridge MAMIT Press
[14] Eberlein T,Kampmeier J, Minderhout V, Moog R S, Platt T, Varma N P and White H B 2008 *Bio.Mol.Biol.Edu.* **36** 262-273
[15] Iswatun I, Mosik M & Subali B *J. Inovasi Pend. IPA* **3** 151-160
[16] National Research Council 2000 *Inquiry and National Science Eeducation Standards.* Washington DC National Academy Press
[17] Koksal E A and Berberoglu G 2014 *International J. Sci. Edu.* **36** 66-78

Acknowledgments
This research was supported by Kemenrisdikti through "Pascasarjana Tim" research grant 2018.

IOP Conf. Series: Journal of Physics: Conf. Series **1156** (2019) 012026 doi:10.1088/1742-6596/1156/1/012026

Analysis of students' scientific literacy in contextual-flipped classroom learning on acid-base topic

M Paristiowati[1,*], T Hadinugrahaningsih[1], A Purwanto[1] and P A Karyadi[1]

[1]Department of Chemistry Education, Faculty of Mathematics and Natural Sciences, Universitas Negeri Jakarta, Jl. Rawamangun Muka, Jakarta 13220, Indonesia

[*] E-mail: maria.paristiowati@unj.ac.id

Abstract. This study aims to determine students scientific literacy skills after learning with a contextual-flipped classroom model. In flipped classroom-contextual learning, students study the lesson material through contextual videos outside the classroom, then conduct discussions in class and lab work. This study used a qualitative method with 36 students in class XI at senior high school in Jakarta. Data collection were carried out through scientific literacy tests, observation, interviews, and reflective journal of students. The students' scientific literacy skills in this study refer to the scientific literacy of PISA. The results showed that 16.7% students had achieved scientific literacy at level 6; 38.9% at level 5; 11.1% at level 4; 27.8% at level 3; 5.6% at level 2 and 1. In addition, the results of the study show that aspects of scientific literacy have also been mastered by students. Mastery of scientific literacy aspects, concerning: aspect of the scientific context is 83.92%, aspect of science knowledge is 84.36%, and aspect of scientific competence is 84.06%. Thus, it can be concluded that acid-base learning with a contextual-flipped classroom model can improve students scientific literacy skills.

1. Introduction

Scientific literacy according to the Program for International Student Assessment (PISA) is the ability to engage in issues related to science and use these abilities to identify questions, acquire new knowledge, explain scientific phenomenon, and draw conclusions based on evidence about science related issues. Miller describe that literacy is the ability to read and write in order to communicate with society, so that scientific literacy is defined as the level of understanding scientific and technological concepts which is used to take the role of a member in the modern industrial society [1].

Based on the results of the 2015 PISA study, the students literacy in Indonesia are still low, and Indonesia is ranked 64th out of 72 countries with an average score that is still far below the International standard, which is 500. This research show that most of the students have not been able to apply scientific knowledge to complex life situations with a high cognitive level, but only able to apply scientific knowledge to life situations with a low cognitive level. Based on this, the teacher have to improve the students scientific literacy through the learning process carried out in school.

The learning model that is currently developing is the flipped classroom, which is a learning model that makes the learning process in the classroom becomes more effective, by "reversing the class". According to Lage, Platt, and Treglia, "flipping classroom" is a learning activity that is doing inside classroom, now it is doing outside classroom [2]. Submission of information can be done using online media, such as reading or watching learning videos made by the teacher. Furthermore, when in the classroom there will be reciprocal activities, such as discussion, problem solving or tests.

Some studies reveal that the use of the flipped classroom model in schools has a positive impact on student achievement when compared to ordinary classes [3]. In chemistry subjects, the flipped classroom learning model can improve student learning outcomes [4]. Flipped Classroom models combined with active learning strategies can improve the performance and attitudes of students [5]. The flipped classroom model combined with active learning, such as inquiry, project based learning, collaborative learning, can improve student learning outcomes [6–8].

In this study, a combination of flipped classroom model with contextual learning was conducted to find out the students' scientific literacy. In contextual learning, context has more meaning than just the connection of certain physical environments at certain times, including the mental and emotional context of each individual, social context and cultural context. Thus, contextual understanding has a broader meaning than applicative learning [9]. So if the flipped classroom model combined with contextual learning, this might develop students' scientific literacy, because the focus of learning is not only on results but on the process of acquiring science itself. The implementation of the flipped classroom-contextual learning model in this study was conducted on acid-base material that has many applications in everyday life.

2. Method

This study aims to determine students' scientific literacy in chemistry learning on acid-base topic using contextual-flipped classroom learning model. Acid-base learning in this study uses 3 videos of learning, i.e acid-base, acid rain, and acid-base experiment. The students' scientific literacy skills refer to the scientific literacy of PISA. The subjects of the study were 36 students of class XI at senior high school in Jakarta. The research method used in this study is a qualitative method. Data collection techniques were carried out through scientific literacy tests, interviews, observation, and reflective journals of students. Then the data were analyzed through stages: data reduction, coding, data presentation, and conclusion.

3. Results and Discussion

The research data obtained from reflective journals, interviews, and observations were reduced and then coded and classified into five parts, i.e students' scientific literacy in the contextual-flipped classroom learning, the activeness of students in contextual-flipped classroom learning, the level of difficulty of the contextual-flipped classroom learning model, strengths and weaknesses of the flipped classroom-contextual learning model.

3.1. Students' Scientific Literacy in Contextual-Flipped Classroom Learning Model

Based on the results of the study, student's scientific literacy scores were obtained which showed the achievement of scientific literacy levels. Students who have reached level 6 are 16.7%, meaning that students can consistently identify, explain and apply science knowledge and knowledge about science in various complex life situations, can connect information sources with different explanations and use scientific facts to make decisions. Students who have reached level 5 are 41.7%, meaning that students can identify scientific components in complex situations, apply both concepts of science knowledge and knowledge of science and can compare, choose, and evaluate scientific evidence that is appropriate response to life situations. Students who reach level 4 are 11.1%, meaning that students can work effectively with situations and problems that may involve explicit phenomena and require students to draw conclusions about the role of science and technology. Students who have reached level 3 are 27.8%, meaning that students are able to identify problems clearly at the level of the scientific context. Students can solve facts and knowledge to explain phenomena and apply simple models or inquiry strategies. Students at this level are able to interpret and use scientific concepts from scientific disciplines and are able to apply them directly. Students are able to form short statements by using facts to make decisions based on scientific knowledge. Students who reach level 2 and level 1 are 2.8%, meaning that students have sufficient scientific knowledge to explain in context or draw conclusions based on simple investigations and students have limited scientific knowledge, so that they can be applied in familiar situations. Figure 1 is students's achievement towards scientific literacy.

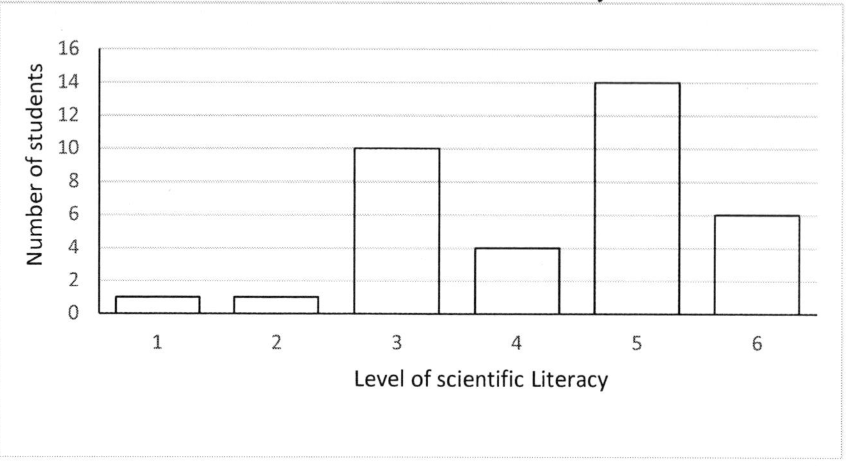

Figure 1. Achievement of Scientific Literacy Level

The data of scientific literacy score is also used to analyse students' ability to answer questions in every aspect of scientific literacy. There are three aspects of scientific literacy, namely aspects of context, knowledge and competence, with results as shown in figure 2.

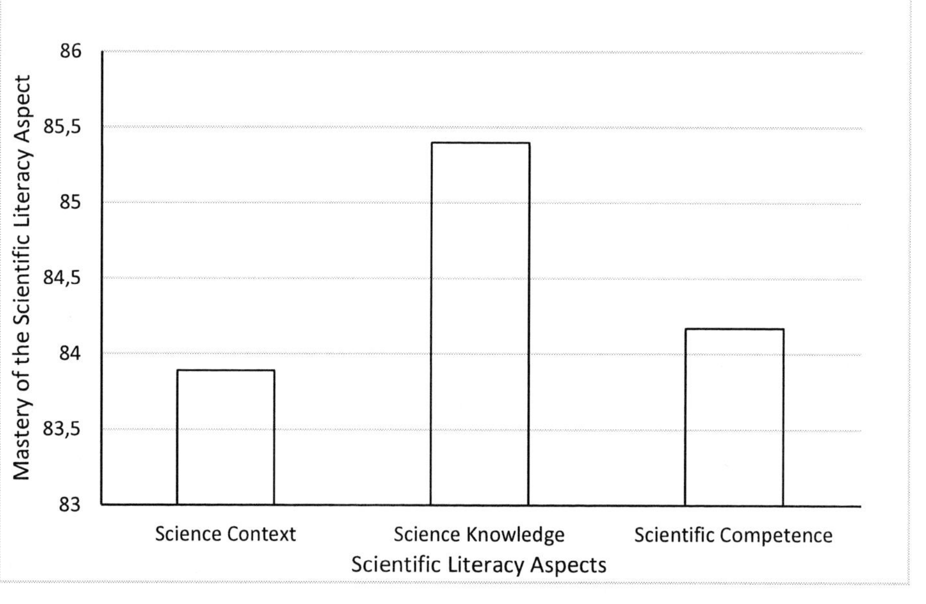

Figure 2. Achievement of Scientific Literacy Level

The scientific literacy test consists of 15 items covering the context of science, scientific knowledge and scientific competence. Description of the achievement of aspects of scientific literacy are shown in table 1.

Table 1. Achievement of the Scientific Literacy Aspect

Scientific Literacy Aspects	Component	Percentage of Knowledge Mastery	Avarage of Knowledge Mastery
Contexts	Global	79.57	83.89
	Personal	89.70	
	Social	82.23	
Knowledge	Knowledge of science	89.68	85.40
	Knowledge about science	81.12	
Competencies	Evaluate and design scientific enquiry	88.12	84.17
	Explain phenomena scientifically	72.49	
	Interpret data and evidence scientifically	91.83	

The data in figure 2 shows that students have mastered 83.89% aspects of the science context, which means that students are able to deal with scientific problems in a personal, social and global context well. The knowledge aspect of science is dominated by 85.40% of students, which means that students are able to conduct scientific investigations and scientific exposures. This is in accordance with the results of observations that show students able to investigate the scientific process and the components contained in the science process when conducting acid-base practicum. During the discussion, it was seen that most of the students were able to explain the results of the science process that had been carried out in the practicum. Thus, students have mastered aspects of science knowledge well.

While the aspect of scientific competence has been mastered by students as much as 84.17%. The scientific literacy description mastered by students in this aspect shows that, students can answer questions about identifying scientific problems, be able to recognize problems, be able to identify keywords to search for science information and students are able to investigate science. In addition, students are also able to answer questions about explanation and predict phenomena scientifically, able to apply scientific knowledge in given situations. Learners are also able to describe and interpret phenomena scientifically and predict their changes, be able to interpret scientific facts, make and communicate the conclusions.

3.2. Activeness of Students in Contextual-Flipped Classroom Learning Model

A contextual-flipped classroom learning model that presents independent learning by watching videos at home, making students actively learn independently at home. When the discussion in class also adds to the activeness of students. Here are some reflective data of student journals.

> *"This learning makes the classroom atmosphere not rigid, pleasant and warm. Using videos and giving quizzes make learning not boring"*
> (Student 5, 30 January 2018)
> *"In learning there is a lot of discussion, so the class atmosphere becomes more alive and active"*
> (Student 9, 30 January 2018)

Students also argue that this learning model makes students who have introvert characters become extrovert, because this model make all students play a role in the learning process to express their thoughts. This learning model makes it easy for students to get information through discussion. Students who are ashamed to ask the teacher, can ask friends through discussion. The teacher acts as a facilitator who facilitates group learning, answers questions and is ready to provide information when students have not gotten information maximally with a group of friends.

3.3. The Difficulty Level of the Contextual-Flipped Classroom Learning model

Based on data obtained from reflective journals and interviews, it was found that some students did not recognize the flipped classroom contextual learning model, so that students needed time to get used to this learning model. However, some students experienced ease in learning with the contextual-flipped classroom learning model. Learners feel that the learning presented through video learning before learning in class makes students' understanding better and supports the readiness of students in learning material in class. Here are some examples of student reflective journal data.

> "*This method is fun, because we learn through video and when in class we also form groups for discussion. It also makes us easier to understand because the learning is not only from printed books or with teachers writing on the board*"
> (Student 15, 30 January 2018)
> "*Because the video uses audio and visual facilities so that it helps students more easily understand the lesson*"
> (Student 21, 30 January 2018)

According to students, this learning model is interesting because it is not as usual, which only focuses on delivering the teacher through verbal and writing on the board. This learning model supports students who have an audio-visual learning style, so that they are more motivated to understand the material. This learning model also supports students who are accustomed to independent learning at home. Through this learning model, students are trained to discuss and collaborate, thus helping students to understand the material more deeply.

3.4. The Advantages of the Contextual-Flipped Classroom Learning Model

Some students are motivated in this learning model, because the discussion models conducted in class make students understand better to the topic, can share knowledge with friends who do not understand and they can ask to their friends who already understand the topic. In this learning model there is interaction between students, exchanging ideas and opinions. When students face difficulties, they can ask the teacher directly so that there is also interaction between students and teachers.

According to students, this learning model provides another environment in the learning process, which usually students just listening to the teacher convey the topic, now they can play an active role directly in the learning process. The advantage of this contextual-inverse classroom learning model is in accordance with Love's research in which learning models make students have basic knowledge to discuss in class and the teacher can maximize the time to examine and pay attention to students' understanding [10].

3.5. Weaknesses of Contextual-Flipped Classroom Learning Models

The contextual-flipped classroom learning model requires students to access and learn the subject matter independently outside the classroom. Students who are used to learn with being guided directly will face difficulties. The use of learning video that can be accessed online through YouTube makes students have to buy enough internet quota, because many students do not have Wi-Fi at home. When learning in class, discussion cannot effectively if not all students watch the video given. This result is in accordance with Brown's research which states that the independent learning at home makes students have more burden than normal learning [11].

4. Conclusion

Based on the results and discussion in this study, it can be concluded that students have achieved scientific literacy from level 1 to level 6. The most scientific literacy mastery is at level 5, which means students can identify scientific components in complex situations, applying both concepts science knowledge and knowledge about science and can compare, choose, and evaluate scientific evidence that appropriate to respond to life situations. Thus, it can be concluded that contextual-flipped classroom Learning model can be used to improve students' scientific literacy.

References

[1] Meinwald J and Hildebrand J G 2010 *Science and the educated American: A core component of liberal education*. American Academy of Arts and Sciences Cambridge MA pp. 241–255.
[2] Lage M J, Platt G J and Treglia M 2000 *J. Econ. Educ.* **31** 30–43.
[3] Zainuddin Z and Halili S H 2016 *Int. Rev. Res. open Distrib Learn* **17** 313–340.
[4] Paristiowati M, Cahyana U and Fitria A 2017 *Indonesian J. Educ. Rev.* **4** 148–152.
[5] Entezari M and Javdan M 2016 A *Int. J. High Educ.* **5** 222–231.
[6] Paristiowati M, Fitriani E and Aldi N H 2017 *Conf. Proc. of ICREAMS* AIP Publishing, p 30006.
[7] Paristiowati M and Nurtanti A 2017 *Proc. of 2017 Int. Conf. on Educ. and E-Learning*. ACM, pp. 22–25.
[8] Foldnes N 2016 *Act. Learn. High Educ.* **17** 39–49.
[9] Siregar E, Hara H and Jamludin 2010 *Teori belajar dan pembelajaran*. Ghalia Indonesia
[10] L Betty, H Angie, C Cynthia and E Dana C 2015 *Primus* **25** 745–762.
[11] Brown C A, Danvers K and Doran D T 2016 *Account Educ.* **25** 256–271.

Acknowledgments

This work was supported by Universitas Negeri Jakarta and Ministry of Research, Technology and Higher Education of the Republic of Indonesia through research fund 2017-2018.

The effect of science-technology-society (STS) model on scientific literacy and scientific attitude of students on the subject of buffer

M G Devi[1*] and N Aznam[2]

[1]Chemistry Education Department, Postgraduate Program, Universitas Negeri Yogyakarta, Indonesia
[2] Chemistry Education Department, Faculty of Mathematics and Natural Sciences, Universitas Negeri Yogyakarta, Indonesia

[*] E-mail: mentari.gayatridevi@gmail.com

Abstract. Scientific literacy and scientific attitudes are two conceptual foundations for students that have to be increased in both developing and developed countries. This research aims to identifythe effect of science-technology-society (STS) model on scientific literacy and scientific attitude of students on the subject of buffer solution. This research was quasi-experimental research using nonequivalent control group design. The sample was determined by cluster random sampling technique. The sample of this study was two classes. One class as an experimental class consisted of 34 students which used STS model. Other class as a control class consisted of 33 students which used direct instruction model. The data analysis used the manova test at the significance level of 5%. The results of the study showed that there was significant effect of STS model on scientific literacy and scientific attitude of students on the subject of buffer solution.

1. Introduction

Indonesia is listed in one of the international evaluation programs which are the Program for International Student Assessment (PISA) organized by the Organization for Economic Cooperation and Development (OECD). The conceptual foundation in PISA emphasizes: scientific literacy. The term scientific literacy was first stated by Paul De Hart, an education expert, in 1958. Hurd defined scientific literacy as a way of understanding science and the application of science in everyday life [1]. Scientific literacy is a person's ability to understand science, communicate science both verbally and in writing and apply scientific knowledge to solve problems so that they have a high attitudes sensitivity towards themselves and their environment in making decisions based on scientific considerations [2]. PISA [3] defines Scientific literacy as:

> "the capacity to use scientific knowledge to identify questions and to draw evidence-based conclusions to understand and help make decisions about the natural world and the changes made to it through human activity"

PISA also explains that there are three competencies used to measure scientific literacy: (a) identifying scientific issues; students recognize scientific issues or problems for the possibility of conducting scientific investigations and recognizing keywords from a scientific investigation, (b) explaining scientific phenomena; students will apply science in certain situations to describe or interpret scientific phenomena and predict change, students also compile descriptions correctly and can provide explanations, (c) using scientific evidence [4]. Students interpret and describe the evidence for drawing conclusions as well as providing assumptions, explanations, and evidence from identified conclusions. Students also consider the implications of science and technology.

By participating in the PISA program, the government gets an overview of the ability of Indonesian students in reading literacy, mathematical literacy, and scientific literacy so that the results can be used as material in the formulation of policies to improve the quality of education in Indonesia. The results shown by PISA show that the ability of scientific literacy of Indonesian students is still deficient. The OECD report shows that Indonesian students' scientific literacy ranked 38th out of 41 countries in 2000, 48th out of 40 countries in 2003, 53rd out of 57 countries in 2006, 38th out of 40 countries in 2009 and ranked 64th out of 59 countries in 2012. From these data, it can be seen that Indonesian students have very low scientific literacy compared to other countries.

In addition to the problem of students' literary abilities, with the rapid advances in science and technology that bring great changes in the quality of community life, new problems arise that are related to global ethics, morals, and issues that threaten human survival [5]. The attitude of students is also one of the problems that must be considered in this case is scientific attitudes. Scientific attitudes is the impulse of feelings and beliefs that arise from within a person to behave or do something to an object by referring to the problem-solving procedure by using the steps of the scientific method [6].

Harlen reveals that the scientific attitudes contain two meanings, attitude toward science that describes a person's attitude to science, and attitude of science that describes the attitude inherent in a person after learning science [7]. PISA also describes scientific attitudes competencies, including indicating interest in science, supporting scientific inquiry, and motivation to act responsibly towards natural resources and the environment.

Based on the problems that have been described, it is necessary to have a learning model that can link scientific issues with technological developments to improve students' literacy skills and their relationship with society. Learning model that aims to make students also care about what is happening in the natural environment and can preserve the environment. One learning model that is considered to meet this need is the Science-Technology-Society learning model (STS).

The National Science Teacher Association (NSTA) defines STS as a learning model that links technology and science in the context of everyday life [8]. NSTA develops STS as a goal of science education. Learning science aims to develop scientific literacy for every individual who understands science, technology, and society and uses his knowledge in making decisions [9]. STS learning emphasizes student involvement in experiences and issues that are directly related to their lives. STS develops students with skills that enable them to become active, and responsible in responding to issues that affect their lives.

Poedjiadi suggests that the purpose of the STS learning model is to form individuals who have scientific and technological literacy and have a concern for the problems of society and the environment [10]. Learning using STS emphasizes student involvement in experiences and issues that are directly related to their lives. STS develops student skills that enable them to be active, and responsible for responding to issues that affect their lives.

Miller states that the STS model can make a substantial contribution in preparing students to face the latest technological issues and changes in science in society [11]. By linking science, technology, and society, it is hoped that this model will not only affect changes in scientific attitudes in a positive direction, but also teaches students to think about the long-term effects of using science.

Based on these data, the research is conduct by implementing the STS model to determine the effect of this model on students' scientific literacy and scientific attitudes of learners on buffer solution.

2. Method

This study is using a quasi-experimental approach to assessing treatment effects in the nonequivalent control group design [12]. The design of this study was chosen based on the purpose of the study to determine the effect of the STS learning model on the science literacy and science attitudes.

This research was conducted at SMA Negeri 4 Kota Ternate. The population in this study is 11th grade students of SMA Negeri 4 Kota Ternate majoring in science with the total population of 173 students. To determine the research sample, cluster random sampling method was used. Both written test and non-test techniques are used as the data collection methods in this study. Tests were conducted to determine the ability of students' scientific literacy before and after the implementation of learning using STS learning models and Direct Instruction (DI) model. Non-test techniques are carried out to measure students' scientific attitudes by questionnaire.

Manova then analyzed the results of the data found in the studymanova. This data analysis method is used to conclude the results of the study based on the purpose of the study to determine the effect of STS learning models and Direct Instruction (DI) models in improving the ability of scientific literacy and scientific attitudes of students. Manova is used to test hypotheses using the SPSS program. Before carrying out the analysis, the test of the normality of the data distribution and the homogeneity of the variants were done first on the data that has been obtained.

The statistical hypothesis test proposed in this study is testing the hypothesis. The hypothesis aims to examine whether or not there is an influence of the STSmodelon the students' scientific literacy and scientific attitudes. Multivariate test statistics in this study used T^2 Hotelling.

3. Results and Discussion

Data on scientific literacy in this study was measured using an essay test instrument consisting of pretest and posttest data on experimental and control classes. Data on the scientific literacy in the STS class and DI class were then analyzed and summarized in table 1.

Table 1. Scientificliteracy ability of STS class and DI class

No.	Description	STS Class		DI Class	
		Pre-test	Post-test	Pre-test	Post-test
1	Mean score	15.99	80.21	13.76	72.35
2	Minimum score	8.33	66.67	8.33	60.42
3	Maximum score	22.92	89.58	20.83	83.33

The scientific attitudes data in this study were obtained from the results of questionnaires filled out by students at the end of the learning process. The questionnaire consists of 18 statement items and filled by the STS class with the number of 34 students as respondents and the DI class with the number of 33 students as respondents. The data was then analyzed and the results of the data analysis are summarized in table 2.

Table2. Results of the science attitude questionnaire

No.	Description	Scientific attitude questionnaire	
		STS class	DI class
1.	Mean score	72.0	66.5
2.	Minimum score	61.0	51.0
3.	Maximum score	84.0	79.0

Hypothesis testing in this study uses manova with the help of the SPSS program.Before testing the research hypothesis, a prerequisite test was done for the normality test and homogeneity test.Test of normality of scientific literacy data and scientific attitude in this research using Kolmogorov-Smirnov test.The results of the normality test of scientific literacy and scientific attitudes obtained the significance value of all variables more than the value of α (0.05) which means that the scientific literacy data and scientific attitudes of the experimental class and control class are normally distributed. The homogeneity test of scientific literacy data and scientific attitude in this research use

IOP Conf. Series: Journal of Physics: Conf. Series **1156** (2019) 012027 doi:10.1088/1742-6596/1156/1/012027

Levene test. The results of the homogeneity of scientific literacy and scientific attitudes obtained the significance value of all variables more than α (0.05) which means that the scientific literacy data and scientific attitudes of the experimental class and the control class have the same variance.

The hypothesis in this study was used to test the effect of STS model on scientific literacy ability and scientific attitude by using manova test. Based on the Multivariate Test table (table 3) obtained a significance value of 0.000 less than α (0.05) which means that there is an influence of STS learning model on the ability of scientific literacy and scientific attitude of students.

Table 3. Multivariate Tests table

Effect		Value	F	Hypothesis df	Error df	Sig.	Partial Eta Squared
Intercept	Pillai's Trace	.997	12542.717^b	2.000	64.000	.000	.997
	Wilks' Lambda	.003	12542.717^b	2.000	64.000	.000	.997
	Hotelling's Trace	391.960	12542.717^b	2.000	64.000	.000	.997
	Roy's Largest Root	391.960	12542.717^b	2.000	64.000	.000	.997
Class	Pillai's Trace	.456	26.800^b	2.000	64.000	.000	.456
	Wilks' Lambda	.544	26.800^b	2.000	64.000	.000	.456
	Hotelling's Trace	.837	26.800^b	2.000	64.000	.000	.456
	Roy's Largest Root	.837	26.800^b	2.000	64.000	.000	.456

This research was conducted on SMA Negeri 4 Kota Ternate in the second semester on the subject of buffer solution. The learning in the experimental class uses the STS model and the learning in the control class using the DI model. The learning process undertaken in this study aims to improve the ability of scientific literacy and scientific attitudes. Learning was carried out as many as 5 meetings for the experimental class and control class.

The influence of STSmodel on scientific literacy skills and scientific attitudes can be seen from the results of the manova test. Manova test results show that the significance value 0.000 is smaller than α (0.05) so it can be concluded that the use of the STS model affects the students' scientific literacy skills and scientific attitudes of learners on buffer solution. This influence is due to the use of STSmodel which according to Poedjiadi can form individuals who have scientific literacy and have concerns about the problems around them. By having good scientific literacy, students are not only able to read and write science but are aware of its impact and care for the social environment and the surrounding environment.

The result can be concluded that there is an influence of the STM learning model on scientific literacy skills. This effect can be seen from the average of scientific literacy skill on the experimental class using the STS model which is higher than the control class using the DI model. The average value of the STS class is 80.21 with a minimum score of 66.67 and the maximum value is 89.58 while the DI class average is 72.35 with a minimum value of 60.42 and the maximum value is 83.33. This difference is in line with the research of Kapici et al. in that students included in the classroom applying the STS were found better able to apply the concept of science learned into everyday life [13].

According to Akgul, STS is a learning model designed to improve students' scientific literacy skills [14]. This is supported by the research of Rahayuni which proves that the STS is better than the problem-based learning model to improve scientific literacy skills, because in STS syntax there are phases where students are asked to associate the concepts that have been learned with the issues or phenomena that occur in daily life and the surrounding environment so students are trained to link science with the surrounding phenomena [15]. The research of Jin & Bierma prove that STS can improve students' scientific literacy skills [16]. This increase is because in the STS contained a method of scientific inquiry wherein the concept formation stage students are directed to find and solve problems related to science, technology, society, the environment including the history and nature of science [17].

The use of STS learning model also affects the students' scientific attitude. This can be seen from the questionnaire that the students filled at the end of the learning. From the results of the questionnaire analysis, the average score of the scientific attitude of students in the experimental class was found higher than the control class. The average scientific attitude of the STS class is 72.0 with a minimum score of 61.0 and a maximum score of 84.0 whereas in the DI class, the average score is 66.5 with a minimum score of 51.0 and a maximum score of 79.0. This result is in line with the research of Mandra; Jumantoro; Kartini et al. that the STM learning model has a significant effect on students' scientific attitudes [18-20].

Akcay & Akcay explained that with the STS model, teachers can create student-centered learning based on the surrounding environment so that they can hone critical thinking skills, problem-solving skills and improve students' scientific attitudes [21]. In addition, after participating in learning by using STS syntax students are more interested in learning science and allows for a career related to chemistry, this can be seen from the students' responses to the questionnaire given. In item 11 which is a description of aspect 3 shows that the students in the experimental class as many as 7 students answered strongly agree, 19 students agree, while 3 students were still hesitant and 5 students answered disagree. This result is in line with Chen et al. which shows that the existence of a positive attitude towards science correlates with students' commitment to science so that it can affect their interest in science and more likely to choose a career in science [22].Relevant research conducted by Chonkaew et al. found that the majority of students studying with STS and environmental model have a positive attitude toward science [23]. Xiao & Sandoval also said that science-centered learning will produce positive results on students' interest in science [24].

4. Conclusion

Based on the results of the research and explanation that has been done, it can be concluded that there was a significant effect of STS model on scientific literacy and scientific attitude of students on the subject of buffer solution. The majority of students studying with the STS model have a positive attitude toward science. With STS students do not only listen to teacher lectures, but students also organize and summarize their knowledge so that students can better understand in depth the principles and theories they learn. The STS model is expected that students are motivated to find more additional information as well as relevant scientific evidence to support it in making decisions related to social-science issues and with the knowledge and the amount of information obtained will affect their attitude to be more responsible to the environment and natural resources.

References

[1] Turiman P, Omar J, Daud A M and Osman K 2012 Fostering the 21st Century Skills through Scientific Literacy and Science Process Skills *Procedia – Social and Behavioral Sciences* 59 pp 110-16

[2] Toharudin U, Hendrawati S, and Rustaman A 2011 *Membangun Literasi Sains Peserta Didik* (Bandung: Humaniora) p 7

[3] PISA 2000 *Preparing Students fo PISA: Science Literacy Teacher's Handbook*

[4] OECD 2007 *PISA 2006 Science Competencies for Tomorrow's World. Volume 1: Analysis* Programme for International Student Assessment

[5] Sri R 2014 Menuju Masyarakat Berliterasi sains: *Prosiding Seminar Nasional Kimia dan Pembelajarannya Jurusan Kimia FMIPA UM*

[6] Dewi W K, Agung G A A and Rati N W 2014 *E-journal MIMBAR PGSD Universitas Pendidikan Ganesha,2(1)*

[7] Bundu P 2006 *Penilaian Keterampilan Proses dan Sikap Ilmiah dalam Pembelajaran Sains di Sekolah Dasar*(Jakarta: Depdiknas) p 140

[8] Yager R E1992 *The Status of Science-Technology-Society Reform Efforts Around the World* (Arlington: ICASE Yearbook) p 2

[9] Yager R E 1996 *Science Tecnology Society as Reform in Science Education* (USA: State University of New York Press) p 4

[10] Poedjiadi A 2010 *Sains Teknologi Masyarakat* (Bandung: PT Remaja Rosdakarya) p 123

[11] Bettencourt C, Velho J L and Almeida P A 2011 Biology Teachers'Perpections About Science-Technology-Society (STS) Education *Procedia Social and Behavioral Sciences* **15** 3148-52

IOP Conf. Series: Journal of Physics: Conf. Series **1156** (2019) 012027 doi:10.1088/1742-6596/1156/1/012027

[12] Sugiyono 2014 *Metode Penelitian Kuantitatif Kualitatif dan R&D* (Bandung: Alfabeta) p 166.
[13] Kapici H O, Akcay H and Yager R E 2017 *International Journal of Progressive Education* ***13****(2)*pp 18-28
[14] Akgul E M 2004 *The Turkish Online Journal of Education Technology* ***3(4)*** 58-61
[15] Rahayuni G 2016 *Jurnal Penelitian dan Pembelajaran IPA* ***2****(2* 131-46
[16] Jin G and Bierma T 2013 *J. Coll. Sci. Teach.* ***42****(6)* 20-26
[17] MacLeod 2013 *EJPE* ***4****(4)*pp 1-12
[18] Mandra I M 2012 *Jurnal Pendidikan IPA Ganesha, 2(1)*
[19] Jumantoro 2012 *Jurnal Pendidikan IPA Ganesha* ***2****(2)*
[20] Kartini N N, Adnyana P B and Jelantik I BS2014 e-*Jurnal Program Pascasarjana Universitas Pendidikan Ganesha* **4**
[21] Akcay B and Akcay H 2015 *International Journal of Education in Mathematics, Science and Technology* ***3****(1)*pp 37-45.
[22] Ying Y L, Chen H T and Hong Z R 2016 *International Journal of Science Education*
[23] Chonkaew P, Sukhummek B and Faikhamta C 2016 *Chemistry Education Research and Practice 1(3)*
[24] Xiao S and Sandoval W A 2017 Sssociations Betwen Attitudes Towards Science and Children's Evaluation of Information About Socioscietific Issues *Springer Sci & Educ.*

IOP Conf. Series: Journal of Physics: Conf. Series **1156** (2019) 012028 doi:10.1088/1742-6596/1156/1/012028

Virtual chemistry laboratory (virtual chem-lab): potential experimental media in hybrid learning

M K Nais[1*], K H Sugiyarto[2] and J Ikhsan[2]

[1]Chemistry Education Program, Graduate School, Universitas Negeri Yogyakarta, Indonesia
[2]Department of Chemistry Education, Faculty of Mathematics and Natural Sciences, Universitas Negeri Yogyakarta, Indonesia

[*] E-mail: miarti.kn@gmail.com

Abstract. A laboratory has a very important role in learning chemistry. It facilitates the experimental activity which supports the students' understanding. The development of information and communication technology has provided an innovation known as the virtual laboratory. This study aims to reveal the quality of virtual chemistry laboratory (virtual chem-lab) integrated into hybrid learning on chemical bonding. This research was a research and development study. The development procedure used in this study consisted of four stages: analyzing, product planning, product developing, and product assessing being adapted from the ADDIE (Analysis, Design, Development, Implementation, and Evaluation) development model. The initial product of the virtual chem-lab was content validated by material and media experts. Product assessment was applied to 28 chemistry teachers, including content, learning quality, and technical quality aspects. The result of this assessment was descriptively analyzed to reveal the quality of the product. The findings of this study reveal that the quality of the virtual chem-lab integrated hybrid learning is in a good category. Thus, the virtual chem-lab is potentially used for further teaching-learning.

1. Introduction

The chemistry learning process is not only limited to studying theories or concepts, but is supported by experiment to prove or develop theories that are studied. Experimental activities are one of the important components in chemistry learning. These activities can generate motivation to learn, develop skills, and support understanding of the material [1]. Unfortunately, experimental activities in schools have not been carried out optimally. Studying chemistry requires a lot of time, because too much materials must be learned [2]. However, the allocation of time provided is very limited. In Indonesia the allocation of time for chemistry subjects in class X is three hours per week, while in class XI and XII there are four lessons per week. Moreover, learning activities are often disrupted by school activities outside learning and commemorating national holidays. This time limitation caused a lack of practical work in schools because the time available was used up to deliver the material.

Unlike other sciences that study macroscopic objects, chemistry tends to study microscopic objects. Most of the material in chemistry subjects is abstract. Objects in chemistry such as atoms, molecules, chemical bonds cannot be seen directly [2] so the imagination ability to understand these objects is essential. This causes students to have difficulty in understanding the chemical materials. Therefore, it is necessary to develop an alternative learning media that can present the visualization of micro-chemical objects to improve the learning outcomes of students. The chemistry materials of grade X are very important material because it is the basis for learning the materials in grade XI and XII. One of the most important materials is chemical bond [3].

The use of information and communication technology is necessary to improve the efficiency and effectiveness of learning. However, it has not optimal, especially to overcome the obstacles faced in learning. In line with increasingly advanced technological developments, an alternative media needs to be developed to overcome the obstacles faced in the implementation of learning and practicum. The virtual chemistry laboratory is an alternative that can be prospectively used to overcome obstacles in the implementation of learning and practicum. This study, therefore, aimed to develop and to know quality of virtual chemistry laboratory (virtual chem-lab) integrated into hybrid learning on chemical bonding.

2. Methods

This study was research and development study. The development procedure used in this study consists of four phases including: analyzing, product planning, product developing, and product assessing which was adapted from the ADDIE development model [4].

2.1. Development Procedure

The analysis phase includes needs assessment, and identifying problems. The needs analysis phase was carried out to gather information about the background of the development of a virtual chemical laboratory. Information gathering was carried out through several steps, the first was reviewing the curriculum to find out the ideal standard of learning implementation. The next step, collecting information about the facts of the implementation of learning in school and the problems that arise in learning activities. Needs analysis was also carried out by analyzing literature related to chemistry learning and virtual chemistry laboratories. The information obtained was then analyzed and used as the basis for the development of a virtual chemical laboratory.

The product planning phase was carried out with the determined and compiled material presented in a virtual laboratory, developed the design of virtual chem-lab and instruments of product assessment. The product development phase was carried out through compiled content from the media in the form of an interactive simulation of practicum on the material of chemical bonding, 3-dimensional models, and tasks that would be published in the media, and developed virtual chem-lab. The virtual chemistry laboratory was developed using several programs namely Panotour Pro 2.5, Blender, Game Maker, and HTML-5. The product virtual chem-lab was packaged in web form. Before assessed by chemistry teacher the virtual chem-lab was content validated by material expert and media expert. The validation was done to review the initial product and provide input for the product. The results of the validation were used to improve the quality of the virtual chem-lab.

The validated virtual chem-lab was assessed by chemistry teachers to find out the quality of product. The assessment was applied to 28 chemistry teachers from public senior high schools and an islamic high school. Assessment results and input from teachers were used as a reference to improve the product so that the initial product was produced. Furthermore, the product was also tested the readability to senior high school students. Readability test was applied to five students to find out whether the media produced could be used by students.

2.2. Data Collection Instrument

Product assessment instrument was in the form of assessment sheet using a 4 points modification of Likert scale (very good to poor). The assessment sheet was validated theoretically by expert. The criteria of quality refers to various quality of computer, internet, and multimedia-based media [5-8]. The aspects measured in the assessment sheet include the content, instructional quality, and technical quality which was described into 20 indicators.

2.3. Data Analysis

To reveal the quality of virtual chem-lab, descriptive-quantitative analysis was used in this study. The result of product assessment were analyzed by categorizing into ideal rating category [9] as shown in Table 1. Meanwhile the suggestion from chemistry teachers was considered to product revision.

Table 1. Ideal Rating Category

Score Range	Category
$Xi + 1.8\ SBi < \chi$	Very Good
$Xi + 0.6\ SBi < \chi \leq Xi + 1.8\ SBi$	Good
$Xi - 0.6\ SBi < \chi \leq Xi + 0.6\ SBi$	Fair
$Xi - 1.8\ SBi < \chi \leq Xi - 0.6\ SBi$	Poor
$\chi \leq Xi - 1.8\ SBi$	Very Poor

Note: Xi = (ideal max score + ideal min score); SBi = (ideal max score - ideal min score); χ = average of assessment score.

3. Result and Discussion

The virtual chemistry laboratory (virtual chem-lab) on chemical bond has been successfully developed through this research development. Virtual chemical laboratory products can be accessed online using laptops, tablets or smartphones. The virtual chem-lab can be used as a substitute or supplement for experimental activities in a real laboratory. The home page of the virtual chem-lab application can be seen in figure 1. The main menu of the virtual chemistry laboratory consists of a menu of basic competencies, experiment tool introduction, material MSDS, experiments, and tasks.

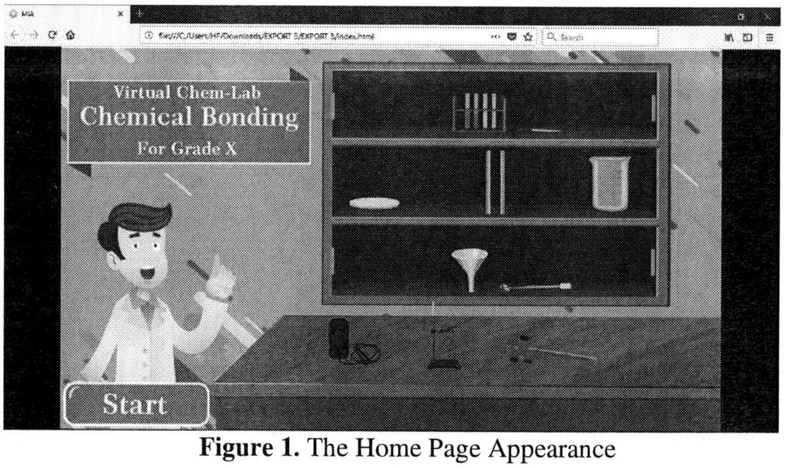

Figure 1. The Home Page Appearance

The basic competency menu contains basic competencies that must be achieved. The introduction menu of the tool contains a picture of laboratory equipment used in the practicum, complete with a description of the name and function of each tool. This menu aims to get students to know the tools used before doing the practicum. The material MSDS menu contains the ingredients used in the experiment supplemented with the MSDS, so the users know the ingredients used before doing the experiment. Thus users can understand the nature of the chemicals used and can be careful of the dangers that may arise from these materials.

The experiment menu contains four experiment, namely the determination of melting point of ionic and covalent compounds, the solubility of ionic and covalent compounds, the conductivity of ionic and covalent compounds, and the polarity of compounds. Each experiment was equipped with an experiment guide that includes objectives, tools, materials, and procedures. The display of practicum instructions can be seen in figure 2. In each practicum users could do an experiment simulation by clicking or moving the tool according to the instructions. While, the task menu contains two task packages that serve to measure students' understanding.

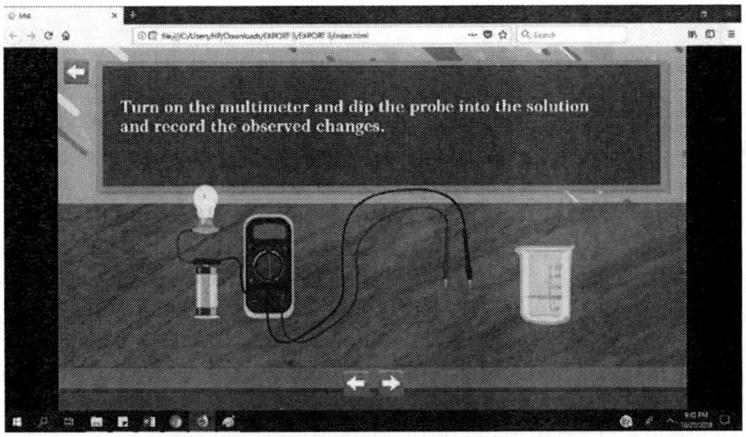

Figure 2. The Experiment Appearance

The initial product of the virtual chem-lab was validated by material experts and media experts. Based on the results of the validation, the virtual chemical laboratory products produced are suitable for use with revisions. Material experts and media experts also provide input to improve the product, as the followings. There were several writing errors that must be corrected. There were also errors in the results of the compound polarity experiment, wherein some compounds should not be deflected. The experts suggested to add more information or instructions and more navigation button so the users could easier to understand and operate. While comments and input from media experts were improving the process of loading from between pages, increase interactivity and add explanations to 3D models. The media expert suggested to fix the navigation buttons, because there is an error navigation button, more over the color composition was less able to adjust again between the background and the image objects.

In addition to being validated by experts, the product was also tested for readability to five students. The five students stated that they could operate a virtual-lab and understand enough the media content. Students were interested in using the media because it resembled a game. The students also provide input for added background music to make it more interesting and added more 3D models.

Product assessment was applied to 28 high school chemistry educators. Assessment covers three aspects, namely content, instruction quality and technical quality. These three aspects were described into 20 indicators. The assessment results for each aspect can be seen in table 2.

Table 2. The Result of Product Assessment.

Aspect	Total Score	Category
Content	16.68 (out of 20)	Good
Instructional Quality	30.16 (out of 36)	Good
Technical Quality	19.82 (out of 24)	Good

The material aspect consists of five indicators, the accuracy of content in a virtual chem-lab, content in a virtual chem-lab contains educational value, the content presented in virtual chem-lab is easily understood by students, the tasks given can be used to measure students' cognitive abilities, content in virtual virtual chem-lab is interrelated to clarify information delivery. In the aspect of content indicator material in the virtual chem-lab contains the value of education included in the very good category. While other indicators are in the good category. The average total aspect score is 16.68 from the maximum score of 20. The score is included in the good category with ideal percentage of

83.4%. Educators consider the content in the virtual chem-lab to be suitable to help form knowledge that resembles practicum in real laboratories. Contents in learning media should be easy to understand [7]. Based on teachers assessment, content in the virtual chem-lab is assumed to be easy to understand.

The quality aspects of learning were consist of nine indicators: compatibility of virtual chem-lab with the needs of current students, the virtual chem-lab can be used to support students in achieving learning objectives, the virtual chem-lab supports learning using hybrid learning communication modes, the virtual chem-lab motivates and attracts students, the virtual chem-lab helps the learning process of students, visual design of virtual chem-lab in accordance with users (teachers or high school students), the virtual chem-lab can stimulate students in practicing, the virtual chem-lab facilitates students to acquire new skills, simulation of experimental activities provides interactive learning opportunities.

All indicators on the quality aspects of learning are in the good category. The average total aspect score is 30.16 from a maximum score of 36. The score was included in good category. Virtual chem-lab were developed according to the needs of students and support the achievement of learning goals. Educators assumed the use of virtual chem-lab will motivate learners to help their learning process. The virtual chem-lab that could be accessed online will make it easier for users to repeat experiment outside of learning so that learning could be done more flexibly.

The third aspect was technical quality. The technical quality aspect consists of six indicators: the virtual chemistry laboratory program has relevance to the technology that is developing at this time, users can easily navigate practicum activities in virtual chem-lab, the combination of forms of visualization can be used to convey information, suitability between objects with the context in the practicum activities, suitability of the 3D model with the context in learning, the illustration shown in the virtual chem-lab is clear

In the quality aspect of the technical indicators the virtual chem-lab program has relevance to the technology that was developing at present in the very good category. While other indicators were in good category. The mean technical aspect score was 19.82 from a maximum score of 24. The score was also included in the good category. Along with the rapid development of technology, virtual chem-lab programs were very relevant to the technology that is developing today. Educators state that students tend to be enthusiastic about using technology-based learning media. Educators could use the product quite easily and the display inside was also quite clear. Objects and 3D models in virtual chem-lab were also in accordance with the context of learning and chemical practicum. Chemistry educators highly appreciate the existence of virtual chem-lab because their use was easy and educators do not need to prepare material tools when they want to practice.

In addition to providing assessments, educators also provide comments and input to improve virtual chem-lab products. Comments and input provided by educators were to added a summary of the material so that it could be used as a source of students' independent learning, color changes need to be displayed when there were two substances that were mixed which indicate a chemical reaction, the pointer button should change to another sign when highlighting clickable objects, added practice questions, we recommend adding videos to make it more interesting, temperature settings using English, should be changed to Indonesian, the display should be made more contrast between the background and objects in the media, there are still errors in writing and practical steps. Comments and input provided by material experts, media experts, educators, and students are taken into consideration in improving the virtual chemical laboratory products.

In general the total score obtained was 66.66 from a maximum of 80. The score was in the good category with an ideal percentage of 83.3%. The virtual chem-lab is potential to be used as experimental media. The virtual chem-lab is an alternative that can be used to overcome obstacles in the implementation of learning and practicum. The results of the Tüysüz [10] study show that virtual laboratories have a positive effect on the achievement and attitudes of students. The virtual chem-lab provides experimental simulations that can be accessed online so students can freely repeat simulations anytime and anywhere. It can enrich their knowledge. 3D models added in a virtual laboratory can visualize microscopic chemical objects, enriching students' knowledge and enthusiasm.

In addition, virtual chemistry laboratories encourage and motivate students to gain deeper and wider knowledge [11] which encourages discussion in the online phase.

The virtual chem-lab is potential for further use. Research conducted by Hawkins and Phelps [12] shows that the use of virtual and traditional laboratories provides relatively similar learning outcomes in electrochemical learning, so that virtual laboratories can be used instead of laboratories. Similar results were also expressed by Tatli and Ayas [13] which stated that virtual chem-lab–applications were as effective as the real labotorium seen from learning outcomes and the ability of students to recognize laboratory tools. According to the results of Arista and Kuswanto [14] virtual chem-lab can even improve students' learning independence and conceptual understanding.

The virtual chem-lab was designed to can be implemented using hybrid learning that combines online and face-to-face learning. Learning that combines face to face and online also influences learning outcomes. In accordance with the results of research by Means, Toyama, Murphy, and Baki [15], hybrid learning has a significant effect on learning outcomes. According to Dale [16], direct and imitation experience of learning will be more effective because more messages can be absorbed. However, several problems made students could not do the experimental activity. The virtual chem-lab can give experimental experience even tough students do not do experimental work directly in the hands-on lab.

4. Conclusion

The virtual chem-lab on chemical bond material has been successfully developed as an experimental media. Based on the assessment of chemistry teachers it was found that the quality of the virtual chem-lab integrating hybrid learning was is in a good category. Thus, the virtual chem-lab is potential for further use.

References

[1] Duit R and Tesch M 2010 *Proc. 7th Int. Conf. Hands-on Sci.* pp 17–30
[2] Middlecamp C and Kean E 1985 *Panduan belajar kimia dasar* (Jakarta: Gramedia)
[3] Sirhan G 2007 *J. Turkish Sci. Educ.* **4** 2–20
[4] Branch R M 2009 *Instructional design: the addie approach* (USA: Springer)
[5] Kustandi C and Sutjipto B 2011 *Media pembelajaran manual dan digital* (Bogor: Ghalia Indonesia)
[6] Nesbit J C and Leacock T L 2009 *Handbook of Research on Learning Design and Learning Objects: Issues, Applications and Technologies* pp 574–588
[7] Oyelekan O S and Olorundare A S 2009 *Int. J. Educ. Develop. Using Inf. Comm. Tech.* **5** 88–104
[8] Squires D and Preece J 1999 *Interact. Comput.* **11** 467–483
[9] Widoyoko E P 2011 *Evaluasi Program Pembelajaran* (Yogyakarta: Pustaka Pelajar) p 238
[10] Tüysüz C 2010 *Int. Online J. Educ. Sci.* **2** 37–53
[11] Chu K C 2000 *Educ. Res. J.* **15** 257-274
[12] Hawkins I and Phelps A J 2013 *Chem. Educ. Res. Prac.* **14** 516–523
[13] Tatli Z and Ayas A *Educ. Tech. Soc.* **16** 159–170
[14] Arista F S and Kuswanto H 2018 *Int. J. Inst.* **11** 1–16
[15] Means B Toyama Y Murphy R and Baki M 2013 *Teach. Coll. Rec.* **115** 1–47
[16] Dale E 1946 *Audio-visual methods in teaching* (New York: Dryden Press)

Acknowledgments

The researcher would like to thank the Ministry of Research and Higher Education, Indonesia for the financial support.

Evaluation of jigsaw puzzles in writing the chemical formula of ionic compounds among the 10th grade students

M I M Damanhuri*, L D P Kumar, M T Borhan, S S Sani and H Taha

Faculty of Science and Mathematics, Sultan Idris Education University, 35900 Tanjong Malim, Perak, Malaysia.

*Email: muhdibrahim@fsmt.upsi.edu.my

Abstract. The objective of this research is to test the effect of using jigsaw puzzles towards students' understanding in writing the chemical formulas of ionic compounds. This quasi-experimental case study design research involves 115 science stream students from two secondary schools in Larut, Matang and Selama district. Achievement test is used to measure students' understanding in writing the chemical formula of ionic compounds. A survey is also conducted on the experimental students' perception on the activity using questionnaires. Descriptive and inferential statistics are used to analyse the data, with the assistance of the Statistical Package for Social Sciences (SPPS) software. The findings proved that the jigsaw puzzles are effective in assisting students to write the chemical formula of ionic compounds. Students also have a positive perception towards the usage of jigsaw puzzles in teaching and learning activity. As conclusion, this jigsaw puzzle activity should be implemented in teaching and learning to assist the students in writing the chemical formula of ionic compounds. This study implies that the elements of game should be infused during the planning of teaching and learning to ensure that learning is more heuristic.

1. Introduction

Chemistry involves describing and explaining chemical reactions using the macroscopic, submicroscopic and symbolic representation [1-3]. Symbolic representation refers to the formula of chemical compound where it contains symbol of elements and subscript numbers referring to the composition of each elements [4]. Mastering the symbolic representation is essential to understand the chemistry concepts and it is fundamental in writing the chemical equations. Students fail to explore chemistry when they face difficulties in writing chemical equation which rises from their inability to write formula of chemical compounds correctly [5]. Studies have shown that the major problem is the poor understanding of the primary knowledge in the background of the formation of ionic compounds and bonding between and among elements [6].

Although students are able to manipulate chemical formula, but they tend to accept it as a mathematical problem without understanding the concept lies behind it [7]. This problem associates with the visualization of chemical representation which involves the cognitive association of conceptual component and visual component [8]. Students could not translate and interpret the chemical representations when they fail to visualize the nature of dynamic and interactive chemical process involved, from the symbol and chemical equation [7-9]. To solve the conceptual problems aroused from the inability in operating the visuospatial representation, Wu and Shah [10] addressed the use of visualising tool in the teaching and learning chemistry. The use of 'eChem', chemistry software, is well known in assisting students' conceptual understanding in chemical representations [8]. In contrast, Rodrigues [11] found that computer assisted visualizing tools acquire the information processing skills, where students need to interpret the logical instruction before understanding the chemistry content. Hence, a suitable technological and pedagogical design is needed in visualizing the concepts involved in writing formula of compounds.

To facilitate the students' conceptual understanding in writing the chemical formula of ionic compounds, game based activities such as 'Rainbow Wheel' and 'Rainbow Matrix', are found to be effective [12]. Further innovation on the teaching and learning strategies introduced 'ChemOkey', a card game to familiarize students with the ionic formulas [13]. Besides that, an interlocking building block was also found to achieve the same intention [11]. Apart of these activities, teacher's intensive focus on the students' ability in chemical representation revealed that students are comfortable in representing the ions with puzzle type pieces [14]. They are able to explain the graphical representation with the formation of ions. Hence, jigsaw puzzles are found to be potential in enhancing the students' ability in visualizing the chemical formulas.

This study proposes the jigsaw puzzles as the visualizing tool in writing the chemical formula of ionic compounds. It is a vital activity to engage the innovative learners in understanding the concept involved in writing the chemical formula of ionic compounds. Generally, jigsaw puzzles do indeed improve the understanding of spatial relations from 3D illustrations, where the implicit guidance of the 3D jigsaw concept requires the user to interact with well-defined objects illustrating a chosen topic [15]. According to Ritter et al., it directs the user's attention to correlations between objects and provides a level of motivation for learning, which is hard to achieve with other concepts.

In this study, some concepts are used as basis and guide to make the research more meaningful. Conceptual framework shown in figure 1 suggests how jigsaw puzzle activity used in teaching and learning increases students' understanding in writing chemical formula of ionic compounds to provoke thoughtful learning.

Generally, constructivism is a theory that explains how students construct their own knowledge based on experience and their existing knowledge. Through constructivism, jigsaw puzzle activity drives the learning process towards experiential learning, in which the jigsaw puzzle activity provides an experience or awareness of the concept under study.

Jigsaw puzzle activity considers the type of stimulation received by the students, the verbal and non-verbal stimuli during the learning process of writing chemical formula of ionic compounds. Jigsaw puzzle activity provides game based learning environment for students to increase their involvement in the learning process. Once students are aware of the concept in writing chemical formula of ionic compounds, then they will interpret and reflect on the concept being studied by visualising the concept. Jigsaw puzzles help students to visualize easily the concept learned. Furthermore, it will help students to evaluate, generalize and apply or test the concept more in detail. Experiential learning environment created by the jigsaw puzzle encourages learning towards thoughtful learning.

The purpose of this study is to investigate the effect of the teaching and learning activity using jigsaw puzzles to write chemical formula of ionic compounds among the form four science students (10th grade) and to evaluate the students' level of understanding in writing the chemical formula after the teaching and learning activity. Students' perception towards the activity is also collected. Consequently, the following research questions were investigated in this study:

1. Is the jigsaw puzzles used as visualizing tool in writing the chemical formula of ionic compounds effective in increasing students' understanding than conventional teaching method?
2. What are the students' perceptions on the usage of jigsaw puzzles in the teaching and learning activity on writing the chemical formula of ionic compounds?

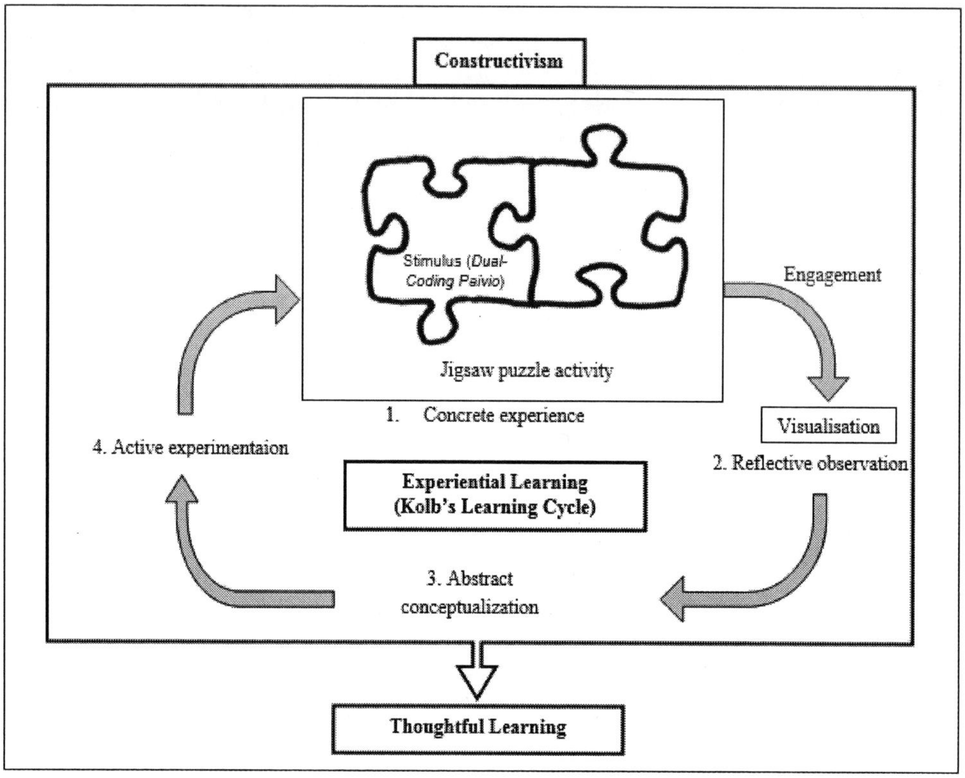

Figure 1. Conceptual framework on how the jigsaw puzzle in teaching and learning increases students understanding in writing chemical formula of ionic compounds

2. Method

A quasi-experimental design case study is employed to research group consisting of 27-30 students. 2 set of controlled and experimental group is selected from 2 different secondary schools in 'Larut, Matang and Selama' district, Perak, where the total of 115 form four students (10th grade) are involved in this study. Table 1 gives brief explanation on the number of participants in each study group.

Table 1. Number of students in each study group

Study Group	School A	School B	Total
Control group	28	30	58
Experimental group	29	28	57
Total	57	58	115

Achievement test on writing chemical formula of ionic compounds are used to evaluate the students' understanding. Pre-test and post-test is executed to compare the mean score. At the end of this study, a survey is also conducted on the experimental students' perception on the activity, using questionnaires. Descriptive statistics (involving mean score and standard deviation) and inferential statistics (involving t-test) are used to analyse the data, with the assistance of the Statistical Package for Social Sciences (SPPS) software.

In this study, the jigsaw puzzles are built to meet the particular need in the selected topic. Puzzles that represent the positive and negative ions are built and the formula of the ions is written on the puzzles. Puzzles for positive ions have loops, where the number of the loops depends on the number of the positive charges of the ions, and the puzzles for the negative ions have heads pointing out. The number of heads shows the number of negative charges of the ions. The design of the puzzles for some of the ions is illustrated in figure 2. The loops in the jigsaw puzzle will give the students the idea of atoms lacking of electrons and the heads as the atoms having excessive electrons. In the jigsaw puzzle game, students will try to solve the puzzle to form a rectangular shape figure. When students combine the puzzles with loops and heads to solve the puzzles, they will visualize the combination of positive and negative ions to form neutral compound.

Students will be able to write the chemical formula of ionic compound based on the number of ions in the solved puzzle. Figure 3 illustrates how students will solve the jigsaw puzzles to write the chemical formula of magnesium hydroxide, as $Mg(OH)_2$.

Achievement test on writing the chemical formula of ionic compounds was administered to both the experimental and comparison groups as a pretest before their study of the chemical formula to ascertain their prior knowledge on the topic. This test contains 25 items and the items were validated by two chemistry lecturers and an 'expert chemistry teacher' to ensure that the relevant content was incorporated in the respective form four (10th grade) chemistry curricula and was covered during instruction. The content validity test is analyzed using the method proposed by Pennfield and Miller [16], which introduces the score confidence interval to improve the assessment of item content relevance. The test was administered again to both groups as a posttest after they had completed studying the topic on chemical formula of ionic compounds. Each test is completed in 45 minutes. The test was developed by the researcher and consisted of open-ended questions. The content of the test was determined from curriculum guidelines, lecture materials and chemistry textbooks, which cover the major concepts on writing the chemical formula of ionic compounds. The concepts that were investigated and the corresponding question numbers in the test are summarized in table 2.

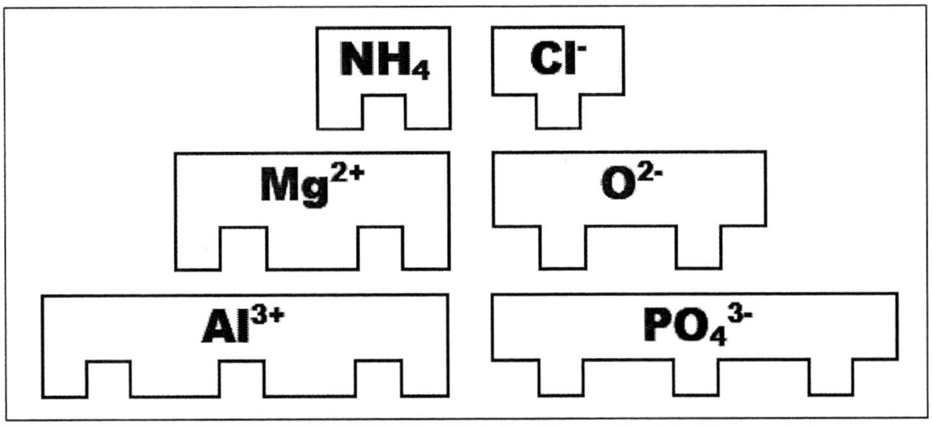

Figure 2. Design of the puzzles for some of the ions

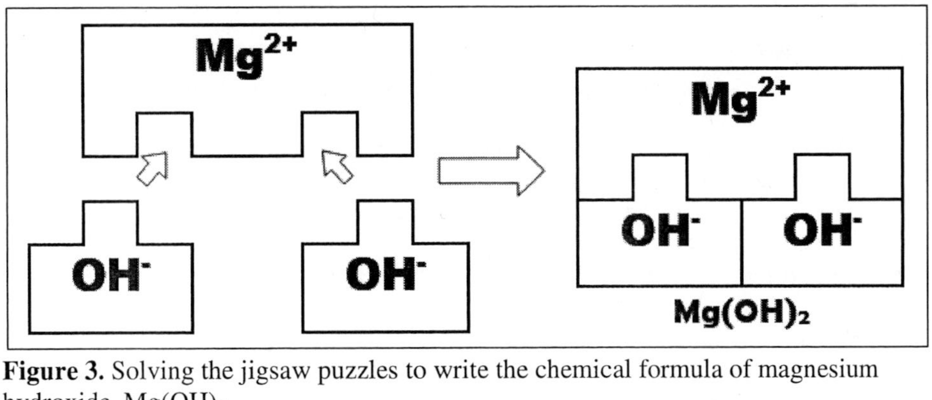

Figure 3. Solving the jigsaw puzzles to write the chemical formula of magnesium hydroxide, $Mg(OH)_2$.

Table 2. Concepts investigated in the achievement test on writing the chemical formula of ionic compounds

No.	Concepts	Question numbers
1	Valency of common ions	1, 5, 9, 13, 17, 19, 24, 25
2	Compounds involving polyatomic ions	2, 6, 10, 14, 20
3	Usage of roman numbers in the IUPAC naming system	3, 7, 11, 15, 18, 21
4	Predicting the formula of new ionic compounds (compounds which are not in the syllabus)	4, 8, 12, 16, 22, 23

Self-evaluation questionnaire on students' perception on the jigsaw puzzle activity was administered to the experimental group on completion of the intervention program. The purpose of this questionnaire was to explore students' perception on the jigsaw puzzle activity in writing chemical formula of ionic compounds. It is adapted from 'My Class Activities' instrument [17] and amended to suit this study. The questionnaire consists of 32 statements to which students responded on a Likert-type scale with five options: 5 = strongly agree, 4 = agree, 3 = not sure, 2 = disagree and 1 = strongly disagree. The items in the test were validated by three chemistry education lecturers to ensure the items are relevant to the activity. Each of the items was explained to students before they responded to the questionnaire to ensure that the original intent of the items was maintained. The aspects evaluated by the questionnaire and the internal consistency reliability for each construct are summarized in table 3. Since the Cronbach's alpha value for each construct is greater than 0.70, hence the instrument is well developed and is reliable [18].

Table 3. Aspects investigated by the self-evaluation questionnaire

No.	Perception aspects	No of items	Item number	Cronbach's alpha
1.	Visualization	4	3, 9, 16, 23	0.703
2.	Interest	7	1, 7, 14, 27, 28, 29, 32	0.844
3.	Understanding	7	4, 10, 11, 17, 18, 25, 26	0.817
4.	Fun	3	21, 24, 31	0.767
5.	Confident	3	5, 12, 19	0.894
6.	Collaboration	3	6, 13, 20	0.782
7.	Challenge	5	2, 8, 15, 22, 30	0.870
	Total	32		

3. Results and discussion

This study was designed to have only two groups: learning using jigsaw puzzle (experimental group) and the conventional learning (control group). The pre-test scores were analysed to determine the equivalence and homogenity of the two groups before the experiment was performed. T-test shows that the pre-test scores for the two research groups did not differ, $t\ (113) = 0.337$, $p = 0.737$. The teaching and learning sessions were conducted since the two groups were similar. The analysis of the students' scores in pre-test and post-test are shown in table 4.

However, the post-test score revealed a significant difference between the two study groups, $t\ (55) = 4.930$, $p = 0.000$. The effect size of the study is 0.92, indicating that the jigsaw puzzle activity is effective in facilitating the students' understanding on how to write chemical formulas of ionic compounds. Thus, visualization is important to understand the concepts behind the chemical representation [8]. The findings prove the potential strength and effectiveness of a new approach in writing the chemical formula of an ionic compound [11-13]. In line with the findings of Waldrip and Prain [14], puzzle-shaped ion representation used in explaining the formation of an ionic compound and the jigsaw puzzle activity can lead the teaching and learning towards high academic achievement.

Table 4. Mean scores and standard error for the experimental group and the control group in School A and B

Research Group	N	Mean Score	
		Pre-test	Post-test
Experimental Group	57	8.56	52.42
Control Group	58	8.07	38.83

The next interest in this study is determining the students' perception on the jigsaw puzzles activity. The experimental group students' perception is investigated using questionnaires. Their perception is tested mainly on visualization, interest, understanding, fun, confident, collaboration and challenge. The results of students' responses to the questionnaires are summarized in table 5 and figure 4. Students responses are presented in three categories: the positive category that included 'strongly agree' (code 5) and 'agree' (code 4), the neutral category that is 'not sure' (code 3) and the negative category that included 'disagree' (code 2) and 'strongly disagree' (code 1).

Table 5. Analysis of results for each component in the self-evaluation questionnaire

Aspects	Category			Mean	Standard Deviation
	Positive (%)	Neutral (%)	Negative (%)		
Visualization	91	0	9	4.34	0.54
Interest	89	4	7	4.27	0.71
Understanding	88	2	88	4.09	0.56
Fun	86	5	9	4.30	0.92
Confident	81	2	18	4.15	0.66
Colaboration	81	0	19	4.18	0.66
Challenge	68	11	21	3.75	0.81

(n=57)

Figure 4 shows that students generally have positive attitude towards the use of the jigsaw puzzle. All the investigated aspects are agreed by more than 60% of the students in the experimental group. Science educators believe that students' interest can be discerned through reports of their beliefs on the level and quality of task engagement and confident in performing a task [10]. Thus, the results of the self-evaluation questionnaire that were generally very positive can be viewed as an indication that the learning environment engendered greater interest among students towards learning science which will lead to effective learning. In general, the students' positive view supports the proposed conceptual framework on how the jigsaw puzzle used in teaching and learning increases the students' understanding in writing the chemical formula of ionic compounds. The high percentage (91%) of students agrees on visualization shows that the jigsaw puzzle activity stimulates verbally and non-verbally during teaching and learning to visualize the combination of ions. This visualization can help them to write the chemical formula of ionic compounds with ease. This finding supports the findings of Wu, Krajcik, and Soloway [19] which shows the ability of jigsaw puzzle in visualizing the concept of chemical representations.

The second highest percentage of students agrees is on the aspect of interest. Many students (89%) claim that the jigsaw puzzle can explore their interests during the process of teaching and learning. Since the jigsaw puzzle activity is a game based activity, so it is capable in encouraging students' interest in the learning process. This aspect has a close relationship with the fun aspect where the percentage of students agreed to this aspect is also high (86%). When students enjoy doing jigsaw puzzles, then this activity is believed to explore their interests. Thus, jigsaw puzzle activity which is an experiential learning activity could provide an active learning environment to increase students' interest and excitement [20]. The results of this study also support Holstermann et al. [21] and Weinberg et al. [8] researches that demonstrated the importance of learning through experience to create interest among students towards learning new concept.

Jigsaw puzzle activity is an experience that can visualize the abstract concept, which in turn give students the understanding in writing chemical formula ionic compounds. This can be seen when 88% of students agree in the aspect of understanding. As proposed in the conceptual framework, high visualization (91%) leads to high understanding (88%). Basically, the jigsaw puzzle activity which uses the constructivism approach gives students the opportunity to develop new knowledge from their visualization in writing chemical formula of ionic compounds. Thus, the innovative approach in writing the chemical formula of ionic compounds is accepted by students. The findings of this study support the findings of Rahayu et al. [22] that revealed the positive impact of new approach in teaching and learning on students' understanding and achievement while increasing their involvement.

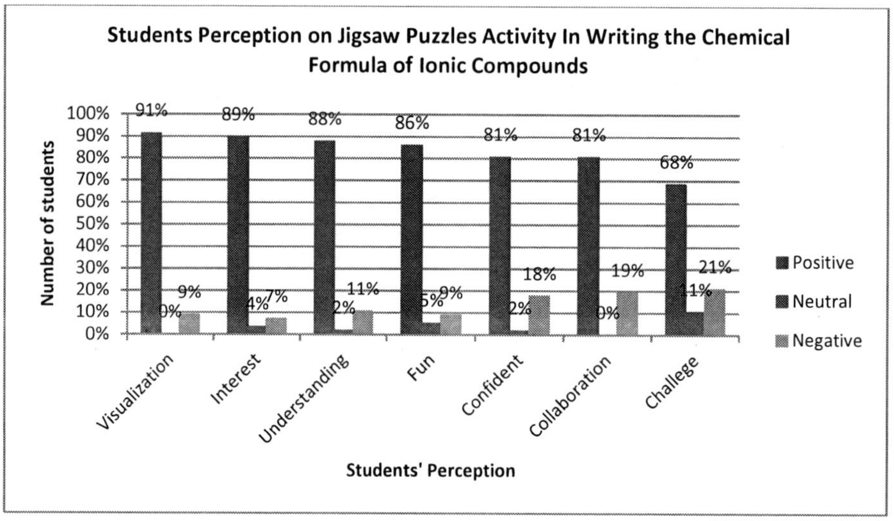

Figure 4. Students' perception on jigsaw puzzle activity in writing the chemical formula of ionic compounds

Students agree that the jigsaw puzzle activity can increase their confidence (81%) and collaboration skills (81%). These aspects are achieved when students explain and discuss the concepts involved in writing chemical formula of ionic compounds, and this occurs when students evaluate, generalize and apply the concepts learned as specified in the conceptual framework. Group activities under focus in the design of teaching and learning strategies [23] succeeded in producing more skilled students. Furthermore, the percentage of students who agreed that the jigsaw puzzle activity challenges the mind and their ability is 68%. This indicates jigsaw puzzle activity, an experiential learning activity can create higher-order thinking skills, HOTS [24] among the students as envisaged. However jigsaw puzzle activity focuses less on higher-order thinking skills, rather it emphasizes students' understanding on how to write chemical formula of ionic compounds. This is one of the factor where there is still a small percentage (33%) of students who disagree and uncertain in the aspect of challenge.

4. Conclusion

As conclusion, the teaching and learning using jigsaw puzzles in the topic of chemical formula assists the early learners in visualizing and understanding the main concept underlies in writing the chemical formula of ionic compounds thus, preventing them from memorizing and accepting blindly formulas. Exploring the chemical formula of ionic compounds will lead to thoughtful learning as proposed in the

Malaysian curriculum. In the Malaysian context, efforts to improve present and future science teachers' attitudes towards inquiry learning as suggested by the new 2012 curriculum are of particular importance and are seen as the precursor of the higher order thinking skills (HOTS) among the students. The suggested teaching innovation is potential to create an active learning environment, where students will be more engaged and experiencing their learning as well as inspiring the HOTS. The jigsaw puzzles activity is also used to capture students' interest and attention during the teaching and learning session. Furthermore, students' positive perception towards the usage of jigsaw puzzles implies that this jigsaw puzzle activity should be implemented in teaching and learning to improve the students' understanding in chemistry. In contrast, the absence of excellent level of understanding in writing chemical formula of ionic compounds among students indicates that teachers need to focus more on chemical representation for in depth learning. Pedagogical element should overcome the game element infused in the jigsaw puzzles during the teaching and learning to ensure the learning objectives are achieved. Teachers should be well comprehended on the design and usage of the jigsaw puzzles as they play the main role of facilitators in teaching and learning. Finally the insights gained from this research are expected to inculcate heuristic as well as thoughtful learning among the young chemistry learners.

References

[1] Gabel D 1998 *International Handbook of Science Education,* eds K T Barry and J Fraser (London: Kluwer Academic) pp 223–248.
[2] Gilbert J K and Treagust D 2009 *Multiple representations in chemical education,* eds D T John and K Gilbert (Netherlands: Springer) pp 1–8.
[3] Johnstone A H 1982 *Sch. Sci. Rev.* **64** 377–379
[4] Chang R 2010 *Chemistry 10th Edition* (New York: McGraw Hill)
[5] Baah R and Ampiah J G 2012 *Int. J. Sci. Res. Educ.* **5** 162–170
[6] Ali T 2012 *SAGE Open* **2** 1-13
[7] Krajcik J 1989 *Meaning of school and communities,* eds M Eisenhart and J G Goets (Washington DC: 88th annual Anthropological Association)
[8] Weinberg A E, Basile C G and Albright L 2011 *RMLE Online* **35** 1-12
[9] Ben-Zvi R, Eylon B-S and Silberstein J 1986 *J. Chem. Educ.* **63** 64
[10] Wu H-K and Shah P 2004 *Sci. Educ.* **88** 465–492
[11] Rodrigues S 2007 *Chem. Educ. Res. Pr.* **8** 1–12
[12] Wulfsberg G P, Sanger M J, Melton T J and Chimeno J S 2006 *J. Chem. Educ.* **83** 651
[13] Kavak N 2012 ChemOkey: A Game To Reinforce Nomenclature *J. Chem. Educ.* **89** 1047–1049
[14] Waldrip B and Prain V 2012 *Int. J. Sci. Math. Educ.* **10** 1191–1213
[15] Ritter F, Berendt B, Fischer B, Richter R and Preim B 2002 *Mensch & Computer,* eds M Herczeg and H Oberquelle (Hamburg, Germany: Teubner) pp. 363–372
[16] Penfield R D and Miller J M 2004 *Appl. Meas. Educ.* **17** 359–370
[17] Pereira N, Peters S J and Gentry M 2010 *J. Adv. Acad.* **21** 568–593
[18] Cronbach L J 1951 *Psychometrika* **16** 297–334
[19] Wu H , Krajcik J S and Soloway E 2001 *J. Res. Sci. Teach.* **38** 821–842
[20] Stupans I, Scutter S and Pearce K 2010 *Innov. High. Educ.* **35** 359–366
[21] Holstermann N, Grube D and Bögeholz S 2010 *Res. Sci. Educ.* **40** 743–757
[22] Rahayu S, Chandrasegaran A L, Treagust D F and Ibnu S 2011 *Int. J. Sci. Math. Educ.* **9** 1439–1458
[23] Dahsah C and Kruatong T 2010 *Int. J. Learn.* **17** 45–60
[24] de Jesus H P, Almeida P, José J T-D and Watts M 2007 *Res. Educ.*, vol. **78** 1–20

Acknowledgment

The authors recognised the sponsorship given by Ministry of Higher Education (MOHE), Malaysia and Universiti Pendidikan Sultan Idris (UPSI).

The effects of scientific approach based jigsaw model on students' self-efficacy and achievement

N M Syawal[1*] and Amanatie[1,2]

[1] Postgraduate of Chemistry Education Program, Universitas Negeri Yogyakarta, Indonesia
[2]Department of Chemistry Education, Faculty of Mathematics and Natural Sciences, Universitas Negeri Yogyakarta, Indonesia

*E-mail: nurlaidymsyawal3@gmail.com

Abstract. This research aims to find out the effect of scientific approach based jigsaw model on self-efficacy and achievement of tenth grade high school student on redox reaction material. This research was a quasy experimental using post-test only design. This study involved 120 tenth grade high school students in Yogyakarta regency, Indonesia. A total of these samples classified into two classes, the experimental class using scientific based Jigsaw model and control class using direct instruction. Data collection in this study used questionnaire to collect students'–self-efficacy and test on redox topic to collect students' achievement data. Data analysis techniques in this study using Multivariate Analysis of Variance (MANOVA) to examine the effect of scientific based jigsaw model on students' self-efficacy and achievement among the experimental and control classes. The results showed that s there was significant effect of scientific based jigsaw model on students' self-efficacy and cognitive achievement. Hence, the teachers are expected to apply scientific based jigsaw model to the chemistry learning because it can increase student self-efficacy that has implications for improving students cognitive achievement.

1. Introduction

Chemical education is an important thing in the development of science and technology, but when associated with learning chemistry in high school has a bad assumption from students. Students think chemistry is the hardest lesson because it is abstract. One of the chemical material that is considered most difficult by students is a redox reaction. The results of the analysis of research from other countries namely Ireland, England and Scotland found that redox was one of the topics considered difficult due to lack of ability mathematics on students[1]. Based on the results of the National Examination obtained from the PAMER application the average value achieved by students was 54.68. It is necessary for teachers to create good and fun learning for students who can overcome these assumptions.

Good and fun learning is student-focused learning [2]. Based on observations, teachers still apply teacher-centred direct learning which has an impact on student activity weaknesses so that it is necessary to apply student-focused learning models. One model that focuses on students is the jigsaw model. The jigsaw model is one type of cooperative learning model. In general, the cooperative learning model is a learning model that encourages students to work together to form small groups and help each other and is a model suitable for all school levels [3]. Based on the results of the research shows achievement. of students after the application of the cooperative model have greater academic achievement, namely the results obtained by the students are higher [4]. Correspondingly, the results of other research also show that the jigsaw cooperative model is effective in improving cognitive achievement compared to other cooperative models because of the acceptance of knowledge from four sources at once namely students themselves, friends of a group of students (team of origin), friends of expert teams and from educators [5]. In addition, the jigsaw model can also be juxtaposed with an approach that is in accordance with the curriculum applied at school. This is what underlies the application of a scientific-based jigsaw model.

One of the predators that influence student success in mastering the concept is self-efficacy. Students must have high efficacy in mastering a learning material because the success of a person depends on the beliefs held in the students [6]. This indicates the importance of self-efficacy towards cognitive learning outcomes. Based on observations obtained in high school researchers showed educators do not instill self-efficacy in students because when the learning process takes place students are just silent and not active in the learning process. The absence of teacher awareness of the importance of self-efficacy for students so that it has an impact on students' cognitive learning outcomes. Based on observations and grades obtained from teachers in the field of chemistry studies showed unsatisfactory results, the researchers offered a scientific-based jigsaw model to improve the efficacy and cognitive achievement. Therefore, this study aims to determine the effect of using scientific-based jigsaw learning models on self-efficacy and cognitive learning outcomes of students.

2. Methods

2.1. Research design and samples

The research design used in this study was quasy experimental using post-test only design. Hence, in this study compare the results of the post-test score of the students' self-efficacy and cognitive achievements among the experimental and control classes after the intervention were applied. Table 1 showed the research design in this study.

Table 1. Research design

No	Class	Treatment	Post-test
1	Experiment	X1	O1, O2
2	Control	X2	O1, O2

Note: X1 = Scientific-based Jigsaw Model; X2 = Direct Instruction model; O1 = self-efficacy questionnaire; O2 = test on redox topic

2.2. Research procedure

The experimental class is a class that receives scientific-based jigsaw learning. The steps in scientific-based jigsaw learning are 1) the teacher outlines learning and learning objectives 2). Students are divided into several groups consisting of 6 students who all serve as experts. 3). Each expert observed the expert assignment sheet after the same material expert group member joined into one group; 4). Faithful experts who found difficulties in learning the material given wrote the difficulties given and then asked fellow expert teams during the expert team discussion; 5). Each expert from each group discussed the same material and collected information from various sources of jam from the student worksheet; 6) The expert group students answer the questions in the expert sheet before returning to the original group and concluding the results of the discussion; 7). Teacher gives quizzes on material in the expert group to check accuracy; 8). Each student returns to the original group to communicate information sharing with one original team; 9). The teacher gave a Quiz about the material distributed to check its accuracy; 10) teachers examine and clarify what students do not seem to understand; 11) teachers give questions to assess individuals using posttest; 12). The teacher re-teaches any material that is not answered in an individual assessment as needed.

The control class is a class that receives Direct Instruction learning. steps in Direct instruction learning are: 1). the teacher determines and explains the learning material by lecture; 2). the teacher asks the students about the material described; 3). the teacher corrects errors related to the material; 4). students are given a quiz to solve questions based on the material that has been taught. Next measure the dependent variable in the experimental class and control class. The dependent variable in the form of self-efficacy was measured using a questionnaire and the dependent variable was cognitive learning outcomes by using oxidation reduction reaction test questions which were all done at the end of the learning process.

2.3. Instruments

The research instrument used in this study was an interview guide to chemistry teachers at public Yogyakarta senior high schools to collect the existing information about the existing teaching-learning instruction, self-efficacy questionnaire to collect the data of students' self-efficacy, and test on redox topics to measure students' cognitive learning outcomes.

In conducting the research, a lesson plan on both experimental and control classes were prepared. All of the instruments including the interview guideline, self-efficacy questionnaire, and test on redox topics were validated by two experts. The two experts confirmed the theoretical validity including the content and face validity.

In addition, the empirical validity of the self-efficacy questionnaire and test on redox topics were carried out with the result of the empirical validity were analyzed using SPSS by comparing R count and R table. The result of the analysis showed that 26 valid questions from 30 questions tested and the self-efficacy questionnaire and 15 valid statements from 15 statements questionnaire given., Moreover, the instrument reliability test was also performed with the results of reliability of both instrument obtained high category

2.4. Data Analysis

The data analysis technique used in the study was Multivariate Analysis of Variance (MANOVA) to determine the effect of using scientific-based jigsaw learning models on self-efficacy and cognitive learning outcomes of students. The analysis of MANOVA technique conducted by the help of SPSS. Manually MANOVA test formulas can be written as follows [7].

$$F = T^2$$

Moreover, perason correlation was used to determine the relationship among the students' self-efficacy toward students' achievement.

3. Result and Discussion

The results of this study were taken using a test and non-test system. The test system is used by the instrument of multiple choice questions on redox topic as many as 26 questions to measure the cognitive learning outcomes of students while the non-test system is used is a questionnaire to measure the self-efficacy of students. The result of MANOVA technique to examine the effect of using scientific-based jigsaw learning models on self-efficacy and cognitive achievement of students showed in the table 2.

Table 2. Data on the results multivariate of self-efficacy and cognitive achievement

Variable	Mean Experiment	Mean control	F value	Sig
Self-efficacy	58,00	54,47	9,401	0,003
Achievement	77,45	69,57	26,942	0,000

Based on the results of the data presented in Table 2 it can be seen that there were differences in students' self-efficacy and achievement among the experimental class and the control class. The average value of the experimental class is higher than the average value of the control class so it is clear that the model applied to the experimental class is better than the model applied to the control class.

The effect of the implementation of scientific-based jigsaw model on self-efficacy and cognitive achievement seen from the results of MANOVA test showed in the Table 2. In the MANOVA test, the significance value is smaller than the 0.05 significance value, so it can be concluded that the significance level of 5% of the implementation of scientific-based jigsaw models can affect the self-efficacy and cognitive achievement of students. In accordance with the research results of several researchers who revealed that cooperative learning can improve self-efficacy and academic achievement because in cooperative learning students work together and help each other so that self-efficacy increases and the ability to achieve achievement will be higher. There is an increase in yields and self-efficacy because the applied learning model is a fun learning for students so that giving effect to motivation this can be seen in the results of the statistical analysis obtained significantly has supported the experimental class on motivation dependent variables [8].

Better experimental class cognitive achievement from the control class indicate that a scientifically based jigsaw model has proven effective in improving students' cognitive achievement. Based on the results of previous studies that the jigsaw model is said to be effective because it shows a significant difference from the experimental class that is better than the control class [9]. This statement is supported by other research which states that jigsaw learning is effective in improving student learning outcomes and the gender of students does not affect student achievement as seen from the results of the quiz analysis of students conducted twice a week [10]. Based on the research results obtained indicate that the jigsaw model has a positive effect in improving student learning outcomes. According to a statement from another study that based on the learning experience of the jigsaw model which is a type of cooperative learning model has a positive impact on improving achievement, mutual respect and attention for each student [11].

Moreover, the Pearson correlation of the students' self-efficacy and achievement was conducted to determine the relationship among these two variables. The result of the analysis showed in the table 3.

Table 3. Pearson correlation among the self-efficacy and achievement

Aspect	Value	Self-efficacy	Achievement
Pearson Correlation	Self-efficacy	1	0.295
Sig (2-tailed)			0.001
Pearson Correlation	Achievement	0.295	1
Sig (2-tailed)		0.001	
N		120	120

In the Pearson correlation test result showed in table 3, it can be seen that there is a correlation between self-efficacy and cognitive learning outcomes of students so that it can be concluded that if the self-efficacy of students increases, learning outcomes also increase. The same results were revealed by previous research that if there is an increase in the students' self-efficacy, the achievement is increases [12]. Other research results also show the results that students who having high self-efficacy tends to have better test results than students who have low efficacy [13]. Self-efficacy is an important thing that can affect the success of students because if the students' self-efficacy is high, the students will not experience fatigue in completing the tasks assigned to them [14]. Students who have high self-efficacy are usually able to think creatively and are able to solve group problems [15]. There are 4 ways to improve self-efficacy, namely: 1). success in achieving results; 2). Learning from experience by observing others achieves results; 3). There is verbal encouragement; 4). Reducing signs of anxiety [16].

Other research also revealed the importance of self-efficacy in high school students, namely the results of research which showed that self-efficacy is an important predator of students' success in completing their first year of study at university so that educators can contribute directly to the development of self-efficacy for the success of students' academic transitions in which academic interest and cognition are directly related to self-efficacy[17] self-efficacy is one of the keys to involving students in learning[18] self-efficacy is one of the keys to involving students in learning. If students have low efficacy, it is ensured that these students will avoid tasks that have made it fail so that the best way is that educators must emphasize the development of self-efficacy that will increase students' motivation in completing school assignment [19] The existence of a link between self-efficacy and learning outcomes is also revealed by several studies which reveal that self-efficacy and motivation are important contributors to performance improvements [20]

4. Conclusion

Based on the results of research and discussion, it can be concluded that. There is an effect of the application of science-based jigsaw learning model on cognitive learning outcomes and self-efficacy on the topic of oxidation reduction reactions in tenth grade high school. this is because in the jigsaw model there is responsibility given to students as experts on a sub topic that is not given to direct learning. the activeness of students in accepting their responsibilities as experts can improve self-efficacy that is rooted directly in improving cognitive achievement. This statement implies that chemistry learning using scientifically based jigsaw models is learning that trains students to be independent, has self-confidence and does not make the teacher as a learning resource so that with a scientific-based jigsaw learning model given a positive impact on improving self-efficacy and results cognitive learning of students.

5. References

[1] Childs P E and Sheehan M 2009 *J Chem. Educ. Res. Pract.*. **10** 204-218
[2] Binggs J 2012 *Higher Education Research & Development* **31** 39–55
[3] Slavin 2015 *Education International Journal of Primary* **43** 5-14
[4] Yamarik S 2010 *Journal of Economic Education* **38** 259-277
[5] Aswirna P 2012 *Journal Al-Ta'lim* **19** 158-165
[6] Schunk D H 2015 *Journal of Educational Psychology* **73** 93-105
[7] Stevens J P 2002 *The Social Science* (London : Lawrence Erlabaum Associates)
[8] Shaaban K 2006 *J. Read. Psychol.* **27** 377–403
[9] Tarhan L and Sesen B A 2013 *Chem. Educ. Res.Pract.* **13** 307–313
[10] Yoruk A 2017 *Journal of Educational Sciences* **15** 377–381
[11] Slavin R E and Karweit N L 2015 *Journal of Experimental Education* **50** 29–35
[12] Bandura A 2007 *Anxiety Research* **177**–98
[13] Jiang Y Song J Lee M and Bong M 2013 *Educational Psychology,* **34** 92–117
[14] Mulki J Lassk F G and Jaramillo F 2008 *Personal Selling & Sales Management* **28** 285-297
[15] Yoon Woo Treagust Chandrasegaran 2012 *Journal of Science Education* **36** 79-102
[16] Jackson, 2002 *Journal of Experimental Education* **70** 243-254
[17] Van R Jansen E and van D 2017 *Research in Post-Compulsory Education* **22** 535–555
[18] Linnenbrink E A and Pintrich P R 2010 *Journal of Educational Psychologist* **19** 119-137
[19] Margolis H and Cabe P 2010 *Education for Children and Youth* **47** 162–169
[20] Vrugt Langereis and Hoogstraten 2010 *Journal of Experimental Education* **66** 61-72

IOP Conf. Series: Journal of Physics: Conf. Series **1156** (2019) 012031 doi:10.1088/1742-6596/1156/1/012031

Effectiveness of using virtual chemistry laboratory integrated hybrid learning to students' learning achievement

R Wijayanti [1,*], K H Sugiyarto [2] and J Ikhsan [2]

[1] Postgraduate of Chemistry Education, Universitas Negeri Yogyakarta, Indonesia
[2] Department of Chemistry Education, Faculty of Mathematics and Natural Sciences, Universitas Negeri Yogyakarta, Indonesia

*E-mail: renniw.rw@gmail.com

Abstract. This research was conducted to analyze the effect of using virtual chemistry laboratory integrated hybrid learning toward students' learning achievement on the subject of thermochemistry. Thus, the research was set as a quasi-experiment research using posttest and non-equivalent control group design. By purpossive sampling technique, the samples from 3 classes composed of 1 public senior high school were selected. The classes were divided into experiment class (EG) using both virtual chemistry laboratory integrated hybrid learning and traditional hands-on laboratory, control class 1 (CG-1) using virtual chemistry laboratory integrated hybrid learning only, and control class 2 (CG-2) using traditional hands-on laboratory only. The corresponding data were collected on the basis of a multiple choice test of chemistry learning achievement. The data were then tested by Anova. The results show that there is a significantly different in students' learning achievement among the three classes EG, CG-1 and CG-2.

1. Introduction

In the era of globalization, the progress of science and technology is growing very rapidly. Delivery of information is quickly needed by all levels of society, including in the world of education. In order to keep up with the development of science and technology, the Government has established policies regarding improving the quality of education in Indonesia, namely by implementing the Curriculum-2013 as a standard for the implementation of national education in Indonesia.

Curriculum-2013 is a scientific approach based curriculum that includes observing, asking questions, gathering information, associating, and communicating for all subjects. The learning process in the Curriculum-2013 is interactive, contextually based, and learner-centered. Learning in the Curriculum-2013 strengthens the learning process and authentic assessment to achieve competencies in attitudes, knowledge, and skills. Authentic assessment includes the readiness of students in following the learning, process, and learning outcomes as a whole [1].

In general, the learning process in senior high school is divided into two, namely the learning process in the classroom and the learning process in the laboratory. In the learning process in the classroom, the teacher places more emphasis on concepts and practice questions so students feel bored. This can be seen from students who are less enthusiastic and interactive during the learning process so that students are less active and do not use the opportunity when the teacher gives the opportunity to ask questions. In addition, because of the large amount of material that must be learned so that the learning activities in the clasroom are not enough. Therefore it's necessary to have additional learning activities that can be done anywhere and anytime.

Learning process in the laboratory, practical activities haven't been effective. This can be seen from the inadequate facilities in the laboratory which are the constraints of chemical practicum activities. Limitations of tools and chemicals that are in accordance with the chemical material that will be delivered are obstacles to the smoothness of chemical practicum activities so that students only do practicums without understanding the essential chemical concepts that can be obtained through the practicum.

Students feel that they have the opportunity to be active with other students as a benefit of the online learning environment [2]. The Indonesian Ministry of Education has realized the importance of ICT (Information and Communication Technologies) in learning, as well as implementing learning with multimedia, as in the HYLITE program (Hybrid Learning for Indonesian Elementary School Teachers) organized by the Consortium of 23 PT Indonesia in 2007 - 2010 which was held in ICT-based hybrids.

The implementation of learning using software for practicum remotely in secondary schools is organized and encouraged by the ministry of national education, which can be adapted for chemistry learning in high school with hybrid learning. The rapid development of ICT has introduced an HTML5 program that has not been used optimally as a learning media for Chemistry. Learning achievement scores of students who take hybrid learning are significantly higher than students who follow conventional learning [3].

Hybrid learning is a learning model that integrates innovation and technological progress through online learning systems with the interaction and participation of traditional learning models. Hybrid learning is a pedagogical approach that combines face-to-face instruction with computer-mediated instructions [4] Hybrid learning method is a combination of face-to-face instructional methods and online learning. The advantage of hybrid learning is that learning can be carried out independently by students, learning can be carried out more effectively and efficiently and able to increase accessibility because with the existence of hybrid learning students are easier to access learning materials.

Technology has potential to change the way teachers teach and learners learn. These systems may influence traditional teaching practices, while also introducing a complex layer of management to teaching programs [2]. Learning management systems such as virtual chemistry laboratory integrated hybrid learning at the forefront of recent technological advances in Higher Education.

Interactive learning environment by using animations and simulations for abstract topic, where students become active in their learning, provide opportunities for students to construct and understand difficult concepts more easily. Therefore, use of virtual chemistry laboratory integrated hybrid learning in labs, overcomes some of the problems faced in traditional hands-on laboratory and make positive contributions in reaching the objectives of an educational system. It is not always possible to see the results of students' studies in a traditional hands-on laboratory, especially in inadequate laboratory conditions. Use of virtual chemistry laboratory integrated hybrid learning can overcome that mistakes occur as a result of such laboratory conditions or misuse of the laboratory.

As part of science, chemistry cannot be separated from the scientific approach and practicum. Therefore it's necessary to have learning media in the form of interactive media so that students can do practical activities whenever and whereever. Virtual chemistry laboratory integrated hybrid learning is a practical solution to improve students' scientific inquiry in the context of senior high school environments [4]. The virtual chemistry laboratory software is at least as effective as real chemistry laboratory [5]. Therefore, a support laboratory such as the Virtual Chemistry Laboratory Integrated Hybrid Learning is needed to overcome the realities of real laboratories.

A virtual chemistry laboratory is a simulation that represents real laboratory experiments in as close a form as possible or a computer simulation that allows important functions of laboratory experiments to be carried out on a computer. Chemical learning materials emphasize the provision of direct learning experiences obtained through practicum activities, but there is abstract material. Visualization of subject matter and abstract objects is needed to improve students' understanding of the material. The benefit of having a virtual chemistry laboratory is that students can visualize and interact with the symptoms that will be experienced when conducting experiments in the laboratory.

Virtual experiments can be used to acquaint students with laboratory techniques and procedures prior to their laboratory sessions, so that they can be better prepared to conduct the same or similar experiments in a real-life chemical laboratory. It also allows students to study the items of apparatus, collecting and assembling items of the apparatus, familiarize themselves with laboratory techniques and procedures. It should be emphasized that virtual chemistry experiments are safe even for novices. Students can carry out such experiments that might be dangerous and expensive in a real laboratory [6].

Hybrid learning is a learning model that integrates innovation and technological progress through online learning systems with the interaction and participation of traditional learning models. Hybrid learning methods are a combination of face-to-face instructional methods and online learning. The intellectual ability of students greatly determines the success of students in gaining achievement. To find out whether or not someone is successful in learning, it is necessary to do an evaluation, the goal is to find out the achievements of students after the learning process takes place. Achievement is defined as the results achieved. Learning is closely related to achievement. Learning outcomes are a process of experience categorized into 3 fields, namely cognitive, affective, and psychomotor fields that appear in behavioural change [7]. Achieving learning outcomes that occur within the student is called learning achievement [5].

2. Methods

2.1 Research Design

An experimental research with a quantitative approach was set in this study. The samples in this study were set into three different classes according to the experimental manipulations, namely Class EG using both virtual chemistry laboratory integrated hybrid learning and traditional hands-on laboratory, Class CG-1 using virtual chemistry laboratory integrated hybrid learning only, and Class CG-2 using traditional hands-on laboratory only. The research design can be seen in table 1.

Table 1: Research Design

Class	Experimental Manipulations	Posttest
Experimental Group (EG)	X1	Y1
Control Group 1 (CG-1)	X2	Y1
Control Group 2 (CG-2)	X3	Y1

Note: X1 = thermochemistry teaching learning mediated by virtual chemistry laboratory integrated hybrid learning and traditional hands-on laboratory; X2 = thermochemistry teaching learning mediated by virtual chemistry laboratory integrated hybrid learning; X3

= thermochemistry teaching learning mediated by traditional hands-on laboratory; Y1 = multiple choiche of chemistry learning achievement.

2.2 Participants

The participants in this study came from public senior high school in Magelang regency, Central Java, Indonesia. School was chosen because has very good school accreditation and has adequate facilities for implementation of hybrid learning such as the internet network, computer, and smartphone. A total of 143 students with 16 years old average were the participants in this study. By cluster random sampling, those 143 participants then were classified into three different classes which were EG of 33 students, CG-1 of 33 students, and an CG-2 group of 33 students.

2.3 Data Collection

According to the objective of this research, the data that would be implemented were obtained from the multiple choice of chemistry learning achievement. The multiple choice of chemistry learning achievement consisted of 40 points of test. Empirical validity was done by testing the multiple test of chemistry learning achievement against the other students that were not used as the samples of this study. Those students used to test have certain criteria. The criteria in this case include having an average age of 16 years old and have studied thermochemistry subject matter. A total of 263 students were obtained to multiple choiche that has been developed. Based on the result of empirical validity analysis there were 40 items of multiple choice does fit with Partial Credit Model (PCM). In addition, the analysis results show an Alpha Cronbach's reliability estimate of 0,82.

2.4 Data Analysis

Data analysis involved normality and homogeneity test and Anova test. The normality test was used to determine the normal distribution of data, while the homogeneity test was used to know the homogeneity of the population. While Anova was used to know whether there was the significant-difference between chemistry learning achievement of students in experimental class and controlled classes.

3. Results and Discussion

The students' chemistry learning achievement data in this study was obtained from the multiple choice of chemistry learning achievement and measured after the experimental manipulations were performed in each class. Results of this study supported several research results reported in the related literature in which it was claimed that virtual laboratories enriched with simulations has many advantages on increasing students' chemistry learning achievement. Although the results of this study supported that the use of virtual chemistry laboratory integrated hybrid learning increased students' chemistry learning achievement levels and made a positive impact on students' attitudes towards chemistry.

Virtual chemistry laboratories have a positive effect on student achievement and attitudes [9]. We do not claim that virtual chemistry laboratory integrated hybrid learning more effective than the traditional hands-on laboratory activities. Instead, we claim that when we are forced not to perform traditional hands-on laboratory activities, due to reasons such as danger of chemical reactions, time concerns, lack of laboratory or equipment, or insufficient lab conditions which limit us to perform a simple laboratory activity, virtual chemistry laboratory integrated hybrid learning can be an alternative. The use of virtual chemistry laboratory integrated hybrid learning has become an increasing issue regarding science laboratories due to the increasing cost of traditional hands-on laboratories, and the increase in distance education [8].

Therefore, instructional materials are developed by using computers can be a solution for schools which has no science laboratory but has a computer cluster or laboratory. It is obvious that instead of making demonstrative experiment in schools where there is lack of laboratory or equipment or where there are over-crowded classes, virtual chemistry laboratory integrated hybrid learning can be an alternative. Another advantage of using virtual chemistry laboratory integrated hybrid learning is that the cost for preparing a science laboratory which is just used for science lessons is higher than for needed to prepare a computer laboratory which can be used for many different lessons.

However, active will of students to participate learning activities plays an important role to lead the learning being effective. Although a visual instruction developed by computerized technology would make a positive contribution to students' chemistry learning achievement in most of the scientific topics and concepts, it is also better to kept in mind that using simulations alone does not solve any problem. To gain a better result from the education, simulations must be supported with appropriate instructional methods and software related to the topics and concepts to be taught in the class. Moreover, a detailed lesson plan in which where simulations will be used, when and what students will do, the parameters related to the topics and concepts that students can change in the system must be defined. It is believed that only after these preparations, an ever-lasting learning can be achieved with simulations based activities in science education.

Learning achievements are the things that have been achieved in the mastery of specific skills in certain learning, where it will change the student's behavior and its change is relatively fixed [5]. The measurement of chemistry learning achievement of the students in this study was carried out in both experimental and controlled classes. The instrument used in this research was learning achievement test of 40 multiple choice questions which have been tested for validity and reliability.

The average number of chemistry learning achievement of students in the experimental class was higher than that of the students in controlled classes. It showed that the teaching-learning use virtual chemistry laboratory integrated hybrid learning and traditional hands-on laboratory was more effective to be applied in the chemistry teaching-learning process on thermochemistry in Senior High School in Magelang Regency grade XI at the academic year 2017/2018. It is because in the teaching-learning use virtual chemistry laboratory integrated hybrid learning and traditional hands-on laboratory, learners can visualize and interact with the symptoms that will be experienced when conducting experiments in hybrid learning laboratories can be done independently by students, learning can be carried out more effectively and efficiently, and able to increase accessibility, because with hybrid learning students increasingly easy to access learning material.

Based on the result, it can be concluded that the teaching-learning using virtual chemistry laboratory integrated hybrid learning on thermochemistry implicates that simulations prepared and used throughout the study had an instructional characteristics with positive contribution to education and improved students' chemistry learning achievement towards lesson at Senior High School in Magelang Regency grade XI at the academic year 2017/2018. Results of this study supported several research results reported in the related literature in which it was claimed that virtual chemimstry laboratory integrated hybrid learning combined traditional hands-on laboratory enriched with simulations has many advantages on increasing students' chemistry learning achievement. Thus, it can be reached a conclusion that the material developed and used in this study increased the students' chemistry learning achievement level by creating an entertaining learning environment.

4. Conclusion

The conclusion of this research was there is a significantly different in students' learning achievement among the three classes EG (class using both chemistry virtual laboratory integrated hybrid learning and

IOP Conf. Series: Journal of Physics: Conf. Series **1156** (2019) 012031 doi:10.1088/1742-6596/1156/1/012031

traditional hands-on laboratory), CG-1 (class using virtual chemistry laboratory integrated hybrid learning only) and CG-2 (class using traditional hands-on laboratory only) to the subject matter of thermochemistry concept. More research needs to be done to determine virtual chemistry laboratory integrated hybrid learning as a suplement for more traditional hands-on laboratory experiences.

References

[1] Kemendikbud 2013 *Elemen Perubahan Kurikulum 2013* Kementerian Pendidikan dan Kebudayaan
[2] Aan H, Susan W, Mallihai T and Denise B 2011 *Aust J. Teach. Educ.* **36**(7) 1-16
[3] Matthew A W, Anne M T, and Elizabeth C A 2016 *Am. J. Pharm. Educ.* **80**(9) 125-154
[4] O'byrne W L & Pytash K E 2015 *J. Adolesc. Adult. Lit.* **59**(2) 137-140
[5] Dermot D, John O, and Oliver M 2013 *Res. Sci. Educ.* **43** 1571–1592
[6] Zeynep T and Alipasa A 2013 *Educ. Tech. Soc.* **16**(1) 159-170
[7] Barney D, Andrea B, and Danny B 2003 *Proc. of the Improving Learning Outcomes Through Flexible Science Teaching The Universty of Sydney* pp 90-95
[8] Nana S 2010 *Dasar-Dasar Proses Belajar Mengajar* Sinar Baru Algesindo Bandung 102
[9] Tüysüz C 2010 *Int. Online J. Educ. Sci.* **2**(1) 37-53
[10] Ian H and Amy J P 2011 *Chem. Educ. Res. Pract.* **14** 516-523

Chemistry students' identity empowerment through etnochemistry in culturally responsive transformative teaching (CRTT)

Y Rahmawati*, A Ridwan, A Rahman and F Kurniadewi

Chemistry Education Study Program, Universitas Negeri Jakarta Rawamangun Muka Street, Jakarta, Indonesia

* E-mail: yrahmawati@unj.ac.id

Abstract. The paper portrays the second year report of a three-year longitudinal study of ethnochemistry integration in Culturally Responsive Transformative Teaching (CRTT) teaching model in secondary schools. The CRTT develop based on Culturally Responsive Teaching (CRT) as an approach that includes students' characteristics and cultural references in all aspects of learning and transformative learning framework. The study developed CRTT teaching model in the first year of study and consists of five phases of self-identification, cultural understanding, collaboration, critical reflections, and transformative construction. The qualitative methodology approach has been employed with observation, interviews, and reflective journals as data collection. The CRTT teaching model was implemented in chemistry classrooms from two secondary schools. The study involved 68 students of year-10 in the topic of electrolyte and non-electrolyte solution. The data was analyzed from the basic principles of cultural identity category. The results showed that the integration of ethnochemistry in CRTT teaching model has engaged students in empowerment their cultural identity and nationalism, cultural differences awareness, and learning identity development. The teaching model is also relevant to the policy of national curricula of nationalism in relation to character and culture identity development.

1. Introduction

Indonesia with its diversity in values, beliefs and practices consists of more than 17,000 islands and more than 300 ethnic groups [1]. In the classroom, cultural background differences affect students' values and ways of thinking and teacher-student interaction. According to [2], teachers need to understand their students' cultural background to engage them in learning experiences. Therefore, it is important to develop 'cultural awareness', not for perpetuating division but for developing understanding of diversity. It is important to equip the students to understand cultural differences as pointed out by [3], students need to have knowledge and skills to engage in differences of age, gender, sexual orientation, class, region, religion, race, ethnicity, ability, and nationality. Meanwhile, the process of globalization and cross-cultural interaction between ethnic groups influence the students' cultural identity and character. [4] stated that education should be able to develop students' competence and maintain their cultural identity. As related to student learning, students' learning stages can be identified as, rote learning, in-depth meaning making, and learning as a cultural phenomenon [5]. However, the study of Indonesia's cultural diversity in relation to chemistry learning for creating meaningful learning experiences has not been widely explored. Therefore, it is important to engage students with their cultural background.

The culturally-based teaching approaches is important in developing social and cultural identity. One of teaching approaches that can integrate culture in the classroom is Culturally Responsive Teaching (CRT) which an approach employs students' cultural references for students' learning engagement[4]. According to Gay [6], CRT applies the cultural knowledge, prior experiences, and performance styles of diverse students to create meaningful learning experiences for these students through their strengths. Multidimensional CRT involves content of curriculum, context of learning, classroom climate, relationships of student-teacher, instructional techniques, and assessments. Students can also participate actively in their own performance evaluations by being better human beings and more successful learners [6]. In the context of learning chemistry from cultural perspectives, ethnochemistry can be integrated. Ethnochemistry as a part of ethnoscience looks at culture from a chemistry perspective which involves a cross-discipline of social sciences and humanities perspectives [7]. In this study, the integration of ethnochemistry in CRT as one way to integrate scientific inquiry within students' cultural background. CRT could empowers students by using cultural referents to develop knowledge, skills, and attitudes [8]. In the culturally responsive teaching, teachers play an important role in engaging students in these ways of knowing. In learning within cultural phenomenon, students will have the opportunity to engage not only Western Science but also their own culture. Students should have abilities to harmonise their own culture (their everyday language) aligns with Western science Learning outcomes and experiences [8]. According to [9], cultural processes involves the dynamic process because of its personal meanings and subjectivity. Therefore, teachers could engage students in their personal worlds by integrating culture in science learning [7].

Chemistry, is one branch of science that studies materials that exist in the universe, the interactions among them and changes in energy-related or caused by the changes of nature [10-12]. In chemistry learning, students are expected to understand chemistry concepts in order to solve problems through building their individual knowledge [13-14]. However, in order to achieve this, students face challenges as many chemistry concepts are considered complex and difficult to contextualize with student lives [15]. In addition to the characteristic of chemistry learning is intact and meaningful if it is associated with the three levels of macroscopic, microscopic and symbolic [16-18]. However, the research shows that generally, students experience difficulties with chemistry because of their inability to visualize the structure and process at the submicroscopic level and connect it with the other chemical representative levels [19].

In the Indonesia context, several problems are evident as related to learning chemistry generally. Linking abstract concepts with students' daily lives, an overloaded curriculum, and rote learning or memorizing all contribute to the problem of effective chemistry education [20]. It is expected that students not only understand chemistry concepts, but also relate the issues to chemistry and play an active role in solving problems, either in the form of ideas or actions. Thus, to achieve these objectives, learning chemistry must be able to be a means of developing students' abilities in social life; to resolve the kinds of problems that require the chemistry teacher to be more creative in applying chemistry to contextual issues in everyday life to create meaningful learning experiences. In relation to chemistry education in Indonesia, one of the main aims in Curricula 2013 is to develop students' scientific inquiry and national character through meaningful chemistry learning experiences. In addition, curriculum 2013 embraces student-centered learning in accordance with their background, characteristics, and a students' prior knowledge and the teacher should act as a facilitator in learning so that students can engaged in their learning experiences [21].

Therefore, this research focuses on the integration of Culturally Responsive Transformative Teaching (CRTT) teaching model which has been developed in the first year of study. The CRTT teaching model is developed based on the principles of CRT and transformative learning consists of five phases of self-identification, cultural understanding, collaboration, critical reflections, and transformative construction. The previous study of CRT approach in chemistry learning has empowered students in chemistry learning [22]. In this paper, the integration of ethnochemistry in CRTT teaching model was explored in the concept of electrolyte and non-electrolyte solution in year-10 chemistry classrooms. In this paper, the implication of CRTT teaching model is explored in relation to students' identity.

2. Methods

The research employed a qualitative method with observation, interviews, reflective journals, and document analysis as data collection. The research was conducted at secondary schools in year 10 in one public school and one private Islamic school, in relation to the concepts of electrolyte and non-electrolyte solution. The participants consist of 68 students from two classrooms in each school. The study consisted of three main phases: 1) exploring classroom and students' cultural background through observations and interviews, 2) CRTT teaching model development, and 3) CRTT implementation by exploring students' identity. The CRTT teaching model involves 5 main steps of self-identification, cultural understanding, collaboration, critical reflective thinking, and transformative construction [22]. In these five steps, students had opportunities to develop self-understanding of their character, their learning style, and their cultural identity. They also learned about others and collaborated as shown in the model below.

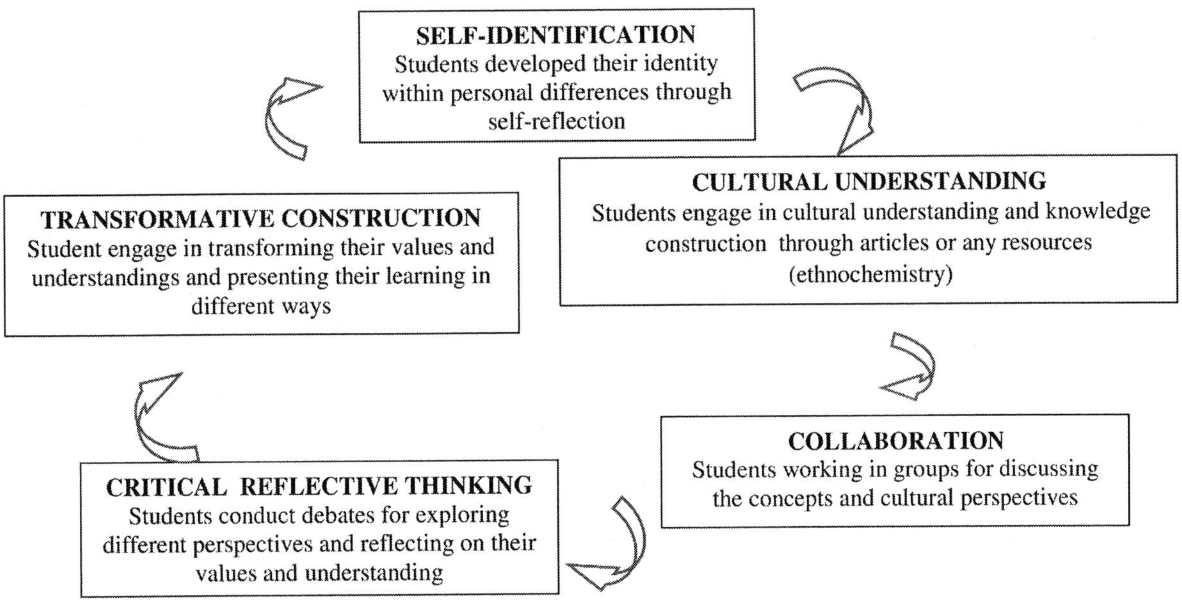

Figure 1. Culturally Responsive Transformative Teaching (CRTT) teaching model

The process above (figure 1.) provides opportunities for teachers and students in exploring their identity and respecting differences. It also involves personal and social learning.

3. Results and Discussion

The study results show the process of integrating ethnochemistry in a CRTT teaching approach in the context of chemistry learning. The researchers worked with the teachers to integrate the model for engaging students. Discussions are explained in two parts according to the model of integration and are followed by the implications that can be made on student empowerment in chemistry learning.

3.1. Culturally Responsive Transformative Teaching (CRTT) Teaching Model

In the first step of **self-identification and cultural understanding** was the process by exploring students' diverse backgrounds through an in-depth interview. It was found that 80% of students were born in Jakarta, however, most of their parents came from Java which is a different ethnic group. The students reflected on their cultural identity within different roles as family members and as students. The researcher conducted classroom observations before implementing the CRT so as to understand classroom culture. **For cultural understanding, the content integration** principle by [23] was employed by providing learning materials associated with the students' cultural background, daily lives, and the chemistry curricula. There were two articles of Jeruk Nipis and Green Coconut Water in relation to the concepts of electrolyte and non-electrolyte solution. Besides, they had to explore their own cultural practices in relation to chemistry concepts, the presented in the classroom. The stories have been developing in different chemistry curricula for engaging students in learning their culture from chemistry perspectives. Students are expected to find the relation of chemistry concepts and of the electrolyte and non-electrolyte solution that they find in the story. Through this content integration, the teacher can develop positive teacher-student relationships through positive relationships between teachers and students. Teachers must believe in their students' ability and provide them with valuable experiences into the classroom [23].

After students have reflected on themselves the next step in the research process was **the collaboration** with teachers and students. In this step, the teachers developed positive student-student interactions by implemented different teaching models. Students engaged in active discussions and collaborations within a group. Teachers provided time for the students to interact with their peers in positive interactions helping them to improve collaboration and communication skills. The next phase, after collaboration, was **critical reflective thinking and transformative construction**. Debates were conducted so that the students could explore different perspectives and reflect on their values and understanding. After reflecting, students engaged in transforming their values and understanding and presenting them in different ways by reflecting on the question *How have the values and perspectives that they hold transformed because of the learning experiences?* The question is relevant to the five ways of being involved in transformative learning, as expressed by [24]: cultural-self knowing (self-realisation), relational knowing (opening to difference), critical knowing, visionary and ethical knowing (over the horizon thinking), and knowing in action (making a difference). These ways of knowing provide opportunities for students to understand themselves and develop the cultural awareness of appreciating differences, especially in collaborating with others.

3.2. Students Identity Empowerment in Chemistry Learning

The CRT approach has impacted on student empowerment in chemistry learning through cultural identity and nationalism, cultural differences awareness, and learning identity development. These aspects were found in each of the different schools. The following case study in each school will be explored in detail other papers.

3.2.1. Cultural identity and nationalism. The ethnochemistry approach not only develops knowledge, but also cultural identity and nationalism. According to [25], identity is a learning process which is constructed with the emotions involvement by the students. The study is relevant according to the definition of cultural identity as development by [26] which describes identity as self-understanding of the social and cultural. When students collaborated with other students they developed their identity was socially constructed in a social/cultural context which included sharing beliefs, relationships, customs, symbol systems (language), knowledge, physical settings, and objects [27]. According to [28], identity involves all factors that constitute our lives, including genetic makeup, our parents, culture, experiences (bad and good, love and suffering). In the following statement students show their cultural identity after it was integrated with the learning culture.

"We learned our own which I realized that it is important to maintain my culture and understand others"
(Student Interview, School A, February 1, 2017)

"In the chemistry learning, the students were engaged and asked the critical questions regarding the cultural practices of Green Coconut Water in the culture
(Observation, School B, January 30, 2017)

"We have to maintain our indigenous knowledge, especially the traditional medicine, so it can compete with modern medicine"
(Students' Reflective Journal, School B, January 30, 2017)

In addition to introducing the idea of bringing their indigenous knowledge into the chemistry classroom, [5] the idea of pluralism in learning science relates to the interaction between students culture and the Western science. Students also face different cultural experiences of their own experiences (and the cultures of science itself. Therefore, learning should recognize this border crossing of cultural experiences. [5] proposes a different idea of enculturation by engaging students through their own identity which recognizes the equity of all students. In addition, it will help teachers to engage students' in their personal worlds [9].

The integration of ethnochemistry was an innovation in the chemistry classrooms in this study. The students felt the learning environment was different to the usual chemistry learning environment where practice in solving the mathematics problems in chemistry was the norm. The CRT approach stimulated their curiosity and motivated them to, not only learn chemistry concepts but also identify different cultural practices. Below is a statement from a student who showed curiosity in the ethnochemistry. The students look more enthusiastic when exploring the chemistry concepts in the cultural practices. Their motivation was stimulated by integrating ethnochemistry in the learning because it is related to everyday thinking. Nanochemistry has stimulated the students learning motivation by describing the chemistry concepts as applied in cultural traditions which make students more interested in chemistry learning. This has led to meaningful learning experiences as recognized by studies of student understanding that support solutions through the use of meaningful learning experiences [29-30]. In this study, students engage in a new learning approach, environment, and in relation to students' daily lives.

"We were engaged with the application of chemistry concepts in everyday life, especially in relation to our culture"
(Students Interview, School B, January 30, 2018)

The CRT approach provides opportunities for students in respecting others' cultures and uses these opportunities for their learning. Culturally responsive teachers develop intellectual, social, emotional, and political learning by "using cultural referents to impart knowledge, skills, and attitudes" [31] explains that culturally responsive teachers not only focus on students' scores, but also cultural identity [3]. In addition to integration, the culture of society and everyday life also relates to nationalism. In relation to this, students stimulated to develop their nationalism, students explained that they become engaged with Indonesia culture. Because of the impact of globalization, they came to understand that they have a limited understanding of their own culture which is considered to be a loss of their cultural identity [28].

3.2.2. Cultural Differences Awareness. The students also learned the different cultures they have to complete the project of exploring their own culture in relation to chemistry concepts. Therefore, at the same time, they have to work with others from different cultures in completing the tasks.

"Its excited that I learned other cultures and worked together to completed the task"
(Student Reflective Journal, School A, February 1, 2017)

" I realized we are different, it should be a problem, but I prefer working with others who care about others and willing to complete the task"
(Student Interview, School B, February 6, 2017)

Based on the above statement from either interviews or reflective journals, the students expressed how they felt when cooperating with each other. They also said they learned to understand and appreciate each other. They realise that it was challenging for working with others who shared differences. They have to learn to collaborate with others, develop empathy and communicate clearly in order to succeed in group working.

Grouping students based on the differences in gender, achievement levels, and ethnic groups helped students develop their empathy with others. Attitudes to listen and respect others' opinions are the characteristic of empathy of accepting another person's perspective, be able to listen to others and be sensitive to the others' feelings. Cultural competence is the ability to successfully teach students who come from different cultures. It helps to develop students' personal and interpersonal awareness, and a set of skills of effective cross-cultural teaching [32]. The teachers tried to understand students' differences and provide different learning experiences. Students developed attitude to be sensitive to others which is demonstrated through the following statement.

"Even though I prefer to work with my close friends as I know them closely, but working with others has helped to understand others and learn new things
(Students Interview, School A, February 6, 2017)

Students indicated that they became more used to talk and getting to know their classmates' character which leads to being good collaborators. In addition, the teaching approach which was implemented helped students to develop their self-confidence. Students appeared confident when discussing in groups by expressing opinions and conveying information held to classmates. In addition to the above, there are suggestions that culturally responsive learning has stimulated students to develop responsibility. The learning process, using cooperative discussions, engaged them in understand the task objectives to be completed.

3.2.3. Learning Identity. Students understand that they have developed their understanding of their identity as the learners. They were realized that they usually learn chemistry from textbook and followed the teachers' instructions. Becoming active learners became challenging since they have to move from their comfort zone from the passive learners to the active learners. In addition to the teachers are empowered to bring culture alongside chemistry concepts helped students develop their knowledge. Teachers should engage students to construct their own knowledge through active processes [33]. In addition [34] stated that the students need to develop their knowledge by interact with their environment.

Students learned to develop their learning identity through their own culture. In this teaching model, students have to collaborate with others in order to construct their understanding of electrolyte and non-electrolyte solution. Even though, the students still focused on learning to the test and giving the right answers. In this stage, the students have been challenged by critical and creative thinking in facing the given problems.

"It is exciting that we had different learning experiences, from practicum, debate, and learning cultures. We are challenged to solve the problem given"
(Student Reflective Journal, School A, February 8, 2017)

" We are not only sitting in the classroom, we are thinking. We feel free to express our ideas"
(Student Interview, School B, February 15, 2017)

"We criticize the problem solution to find out the best, it becomes important to argue in discussion"
(Student Reflective Journal, School A, February 1, 2017)

The statements demonstrate that the students showed their engagement while solving a problem. Students stated that during the lesson the teacher assigns students to develop their critical voices in the debate. The above statements show that it is important for students to build an effective understanding of chemistry concepts as according to [35], it is important for students to retain new knowledge with their critical thinking which required them to be active learners.

Finally, integrating cultural ethics and values into the chemistry learning should be developed in relation to empowering students' identity development within character education.The problem is not in the usefulness of knowledge but in how to help them in using their knowledge for their future roles. The previous study showed that the students faced the challenges in shifting the learning from focusing into exams [36]. Teachers should facilitate students in understanding their identity by approaching with the contemporary context of problems, interests, issues, and concerns. However, both teachers and students need to shift their values and beliefs in their perspectives of teaching and learning.

4. Conclusion

The CRTT teaching model has been implemented in this study which consists of the five main steps of self-identification, cultural understanding, collaboration, critical reflective thinking, and transformative construction. The students showed empowerment in developing their cultural identity and nationalism, cultural differences awareness, and learning identity. The learning experiences have stimulated their learning motivation and curiosity in order to develop their knowledge of chemistry and culture. The implementation of CRTT stimulated students' awareness of the role of chemistry in their everyday lives, especially in their culture that also affected their cultural understanding. In applying this approach, the teacher should develop their creativity and ability to prepare the learning resources for students, especially in ethnochemistry topics. This approach also provides the opportunity for parents to participate in students learning as part of the learning community.

References

[1] Blackley S, Rahmawati Y, Fitriani E, Sheffield R and Koul R 2018 *Issues in Educ. Res.* **28** 18-42
[2] Fitzpatrick KR 2012 *Music Educ. J.* **98** 53-9
[3] Griffin RA and Jackson NR 2011 *Communication Teacher* **25** 1-6
[4] Ladson-Billings G 1995 *Am. Educ. Res. J.* **32** 465-91
[5] Aikenhead G S 2000 Improving Science Education, Eds In R Millar, J Leach, and J Osborne, (Philadelphia: University Press) pp 245 – 64
[6] Gay G 2000 *Culturally Responsive Teaching: Theory, Practice and Research* Ed Banks James A (New York: Teachers College Press) pp 1-21
[7] Ingold T 2000 *The Perception of the Environment: Essays on Livelihood, Dwelling and Skill.* (London: Psychology Press) pp 145-60
[8] Lee JS and Anderson KT 2009 *Rev. Educ. Res.* **33** 181-211
[9] Hollan D 2000 *Am. Anthropol.* New Series **102** 538-50
[10] Oxtoby DW, Nachtrieb NH and Gillis HP 2001 Principles of Modern Chemistry, 4th Edition (London: Brooks Cole) pp 392-473
[11] Goldberg DE 2005 *Schaum's Outline of Theory and Problems of Beginning Chemistry* (Amerika: The McGraw-Hill Companies, Inc) pp 1-6
[12] Brady J E 2000 Chemistry: Matter and Its Change (New Jersey: John Wiley & Sons) pp 5-8
[13] Taber K 2002 *Chemical Misconceptions: Prevention, Diagnosis and Cure* Ed Osborn C and Pack M (London: The Royal Society of Chemistry) p11
[14] Koballa 2010 *Int. J. Sci. Educ.* **22** 215-17
[15] Pfundt H and Duit R 2000 *Bibliography: Student's Alternative Frameworks and Science Education* (Kiel, Germany: University of Kiel) pp 65-8
[16] Johnstone A H 1991 *J. Comp. Assisted Learning* **7** 75-83

[17] Gabel D L, Briner D and Haines D 1992 *Sci. Teach.* **59** 58-63
[18] Garnet P J, Garnet P J and Hackling M W 1995 *Stud. Sci. Educ.* **25** 69-95
[19] Chittleborough G and Treagust D F 2007 *Chem. Educ. Res. Pract.* **8** 274-92
[20] Rahmawati Y 2013 *Revealing and Reconceptualising Teaching Identity Through the Landscapes of Culture, Religion, Transformative Learning, and Sustainability Education: A Transformation Journey of a Science Educator* Doctoral thesis (Australia: Curtin University) p116
[21] Ministry of Education and Culture 2013 *Pengembang Pendidikan Budaya dan Karakter Bangsa.* (Jakarta: Ministry of Education and Culture) pp 11-9
[22] Rahmawati Y and Ridwan A 2017 *Chemistry: Bulg. J. Sci. Educ.* **26** 813-30
[23] Hernandez C M, Amanda R M and Shroyer M G 2013 *Cult. Stud. Sci. Educ.* **8** 803-20
[24] Taylor P C 2013 *Meaning-Centered Education International Perspectives and Explorations in Higher Education* Ed O Kovbasyuk and P Blessinger (New York: Routledge) pp 168-85
[25] Geijsel F and Meijers F 2000 *Educ. Stud.* **31** 419-30
[26] Erickson M 2010 *Hist. Hum. Sci.* **23** 68-91
[27] Vygotsky L S 1978 Mind in society Ed M Cole, V John-Steiner, S Scribner and E Souberman (Cambridge, MA: Harvard University Press) pp 79-91
[28] Palmer P J 2004 *A Hidden Wholeness: The Journey Toward an Undivided Life* (San Fransisco: Jossey-Bass-A Wiley Imprint) pp 31-50
[29] Rickey D and Stacy A M 2000 *J. Chem. Educ.* **77** 195-211
[30] Treagust D F, Chittleborough G D and Mamiala L T 2004 *Res. Sci. Educ.* **34** 1-20
[31] Ladson-Billings G 1992 *Theor. Pract.* **31** 312-20
[32] Diller J V and Moule J 2005 *Cultural Competence: A Primer for Educators* Ed. Kerr M D, Dashiell R and Allen G (Belmont, CA: Wadsworth) pp 225-30
[33] Henderson J G and Kesson K R 2004 *Curriculum Wisdom* (New Jersey: Pearson Education) p79
[34] Rahmawati Y and Koul R 2016 *Issues in Educ. Res.* **26** 147-64
[35] Bekoff M 2000 *Hum. Ecol. Rev.* **7** 59-67
[36] Rahmawati Y and Ridwan A 2017 *AIP. Conf. Proc.* **1868** (1), 030009

Acknowledgement
We would like to thank Ministry of Research, Technology, and Higher Education, Indonesia for the research grant, Universitas Negeri Jakarta, SMA 51 Jakarta, and SMA IT Nurul Fikri for their support.

Developing critical and creative thinking skills through STEAM integration in chemistry learning

Y Rahmawati[1*], A Ridwan[1], T Hadinugrahaningsih[1] and Soeprijanto[2]

[1]Chemistry Education Study Program, Universitas Negeri Jakarta, Rawamangun Muka Street, Jakarta, Indonesia
[2]Electrical Engineering Vocational Education Study Program, Universitas Negeri Jakarta, Rawamangun Muka Street, Jakarta, Indonesia

[*]E-mail: yrahmawati@unj.ac.id

Abstract. This paper describes a second year of two-year of longitudinal study of the integration of Science, Technology, Engineering, Art, and Mathematics (STEAM) in chemistry learning in secondary schools in which the importance of critical and creative thinking skills were explored. The study involved 76 students of Year 11 from two secondary schools in Indonesia. The STEAM integration was implemented in the teaching of acid and base through a project-based learning model. The research employed a qualitative approach to explore the integration of STEAM in chemistry, students' critical and creative thinking skills development. Data were collected through observations, interviews, reflective journals, critical and creative thinking skills assessment. The students develop their STEAM projects by integrating chemistry concepts with STEAM principles. The study results were analyzed based on the pattern themes found in different data sources. It showed that students developed creativity and critical thinking skills. The study faced challenges of STEAM integration with chemistry concepts, empowering teachers, engaging students, and managing time and resources. The integration of STEAM provided the opportunities for students in chemistry learning engagement and several areas of thinking skills development in relation to the current curricula objectives.

1. Introduction

The science and technology development in various aspects of life has an impact on the demands of graduate competencies. Therefore, the learning reform needs to be conducting in implementing education in 21st century era. Various efforts for preparing the 21st century generation have been conducted. The aims of education in secondary school should ensure the students to have competences in facing the challenge in college and in the workplace [1]. However, the stakeholders of the companies claim that the secondary schools graduates don't have the necessary basic skills that they need. In Indonesia, the secondary schools consist of the senior high schools and vocational schools which have different aims of preparing graduates for continue the study and work. However as stated by [2] there is still debate in postsecondary institutions in defining "readiness".

The 21st-century learning has been discussed for facing the challenges in revolution industry 4.0. The term 21st-century skills is generally refer to core competencies of collaboration, digital literacy, critical thinking, and problem-solving in facing globalisation world [3]. In addition to the Partnership for 21st Century Skills also believes that graduates should have learning skills in information, the communication, critical thinking and problem solving [4]. Therefore, it is important to teach the higher order thinking skills, especially problems solving in real-world problems [5]. Rich stated that 21st-century learning refers to students should have competences in producing, synthesizing, and evaluating information within different cultures [6]. As the results, they demonstrate digital literacy and civic responsibility. Virtual tools and open-source software create borderless of students' learning [7]. The design of classroom learning experiences should relate to students' competencies development within this 21st-century learning. Besides, these learning experiences should empower students as individual and citizen [8].

Chemistry is the study of matter and the changes, interactions, and rules that underlie its change [9]. Chemistry education focuses on students' learning to understand the chemistry concepts, however, chemistry with its abstract concepts is considered difficult for students to connect it with the real situations [10]. According to [11], chemistry education research should focus to improve chemistry learning engagement. In this paper, not only focused on the fundamental concepts, but also 21st-century skills development [12].

[13] states that the STEAM approach is an evolving approach of integrating art into the STEM. It plays the important role to enhance understanding of the structure of science, technology, art, and mathematics. In this study, the STEAM has been integrated with project-based learning which focusing on student-centered within dynamic classroom approach. The integration of art in STEM learning can make learners more creative and innovative as students involved in exploration of real-world challenges and problems [14-16] suggest that STEAM as an integrated approach to encourage creativity. In STEAM, learners are taught to think comprehensively. STEAM is integrated into the curriculum as a whole as a supporter of project-based learning methods. Learners will learn, solve and analyze problems using technology tools and collaborative learning strategies across the STEAM curriculum [17]. Each learner will be given the material cross-science by putting forward work collaboration and high social capabilities. This educational system is believed to trigger learners into individuals full of responsibility, leader, and entrepreneur [18].

This STEAM approach leads learners to develop problem solving, critical thinking, and collaboration skills [19]. The previous study by [2] stated that the STEAM approach has engaged students in developing 21st century skills. The main aim of education is not only to teach basic knowledge, but to use thinking skills such as creative thinking skills [20], problem-solving skills [21], scientific and technological abilities [22-23] as these are skills needed for sustainability and lifelong education. Starko made a discussion about the importance of creativity in the classroom, so creative students can learn more in the classroom [24]. In addition, the National Center for Education and the Economy [25] states that for the curriculum reform in education, creativity and innovation are essential. In addition to [26] also predicts that in the future, the world will be increasingly complex with problems that require new solutions, therefore creative and critical thinking are important. In addition to critical thinking allows learners to evaluate the evidence, assumptions, logic, and language that underlies others' statements [27]. According [28], critical thinking is a well-organized mental process in decision-making to solve problems by analyzing and interpreting data in scientific inquiry activities. It employs evidence, logic, and understanding to solve the problems.

In Indonesia, the current curriculum of Curriculum 2013 embraces the pattern of students' character development and the 21st-century learning principles. Chemistry in context should be taught, so that students can learn to solve real lives problems by using their chemistry knowledge. In this study, the innovation of integration of STEAM in chemistry learning within Indonesia context has been explored. According to [15], STEAM approach can develop students' understanding and creativity which has been implemented in Korea and 17 other countries. In addition to enrichment of chemistry learning within perspectives of different disciplines. Therefore, the research focusing on the implementation of STEAM and exploring the implications on students' learning.

2. Method

This qualitative research was employed the with data collection of interview, observations, reflective journal, and critical and creative thinking assessment. The quantitative results of test were not explored in this paper. The activities carried out through the phase of relating, planning, developing, cooperating, and transferring [30]. The focus on developing the STEAM project with employed acid base chemistry concepts as the principles through the hydroponics plants project. The steps of teaching methods in STEAM approach shown below.

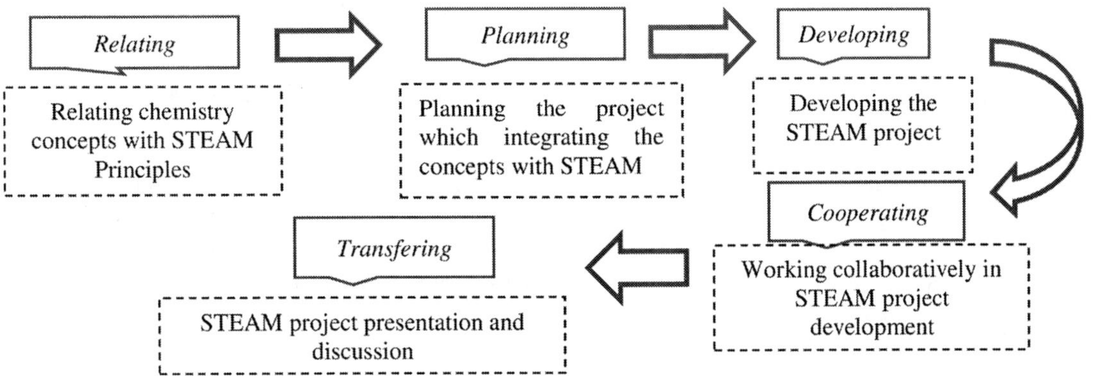

Figure 1. Project Based Learning in STEAM Project [30]

In the project, students had to find out the pH that the plan can grow well. They have tried different acid and base solution to get the best condition for the plants. In addition to they have to learn in constructing the hydroponic plants the challenges were integrating the STEAM in chemistry learning as chemistry is studied as a discipline, not as an interdisciplinary subject. Therefore, relating step becomes the main challenge in the project development. In each process, teachers play important roles as facilitators, then students' working collaboratively in completing the project.

3. Results and Discussion

In this initial study of the implication of STEAM project in students' learning has been explored which focusing on critical and creative thinking skills. As stated before there are challenges in integrating chemistry subject with different disciplines. The integration of STEAM has encouraged students to exploring daily lives problems by using their chemistry knowledge. STEAM project was conducted with the group working. The STEAM approach which was integrated through project-based learning of hydroponics plants has helped students to develop knowledge and skills by working for certain period of time. In the project, students have integrated the concept of STEAM (figure 1). In science, they have to find out the pH of suitable for Kangkung to grow up which is related to acid and bases concept. In technology, engineering, and mathematics, the students developed and designed the hydroponics pot structure to optimize the plants' growth, including using the aerator. Then in art, the students engaged in creating an aesthetic structure of hydroponics plants.

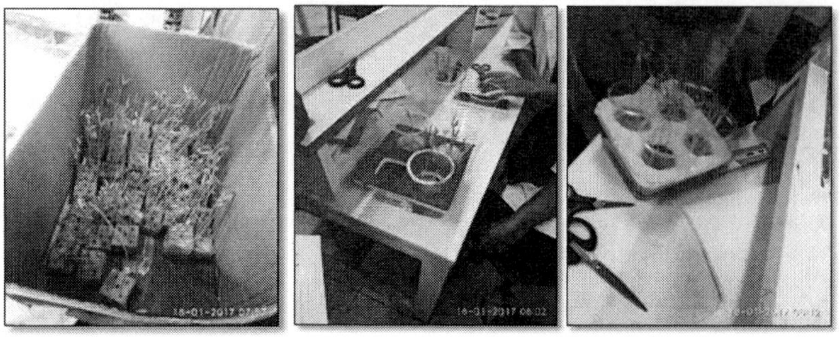

Figure 2. Hydroponics STEAM Project

In the process of a STEAM project, the students are stimulated to develop their critical and creative thinking skills as stated below.

The student asked if the aerator is not used, do the plants die? Should it have certain pH?
(Observer of Students 1, January 18, 2017)

after continuous observation, we found that the kangkung died at pH 12 because the solution's pH is too high, and not suitable with suitable pH from literature which states 6.5- 8.4
(Student Interview, Student 29, January 18, 2017)

Our group hydroponics plants have a pH of 8 when was checked using universal indicators and pH meters. It is because our hydroponics nutrition solution consists of 100 mL of HCl solution and 300 mL of KOH solution so it tends to be alkaline
(Student Observation Sheets, Student 11, January 30, 2017)

The students also found out and solve the problems by found out different learning resources, including internet and books. The students explored different resources and asking critical questions in completing the project.

We concluded that the aerator used should not be exposed to water because if the battery is exposed to water it will die
(Student Observation Sheets, Student 33, January 11, 2017)

I recorded the color change after solution added natural indicators and matching observations with reference to books and on the Internet
(Student Interview, Student 11, January 18, 2017)

Ruby planted crops with hydroponics media cannot grow properly because Rubi plants kangkung at pH 3 and 8, where the pH is not a suitable pH. pH suitable for kangkung that is 6,5-7. "
(Test Results, answer of Student 2, January 25, 2017)

The students learned to develop different ways of learning by asking critical questions and explored different resources in the project. According to [29], the students had opportunities to express their arguments which was different from their learning experiences. According to [19-30], the integration between STEAM learning and project-based learning not only keeps students learning basic theories in project completion techniques but learners will also learn problem-solving skills, critical thinking, and collaboration skills. Project-based learning enables learners to go through a lengthy process of inquiry, responding to questions of complex problems, or challenges, practicing the skills demanded in the 21st century, one of which critical thinking is [3]. Students' creativity and critical thinking have been stimulated in this STEAM project as students have to find and solve the problems in hydroponic plant within their understanding of acid-base concepts.

In this study, students can continuously develop their critical thinking by asking questions, defining a problem, examining evidence, analyzing assumptions, and considering other interpretations [31]. The result shows the opportunities to further development of higher order thinking skills in STEAM integration in chemistry learning. However, teachers play important roles to stimulate the process. In globalization world, information, media, and technology skills are important, thus in a STEAM project, students learn to find out the information for the project development and the problems solving of determining the best pH conditions.

Students are able to use various creative ways to get ideas, such as found out different answers and ideas to solve the problems. Students were able to create new ideas by using their experiences. Therefore, when the learning process takes place, it is important to stimulate the students to provide different ways or suggestions to do various things.

I felt my creative ideas were developed because I found different new things in the project
(Student Reflective Journal, Student 11, January 23, 2017)

My creative idea is flourished, I knew new things, such as the pH solution for hydroponics' plants
(Student Reflective Journal, Student 11, January 13, 2017)

Their answers are the indicators of the ability to think fluently which when students can generate ideas, questions, and answers in varied ways. Students can see a problem from different perspectives. As a result, when they face different problems, the students can change the way of thinking to provide various alternative solutions to solve the problems.

Student's (number 24) way of thinking was developed as they were able to create questions from different perspectives. Other students focused on the plant's stem, while student 24 asked about the roots of plants
(Observation of, Student 24, January 23, 2017)

Learners 24 have a good flexibility in thinking skills, this can be seen through the observation results when the students 24 answered the question when the discussion. In addition, the students also keep asking questions

Our plants were wilted because it is not exposed to sunlight as we put the plants on the table
(Student Observation Sheets, Student 9, January 11, 2017) "

"What is this technology called? What does it do for? Later if the batteries run out how the effect on our plants?
(Observation of Student 34, January 18, 2017) "

Students have generated the ideas into designing a product and carrying out activities. Therefore, learners become accustomed and continue to develop their creative thinking skills.

The cause of one wilted plant is because the nutrient solution is alkaline, while the cause of the three wilted plant is due to the acidic nutritional solution. Plant 2 does not wilt because it is neutral, then the need for plant minerals is sufficient.
(Test Results, answer of Student 9, January 9, 2017)

Based on the answers given by learners 9, overall it is close enough to the indicators and test instructions. However, learners 9 have not been able to explain in detail the reasons. Students have faced the challenges in looking different information to solve the problems in the project. They understood that the information widely provided on the internet, however sometimes it is not considered as valid information. Therefore, they not only learned to explore the information but also to evaluate it.

Jackson and Shaw discuss strategies for developing students' creativity and designing learning systems to help students to develop their creativity [32]. For example, providing opportunities for learners to experience and practice their own creativity through stimulating, relevant, and authentic learning experiences in the field of their subject matter. In addition, it also gives students with the problems in the project which allow them to practice for developing creativity and active inquiring. In addition, individuals who are creative have a positive attitude toward creative thinking [32]. Therefore, creative thinking as a logical and divergent thinking to produce something new. This is especially important in the Curriculum 2013 that requires the development of the holistic students. Chemistry learning is also expected to educate graduates who are able to interact socially with the environment, school environment, community, and environment.

Finally, the STEAM approach has helped students to relate the chemistry concepts and different disciplines of STEAM through an application of knowledge into everyday life. The STEAM approach provides the opportunity for helping students to understand working in real-life projects through collaborative working and use their knowledge and skills from different subjects, beside the challenges of relating the chemistry concepts, time and resources management.

4. Conclusion

The study explored that STEAM approach can be integrated into chemistry learning within real-lives problems in Indonesia context. The important elements of critical and creative thinking skills have been stimulated in the study by active self-reflection and questioning followed by defining the problem, examining evidence, analyzing assumptions, and considering other interpretations. Students learnt to implement their chemistry knowledge in solving the problems within different ways of knowing. They learnt to think fluently by exploring problem solution from different perspectives. In Indonesia context, the STEAM integration has provided the opportunity to develop students' competencies in relation to the curriculum 2013. The challenges of integrating STEAM within the chemistry curricula, empowering students, and managing the teaching and time resources need to be explored. However, the further research will be conducting in exploring the challenges of implementing the STEAM in chemistry learning.

References

[1] O'Sullivan and Dallas K B 2010 *Education Libraries* **33** 187
[2] Olson L 2006 *Education Week* **25** 201
[3] Partnership 21 2008 *21st Century Skills, Education and Competitiveness: A Resource and Policy Guide* (Tucson: Partnership 21) p 13
[4] Jobs for the Future 2005 *Education and Skills for the 21st Century: An Agenda for Action.* (Boston: Jobs for the Future) p 273
[5] Davies H A, Csette J and Poon L K 1999 *Int. J. of Eng. Ed.* **15** 191-98
[6] Rich E 2010 *Education Week Teacher PD Source Book* **04** 32-35
[7] Scott C L 2010 *The futures of learning 2: What kind of learning for the 21st century?* ERF Working Papers Series, No. 14 (Paris: UNESCO Education Research and Foresight)
[8] Rotherham A J and Willingham D 2009 *Educ. Leadersh.* **67** 16-21
[9] Chang R 2005 *Kimia dasar konsep-konsep inti* (Jakarta:Erlangga) pp 683-89
[10] Sirhan G 2007 *J. Turkish Sci. Ed.* **4** 2-20
[11] Gabel D 1999 *J. Chem. Ed.* **76** 548-54

[12] Treagust D F, Chittleborough G and Mamiala T 2015 *Int. J. Sci. Ed.* **25** 1353-68
[13] Crayton J and Svihla V 2015 *The STEAM Journal* **2**
[14] Blackley S, Rahmawati Y, Fitriani, E, Sheffield R, and Koul, R 2018 *Issues in Educ. Res.* **28** 18-42
[15] Yakman G and Hyongyong L 2012 *J. Korea Association Science Education* **32**
[16] Guy B 2012 *the 63rd International Astronautical Congress* (Naples: Italy) p 9995
[17] Blumenfeld PC 2011 *J. Ed. Psy.* **26** 369-98
[18] Jeon S and Lee Y J A 2014 *Society for Information Technology & Teacher Education International Conference* (Chesapeake, VA)
[19] Messier N 2015 *The how's and why's of going 'full steam ahead' in your classroom* (Burlingame: Edsurge)
[20] Runco, MA 2015 Creativity theories and themes: *Research, development and practice* (USA: Elsevier Inc.) pp. 230-35
[21] Segal J W, Chipman S F and Galser R 1985 *Thinking and learning skills: Relating instruction to research.* (New York: Routledge) pp. 327-30
[22] Tortop H S 2013 *J. for the Education of the Young Scientist and Giftedness* **2** 21-31
[23] Lawless K A and Brown S W 2015 *The Curriculum J.* **26** 268-89
[24] Starko A J 2012 *Creativity in the classroom* (New York: Taylor & Francis) pp. 273-81
[25] NCEE 2007 *Executive Summary* (NCCE:Washington)
[26] Cramond B 1999 (San Diego-Academic Press) pp. 84-89
[27] Larmer J and Mergendoller J 2010 *Educ. Leadership* **68** 34-37
[28] Johnson 2007 *Berpikir Kritis dan Kreatif* (Bandung: MLL) pp. 183-86
[29] Rahmawati Y and Koul R 2016 *Issues in Educ. Res.* **26** 147-64
[30] Hadinugrahaningsih T, Rahmawati Y and Ridwan A 2017 Proc. Of *ICRIEMS* (AIP Conf. Pro., 1868, 030008. http://dx.doi.org/10.1063/1.4995107)
[31] Wade C 1995 *Teach. Psychol.* **22** 24-28
[32] Jackson N and Shaw M 2006 *Developing subject perspectives on creativity in higher education, Ed* N Jackson, M Oliver, M Shaw and J Wisdom, (New York: Routledge) pp 89-108

Acknowledgment
The authors thank the Ministry of Research, Technology, and Higher Education for the research funding, Faculty of Mathematics and Natural Science, Universitas Negeri Jakarta, and SMA 51 and SMA 107 Jakarta for the supports and collaborations.

IOP Conf. Series: Journal of Physics: Conf. Series **1156** (2019) 012034 doi:10.1088/1742-6596/1156/1/012034

Effect of virtual chemistry laboratory toward cognitive learning achievement

Z Latifah[1*], J Ikhsan[2] and K H Sugiyarto[2]

[1]Chemistry Education Program, Universitas Negeri Yogyakarta, Indonesia 55281
[2]Department of Chemistry Education, Faculty of Mathematics and Natural Sciences, Universitas Negeri Yogyakarta, Indonesia 55281

[*] E-mail: zuyinatullatiffah@gmail.com

Abstract. The aim of the study was to determine the effect of using a virtual chemistry laboratory on students' cognitive learning achievement. The study sample was 163 eleventh grade students from two high school in Cilacap regency, Central Java, Indonesia and selected randomly. The research sample was divided into three groups, where each group had different learning activities. Learning activities in the CG were chemistry learning with the practicum in the traditional laboratory, EG-1 were chemistry learning with the practicum in the virtual chemistry laboratory, and EG-2 were chemistry learning with the practicum in traditional laboratory and virtual chemistry laboratory. The data were collected by multiple-choice test questions with 30 items and analyzed using ANOVA (Analysis of Variance). The result showed that the cognitive learning achievement of students in the EG was higher than the CG. This result indicates that there is an effect of using virtual chemistry laboratory on student's cognitive learning achievement. This result of the study can be used as a reference for chemical educators to be more creative and innovative in providing facilities to support the achievement of cognitive learning achievement.

1. Introduction

Virtual Chemistry Laboratory (VCL) is defined as a virtual learning environment that contains a simulation of chemical practicum activities. Virtual laboratories are also interpreted as a form of interaction between students and practical tools through a computer or smartphone [1]. The advantage of virtual laboratories is that it is safe, practicum time is more efficient, laboratory staff is not too necessary, laboratory tools do not have any risk to break down or to lose, and practicum cost is more saving. It is because all tools and substance of this practicum tend illusion [2]. The virtual laboratory acts as a supporter of conventional practicum and forms of e-learning promotion [3]. Laboratory experience is important and challenging. So, the virtual laboratory needed as a potential facility to mold interactive learning [4].

One characteristic of interactive learning is that students are involved more actively in the learning process than teachers. The role of the teacher is only as a facilitator and students as the main actors. Potential tools for achieving interactive learning through virtual laboratories are multimedia. Multimedia elements include text (print or e-book), sound, images, animation, videos, tables and graphs [5]. Various studies have reported the impact of adding multimedia to VCL. The results of the first study showed that the use of images, sound, and animation proved to have an effect on increasing knowledge retention tests [6]. The results of the second study indicate that simultaneous video and image presentation can strengthen memory in long-term memory [7]. The results of subsequent studies state that the use of multimedia in learning can reduce the cognitive burden of students [8]. Finally, the overall effect on improving cognitive learning achievement [9].

The addition of multimedia in VCL needs to be adjusted to the characteristics of students so that cognitive learning achievement can be improved [5]. A study shows that students are more interested in using multimedia-based technology [10]. However, another fact is found that most educators have difficulty developing technology-based multimedia because of limited facilities and capabilities. This is predicted to inhibit the construction of students' knowledge and have an impact on cognitive learning achievement. Therefore, it is necessary to examine the effect of VCL which is equipped with multimedia on the cognitive learning achievement of students. Based on the results of the study it can also be known that multimedia standards are good and in accordance with the learning characteristics of students.

2. Material and methods

VCL influence testing toward cognitive learning achievement was done through quasi-experiment with post-test only design. The study began with the development of questions on cognitive learning achievement tests specifically for petroleum materials. Test questions that have been developed are then theoretically validated by instrument experts and organic chemists. The results of theoretical validation from experts in the form of judgment are complemented by criticism and suggestions as a reference for revisions (summarized in table 1).

Table 1. Details of development and results of the validation of test questions

Indicators of Competence Achievement	Problem form	Number of questions	Validation result instrument expert	Validation result organic chemist
Explain the process of oil and natural gas formation.	Multiple choices (5 choices)	10	• Written word error. • The answer choices are not equivalent. • Answer choices are directed at the cognitive level of analysis.	• Writing numbers that indicate the number of atoms C is not right.
Explain the separation technique using multilevel distillation.		9		
Analyze the impact of burning fuel on the environment and health.		11		

After theoretical validation by instrument experts and organic chemistry, test questions also need to be empirically validated by 109 students outside the research subject. Empirical validation is used to investigate the accuracy of each question in the test question. Each question is stated to be accurate if the students' understanding of consistency is obtained [11]. Empirical validation was analysis through Quest application until it was obtained reliability score of estimate and item fit (summarized in table 2).

Table 2. Reliability of estimate and item from Quest application

Instrument	Coefficient reliability of the estimate	Item amount (initial)	Item fit (end)
cognitive achievement tests	0.91	30	30

Reliability is measurement consistency; it means that measure was declared reliable if consistency score in a situation was the same. Inconsistency Measurement result was represented with reliability coefficient 0.00 [12]. while, consistency of measurement result is shown by the reliability coefficient approaching 1.00 [13]. Consistency item of cognitive tasks on table 3 was declared good because it is obtained reliability result of estimate approaching 1.0. Then, the amount of cognitive tasks of item fit for post-test is 30, it means that all item is valid (there is no failed item) and it can be declared appropriated with the model. Item compatibility with the model can also be seen through INFT MNSQ value and OUTFIT MNSQ value about one, as well as INFT t value and OUTFIT t value about zero [14]. In this context, it is proven that it is obtained INFT MNSQ value and OUTFIT MNSQ value about one as well as INFT t value and OUTFIT t value about zero (See Figure. 1). it is clear on item fit map, where star sign is on two vertical lines dotted with the mean square scale 0.56-1.80 (See figure. 2).

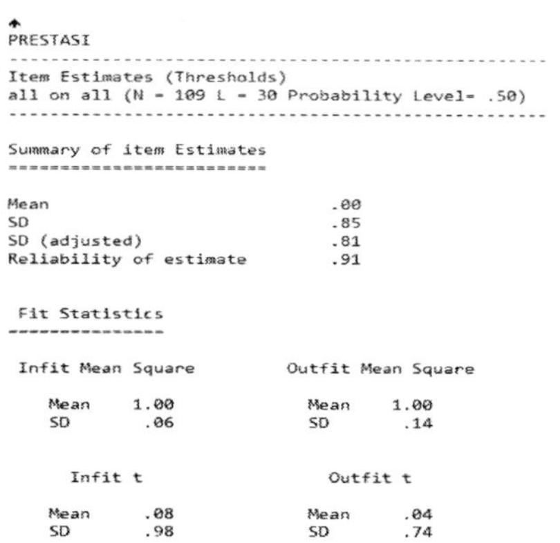

```
PRESTASI
-----------------------------------------------------
Item Estimates (Thresholds)
all on all (N = 109 L = 30 Probability Level= .50)
-----------------------------------------------------

Summary of item Estimates
=========================

Mean                          .00
SD                            .85
SD (adjusted)                 .81
Reliability of estimate       .91

Fit Statistics
===============

Infit Mean Square          Outfit Mean Square

   Mean     1.00              Mean     1.00
   SD        .06              SD        .14

     Infit t                   Outfit t

   Mean      .08              Mean      .04
   SD        .98              SD        .74
```

Figure 1. INFIT and OUTFIT value mean square

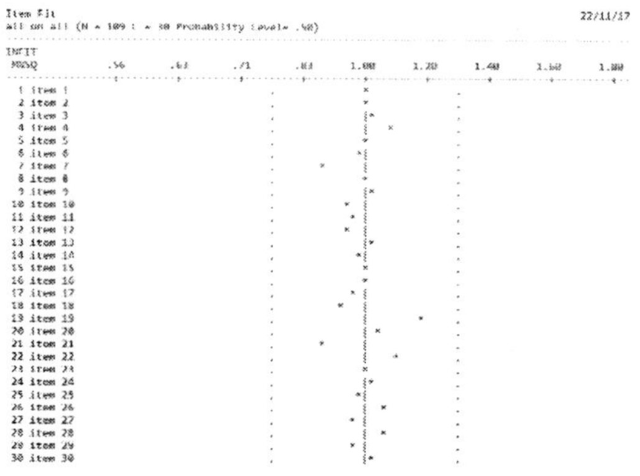

Figure 2. INFIT Value and OUTFIT t

Research subject involved is 163 eleventh learners from two senior high schools. Research subject was divided into three independent group, where every sample consisted of two classes. Every sample was treated randomly by different practicum model. It is based on normality test result where all classes are distributed normally with significances value bigger than 0.05 (summarized in table 3).

Table 3. Normality test for each class

Senior high school	Class	Kolmogorov-Smirnov Significance	Shapiro-Wilk Significance
A	1-XI (Natural science 1)	0.200	0.399
	1-XI (Natural science – 2)	0.190	0.098
	1-XI (Natural science – 3)	0.122	0.055
B	2-XI (Natural science – 1)	0.200	0.726
	2-XI (Natural science – 4)	0.200	0.876
	2-XI (Natural science – 5)	0.200	0.193

Next, six classes are grouped randomly into three independent group (summarized in table 4).

Table 4. Independent group are grouped

Independent group	Class	Amount (N)
CG	1-XI (Natural science – 3) 2-XI (Natural science – 5)	59
EG-1	1-XI (Natural science – 2) 2-XI (Natural science – 4)	51
EG-2	1-XI (Natural science – 1) 2-XI (Natural science – 1)	53

Each independent group has a different form of practicum (summarized in table 5).

Table 5. The form of practicum activities in each independent group

Group	Activity
CG	Practicum activities in conventional laboratories
EG-1	Practicum activities at VCL
EG-2	Practicum activities in conventional laboratories and VCL

Influence of VCL implementation toward cognitive learning achievement was analyzed through one way ANOVA helped by SPSS (*Statistical Product and Service Solutions*) *version* 21. However, before one way ANOVA analysis was done, two assumptions must be fulfilled as well as dependent variable every group distributed normal and homogeneity population variances [15]. In this research, first one way ANOVA assumption had been fulfilled because of significances value obtained bigger than 0.05 (summarized in table 6).

Table 6. Normality test result for each independent group

Group	Amount (N)	Shapiro-Wilk Significance	Lillifors significance
CG	59	0.109	0.100
EG-1	51	0.095	0.174
EG-2	53	0.091	0.072

While the third assumption related to Homogeneity population variances had also been fulfilled, where it is shown by significance value 0.753 or bigger than 0.05 (summarized in table 7).

Table 7. Homogeneity test result of population variances

Levene statistic	df1	df2	Significance
0.285	2	160	0.753

Overall, it can be concluded that the three assumptions above had been fulfilled, so analysis through ANOVA can be done.

3. Result and discussion

The results of the one-way ANOVA test were shown through the existence of statistically significant differences in cognitive learning achievement between students in the three independent groups (p-vale < 0.05) (summarized in table 8). This was proven through the acquisition of significantly smaller than 0.05, which is 0.00.

Table 8. The significant difference of every independent sample

	Sum of squares	Df	Mean square	F	Significance
Between groups	1308.408	2	654.204	8.301	0.000
Within groups	12609.901	160	78.812		
Total	13918.309	162			

Other findings indicate differences in the average cognitive learning achievement among students in three independent groups (see figure 3).

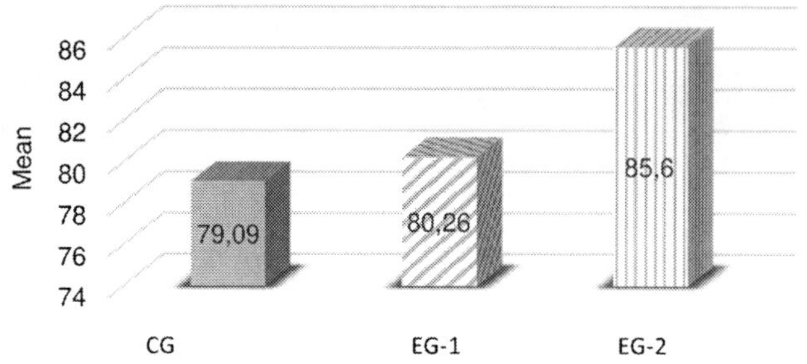

Figure 3. Differences in the average cognitive learning achievement in each independent group

Based on figure 3, it can be seen that students and the experimental group 2 had an average cognitive learning achievement higher than others. Some speculation arises relating to the reason of acquisition the highest average value on experimental group two. The first speculation relates to chemistry learning supported by practicum activity on conventional laboratory and VCL. A combination between virtual laboratory and remote laboratory had been done and impacted on understanding enhancement [16]. Virtual RTD laboratory application as support of conventional laboratory was also proven effective because it is produced the better concept of understanding [3]. The higher cognitive level was declared to be reached when laboratory technique and kinesthetic details were applied simultaneously [17]. Enhancement of understanding concept and skills also happen when the virtual instrument was combined with real practicum [18]. Practicum combination becomes tools to achieve high understanding and cognitive value because it is fulfilled representational competence [19].

Second speculation related to multimedia from practicum activity on conventional laboratory and VCL. In this context, the conventional laboratory has a role in giving media in the form of direct experience such as substance introduction and tools utilization, real practicum activity. While VCL contributed in giving media in the form of animation video, sound, virtual practicum simulation, learners' worksheet (PDF), and an introduction of 2D/3D practicum tools. Multimodal utilization was proven effective as detail information giver about physic/chemistry characteristic of substance as well as becoming chemistry theory ability enhancer [20]. Multimedia in the virtual laboratory was also reported impact on achievement enhancement [2]. Multimedia can also act as abstract chemistry material explainer, so it is obtained concept understanding as well as final value enhancement [21]. Another reason was conveyed through the results of research, where multimedia can play a role as a provider of detailed information about the physical or chemical nature of the material. Multimedia is also stated to be useful as an explanatory abstract chemical material, so as to gain an understanding of concepts and increase in the final value. Based on various research results above, so it can be concluded that better understanding can be reached through multimedia such as simulation; interactive animation; video lectures; and lab view video [22]. In this context, multimedia must interact and be interrelated with each other in order that experiment activity run more effective.

4. Conclusion

Statistically, it can be concluded that there are significant differences in cognitive learning achievement between students who conduct practicum activities in conventional laboratories, VCL and practicum activities in conventional laboratories and VCL has higher cognitive learning achievement than others. This finding can be the basis of subsequent research, especially related to the combination of practicum activities and a more varied presentation of multimedia so that higher cognitive learning achievement is achieved. In the next study also needs to examine the effect of a combination of practicum activities on other academic performances such as learning motivation, creativity or scientific attitude.

IOP Conf. Series: Journal of Physics: Conf. Series **1156** (2019) 012034 doi:10.1088/1742-6596/1156/1/012034

References

[1] Jezierska K, Podraza W, Domek H and Szwed J 2016 *Nat. Acad. Sci Lett.* **39** 295–99
[2] Tatli Z and Ayas A 2013 *Educ. Technol. Soc.* **16** 159–170
[3] Domingues L, Rocha I, Dourado F, Alves M, and Ferreira E C 2010 *Educ. Chem. Eng.* **5** 22–27
[4] Polly P, Marcus N, Maguire D and Belinson Z V 2014 *BMC Medic. Educ.* **14** 1–9
[5] Ercan O 2014 *J. Baltic. Sci. Educ.* **13** 608–21
[6] Mayer R E and Moreno R 1998 *J. Educ. Psychol.* **90** 312–20
[7] Watson S L, Loizzo J, Watson W R, Mueller C, Lim J and Ertmer P A 2016 *Educ. Technol. Res. Dev.* **64** 1273–1300
[8] de Oliveira Neto J D, Huang W D and de Azevedo Melli N C 2015 *Educ. Technol. Res. Develop.* **63** 555–573
[9] Aloraini S 2012 *J. King Saud Univ - Lang. Transl.* **24** 75–82
[10] Su K D 2008 *Int. J. Sci. Math. Educ.* **6** 225–249
[11] Adams W K and Wieman C E 2010 *Int. J. Sci. Educ.* p 1–24
[12] Yang Y, Oosterhof A and Xia Y 2015 *J. Educ. Res.* **108** 465–79
[13] Gliem J and Gliem R R 2003 *Midwest Res. Pract. Conf. in Adult, Continuing Com. Educ.* 82–88
[14] Adams R J and Khoo S T 1996 *ACER quest: The interactive test analysis system* (Victoria: Australian Council for Educational Research)
[15] Stevens J 2002 *Applied multivariate statistics for the social sciences* (New Jersey: Lawrence Erlbaum Associates)
[16] Chaos D, Chacón J, Lopez-Orozco, J A and Dormido S 2013 *Sensors (Switzerland).* **13** 2595–2612
[17] Woodfield B F, Andrus M B, Andersen T, Miller J, Simmons B, Stanger R, Waddoups G L, Moore M S, Swan R, Allen R and Bodily G 2005 *J. Chem. Educ.***82** 1728–35
[18] Marques M A, Viegas M C, Costa-Lobo M C, Fidalgo A V, Alves G R, Rocha J. S and Gustavsson I 2014 *IEEE Transac on Educ.* **57** 151–59
[19] Plass J L, Milne C, Homer B D, Schwartz R N, Hayward E O, Jordan T, Verkuilen J, Ng F, Wang Y and Barrientos J 2012 *J. Res. Sci Teach* **49** 394–419
[20] Osman K and Lee T T 2013 *Int. J. Sci. Math. Educ.* **12** 395–421
[21] Agrawal A, Uppaluri R and Verma A 2013 *CSI Trans. ICT.* **1** 75–90

Acknowledgments

Thanks to Directorate General of Affirmation and Enhancement Research Ministry of Technology and High Education Indonesia which help in giving fund for this research.